Herpesviruses and Immunity

INFECTIOUS AGENTS AND PATHOGENESIS

Series Editors: Mauro Bendinelli, *University of Pisa*
Herman Friedman, *University of South Florida College of Medicine*

Recent volumes in the series:

DNA TUMOR VIRUSES
Oncogenic Mechanisms
Edited by Giuseppe Barbanti-Brodano, Mauro Bendinelli, and Herman Friedman

ENTERIC INFECTIONS AND IMMUNITY
Edited by Lois J. Paradise, Mauro Bendinelli, and Herman Friedman

FUNGAL INFECTIONS AND IMMUNE RESPONSES
Edited by Juneann W. Murphy, Herman Friedman, and Mauro Bendinelli

HERPESVIRUSES AND IMMUNITY
Edited by Peter G. Medveczky, Herman Friedman, and Mauro Bendinelli

MICROORGANISMS AND AUTOIMMUNE DISEASES
Edited by Herman Friedman, Noel R. Rose, and Mauro Bendinelli

NEUROPATHOGENIC VIRUSES AND IMMUNITY
Edited by Steven Specter, Mauro Bendinelli, and Herman Friedman

PSEUDOMONAS AERUGINOSA AS AN OPPORTUNISTIC PATHOGEN
Edited by Mario Campa, Mauro Bendinelli, and Herman Friedman

PULMONARY INFECTIONS AND IMMUNITY
Edited by Herman Chmel, Mauro Bendinelli, and Herman Friedman

RAPID DETECTION OF INFECTIOUS AGENTS
Edited by Steven Specter, Mauro Bendinelli, and Herman Friedman

RICKETTSIAL INFECTION AND IMMUNITY
Edited by Burt Anderson, Herman Friedman, and Mauro Bendinelli

VIRUS-INDUCED IMMUNOSUPPRESSION
Edited by Steven Specter, Mauro Bendinelli, and Herman Friedman

---

Herpesviruses and Immunity

Edited by

Peter G. Medveczky and
Herman Friedman
University of South Florida College of Medicine
Tampa, Florida

and

Mauro Bendinelli
University of Pisa
Pisa, Italy

Plenum Press • New York and London

Library of Congress Cataloging-in-Publication Data

Herpesviruses and immunity / edited by Peter G. Medveczky and Herman Friedman, and Mauro Bendinelli.
p. cm. -- (Infectious agents and pathogenesis)
Includes bibliographical references and index.
ISBN 0-306-45890-X
1. Herpesvirus diseases--Immunological aspects. I. Medveczky, Peter G. II. Friedman, Herman, 1931- . III. Bendinelli, Mauro. IV. Series.
[DNLM: 1. Herpesviridae--immunology. 2. Herpesviridae--genetics. 3. Herpesviridae Infections--immunology. QW 165.5.H3 H5638 1998]
QR201.H48H48 1998
579.2'434--dc21
DNLM/DLC
for Library of Congress 98-42559
CIP

ISBN 0-306-45890-X

A Division of Plenum Publishing Corporation
233 Spring Street, New York, N.Y. 10013

http://www.plenum.com

10 9 8 7 6 5 4 3 2 1

Printed in the United States of America

Contributors

JENS-CHRISTIAN ALBRECHT • Institut für Klinische und Molekulare Virologie, Friedrich-Alexander Universität Erlangen-Nürnberg, 91054 Erlangen, Germany

ANN M. ARVIN • Department of Pediatrics and Microbiology/Immunology, Stanford University School of Medicine, Stanford, California 94305

SALLY S. ATHERTON • Department of Cellular and Structural Biology, University of Texas Health Sciences Center at San Antonio, San Antonio, Texas 78284-7762

DAVID C. BLOOM • Department of Microbiology, Arizona State University, Tempe, Arizona 85287-2701

BERNADETTE DUTIA • Department of Veterinary Pathology, The University of Edinburgh, Summerhall, Edinburgh EH9 1QH, Scotland

BERNHARD FLECKENSTEIN • Institut für Klinische und Molekulare Virologie, Friedrich-Alexander Universität Erlangen-Nürnberg, 91054 Erlangen, Germany

PETER GECK • Department of Anatomy and Cellular Biology, Tufts University Health Science Schools, Boston, Massachusetts 02111

HARTMUT HENGEL • Max von Pettenkofer-Institut für Hygiene und Medizinische Mikrobiologie, Lehrstuhl Virologie, Ludwig-Maximilians-Universität München, D-80336 München, Germany

GEORGE KLEIN • Microbiology and Tumor Biology Center (MTC), Karolinska Institutet, S-171 77 Stockholm, Sweden

ULRICH H. KOSZINOWSKI • Max von Pettenkofer-Institut für Hygiene und Medizinische Mikrobiologie, Lehrstuhl Virologie, Ludwig-Maximilians-Universität München, D-80336 München, Germany

TAKESHI KURATA • Department of Pathology, National Institute of Infectious Diseases, Tokyo 162, Japan

PATRICK S. MOORE • Division of Epidemiology, Columbia University School of Public Health, New York, New York 10032

ANDREW J. MORGAN • Department of Pathology and Microbiology, School of Medical Science, University of Bristol, Bristol BS8 1TD, England

ANTHONY A. NASH • Department of Veterinary Pathology, The University of Edinburgh, Summerhall, Edinburgh EH9 1QH, Scotland

FRANK NEIPEL • Institut für Klinische und Molekulare Virologie, Friedrich-Alexander Universität Erlangen-Nürnberg, 91054 Erlangen, Germany

SONJA J. OLSEN • Division of Epidemiology, Columbia University School of Public Health, New York, New York 10032

BERNARD ROIZMAN • The Marjorie B. Kovler Viral Oncology Laboratories, The University of Chicago, Chicago, Illinois 60637

BARRY T. ROUSE • Department of Microbiology, University of Tennessee, Knoxville, Tennessee 37996-0845

TAKESHI SAIRENJI • Department of Biosignaling, School of Life Science, Faculty of Medicine, Tottori University, Yonago 683; and Department of Pathology, National Institute of Infectious Diseases, Tokyo 162, Japan

JAMES P. STEWART • Department of Veterinary Pathology, The University of Edinburgh, Summerhall, Edinburgh EH9 1QH, Scotland

DAVID A. THORLEY-LAWSON • Department of Pathology, Tufts University School of Medicine, Boston, Massachusetts 02111

EDWARD J. USHERWOOD • Department of Veterinary Pathology, The University of Edinburgh, Summerhall, Edinburgh EH9 1QH, Scotland; *present address:* Department of Immunology, St. Jude Children's Research Hospital, Memphis, Tennessee 38105

EDWARD K. WAGNER • Department of Molecular Biology and Biochemistry, University of California, Irvine, Irvine, California 92697-3900

PATRICIA L. WARD • The Marjorie B. Kovler Viral Oncology Laboratories, The University of Chicago, Chicago, Illinois 60637

HANS J. WOLF • Institut für Mikrobiologie und Hygiene der Medizinischen, Fakultät der Universität Regensberg, D-93053 Regensberg, Germany

Preface to the Series

The mechanisms of disease production by infectious agents are presently the focus of an unprecedented flowering of studies. The field has undoubtedly received impetus from the considerable advances recently made in the understanding of the structure, biochemistry, and biology of viruses, bacteria, fungi, and other parasites. Another contributing factor is our improved knowledge of immune responses and other adaptive or constitutive mechanisms by which hosts react to infection. Furthermore, recombinant DNA technology, monoclonal antibodies, and other newer methodologies have provided the technical tools for examining questions previously considered too complex to be successfully tackled. The most important incentive of all is probably the regenerated idea that infection might be the initiating event in many clinical entities presently classified as idiopathic or of uncertain origin.

Infectious pathogenesis research holds great promise. As more information is uncovered, it is becoming increasingly apparent that our present knowledge of the pathogenic potential of infectious agents is often limited to the most noticeable effects, which sometimes represent only the tip of the iceberg. For example, it is now well appreciated that pathologic processes caused by infectious agents may emerge clinically after an incubation of decades and may result from genetic, immunologic, and other indirect routes more than from the infecting agent in itself. Thus, there is a general expectation that continued investigation will lead to the isolation of new agents of infection, the identification of hitherto unsuspected etiologic correlations, and, eventually, more effective approaches to prevention and therapy.

Studies on the mechanisms of disease caused by infectious agents demand a breadth of understanding across many specialized areas, as well as much cooperation between clinicians and experimentalists. The series *Infectious Agents and Pathogenesis* is intended not only to document the state of the art in this fascinating and challenging field but also to help lay bridges among diverse areas and people.

M. Bendinelli
H. Friedman

Preface

Although virology and immunology are now considered separate disciplines, history shows that these areas of investigation always overlapped and one cannot really exist without the other. This trend has become particularly significant and fruitful in the past few years in the area of herpesvirus research. The genomes of the most important herpesviruses have been sequenced, a significant portion of their genes have been identified, and many secrets of regulation of gene expression have been unraveled. Now this progress sets the stage for a true revolution in herpesvirus research: analysis of interactions between the host and the virus. Because herpesviruses can induce, suppress, and fool the immune system, the most productive herpesvirologists are also expert immunologists, and the current results of this interdisciplinary effort are truly remarkable.

Because herpesviruses cause many important human diseases, the development of vaccines against these agents is a very significant goal. This effort is also very challenging because of the complexity of herpesviruses and the lack of sufficient information about immune responses.

The remarkable ability of herpesviruses to escape immune responses is another feature that brings immunology and virology together. Herpesviruses encode many proteins that interact with and down-regulate some key elements of the immune system. This property of herpesviruses represents a major challenge in developing strategies against these viruses. On the positive side, these viral proteins also provide novel tools for analyzing specific immune reactions and molecular mechanisms.

The gamma subgroup of herpesviruses have two additional links to immunology. These viruses undergo latency in immune cells and, in some instances, participate in oncogenic transformation of T or B cells. Moreover, these viruses encode several genes transduced from the host, such as lymphokines, cellular receptors, and complement control genes. Herpesviruses probably use these pirated protein products of the immune system to aid their survival in the infected host.

The aim of this book is to present reviews by internationally renowned herpesvirus virologists and immunologists summarizing their work and some of the exciting newer developments in the field of herpes virology. The book is not intended to be a comprehensive work on herpesviruses but is designed to provide an in-depth analysis of "cutting-edge" areas of herpesvirus research. For example, the first chapter by Drs. Patricia Ward and Bernard Roizman of The University of Chicago describes evasive strategies of human herpesviruses against host defenses. The immunopathology of herpesvirus infection is discussed by Drs. Barry Rouse and Sally Atherton. Then Drs. Edward Wagner and David Bloom discuss gene expression by herpesviruses during latent infections and reactivation. Dr. Peter Geck (Chapter 4) and Drs. Sonja Olsen and Patrick Moore (Chapter 5) describe Kaposi's sarcoma-associated herpesvirus and *herpesvirus saimiri* effects on T lymphocytes and lymphokines. There is then a review of the immunobiology of murine gamma herpesvirus by Drs. Stewart, Usherwood, Dutia, and Nash.

Dr. George Klein (Chapter 7) and Drs. Takeshi Sairenji and Takeshi Karada (Chapter 8) discuss the Epstein–Barr virus infection and discuss the role of the immune response and its role in induction of tumors. Dr. David A. Thorley-Lawson discusses EB virus persistence *in vivo* and to evasion of the immune response. Drs. Wolf and Morgan provide a summary regarding EBV vaccine development. Drs. Hartmut Hengel and Ulrich H. Koszinowski discuss cytomegalovirus effects on MHC class I functions. Dr. Ann Arvin discusses cellular immunity and its role in controlling varicella-zoster virus infection. Drs. Jens-Christian Albrecht, Frank Neipel, and Bernhard Fleckenstein discuss the effects of rhadinoviruses on the complement system of the immune response.

It is anticipated by the editors of this volume that the readers of the book will include both immunologists and virologists as well as pre- and postdoctoral students and their mentors. The editors are confident that newer developments in the field of herpesvirus immunology and pathogenesis will continue to stimulate interest in the field of virology and immunology. It is anticipated that the discovery of newer knowledge will occur as rapidly in the future as it has in the past. The editors of the volume are grateful for the time and effort of all contributors. The editors also thank Ms. Ilona M. Friedman for invaluable assistance as editorial coordinator and managing editor for the preparation of this volume. We also thank the editorial staff of Plenum Publishing for their invaluable assistance.

Peter G. Medveczky
Herman Friedman
Mauro Bendinelli

Contents

3. **HSV Gene Expression during Latent Infection and Reactivation**

EDWARD K. WAGNER AND DAVID C. BLOOM

4. T Cell Activation and Lymphokine Induction in *Herpesvirus saimiri* Immortalized Cells

PETER GECK

7. EBV and B Cell Lymphomas

GEORGE KLEIN

8. Immune Responses to Epstein–Barr Viral Infection

TAKESHI SAIRENJI AND TAKESHI KURATA

9. **EBV Persistence *in Vivo*: Invading and Avoiding the Immune Response**

DAVID A. THORLEY-LAWSON

10. **Epstein–Barr Virus Vaccines**

HANS J. WOLF AND ANDREW J. MORGAN

Herpesviruses and Immunity

1

Evasion and Obstruction

The Central Strategy of the Interaction of Human Herpesviruses with Host Defenses

PATRICIA L. WARD and BERNARD ROIZMAN

Whatever Nature has in store for mankind,
unpleasant as it may be, men must accept,
for ignorance is never better than knowledge.
Enrico Fermi, in Laura Fermi,
Atoms in the Family
(University of Chicago Press, 1954)

1. INTRODUCTION

One of the fascinating aspects of the biology of the human herpesviruses is their relationship with their hosts. Herpesviruses have evolved strategies that enable them to persist and disseminate widely throughout the human population, usually without dire consequences to their hosts.[1] In the absence of effective host defense mechanisms, however, these viruses can and do cause severe disease. The mechanisms by which these viruses replicate and persist in the face of the host's immune response reflects a complex interplay between the virus and the host. An understanding of these phenomena must take into account both the shared and unique features of each of the members of the herpesvirus family infecting humans.

PATRICIA L. WARD and BERNARD ROIZMAN • The Marjorie B. Kovler Viral Oncology Laboratories, The University of Chicago, Chicago, Illinois 60637.

Herpesviruses and Immunity, edited by Medveczky *et al.* Plenum Press, New York, 1998.

Herpesviruses are a diverse group of large double-stranded DNA viruses that have a common virion structure, a few highly conserved genes, and can establish both productive and latent infections.[2] The wide variety of mechanisms by which the human herpesviruses interfere at the molecular level with the host immune system reflect both the unique biological features of each virus and the nature of the infection. All of the human herpesviruses establish latent infections which allow viral persistence to go largely unnoticed by the immune system.

A common strategy for immune evasion among pathogenic organisms, including many RNA viruses, is to alter the antigenic profile by mutation. Although mutants of the species flourish, the cost in terms of immune destruction of non-mutated viruses is very high. DNA viruses have evolved alternate mechanisms to evade or reduce the impact of specific host defense processes. In this chapter, we focus on evasive mechanisms with particular emphasis on herpesvirus proteins that interact with host defense systems resulting in inhibition, interference, or modulation of the host response to infection. Ultimately, this results in evasion of the host defense system by the virus and a long term association with its host.

2. IMMUNE EVASION DURING LATENT INFECTION

The hallmark of the biology of human herpesviruses is their ability to remain latent for the lifetime of the host. For the purposes of this chapter, we shall define latency as the lifetime maintenance of the viral genome in the infected cells without full expression of viral genes characteristic of productive infection. Latent infection reflects a complex interaction between the virus and the host cell that is unique for each virus and is characterized by (i) the nature of the cell harboring the virus and (ii) viral gene expression in the cell harboring a latent virus. The host cell defines the relationship of the virus with the host immune defense system. Thus, not surprisingly, cells that harbor a latent virus lack certain features necessary to become targets of a full-scale immune attack. The extent and nature of gene expression defines a second key parameter because a latent virus must either restrict viral gene expression, thus remaining antigenically silent, or suppress presentation of antigenic peptides to avoid attack by its host. The ability of human herpesviruses to remain antigenically silent and/or hidden from the host's immune system enables these viruses to form a reservoir available for dissemination upon reactivation and propitious contacts between infected and uninfected individuals.

The mechanism by which viruses remain hidden from host defense systems depends on the types of cells in which they establish a latent infection. Current knowledge of herpesvirus latent infection suggests that these viruses establish latency in cells that are at least largely immune privileged or that are components of the host's immune system, allowing for direct modulation of host immune responses by the virus.

The human members of the *Alpha herpesvirinae* subfamily of the *herpesviridae* family, herpes simplex virus -1 and 2 (HSV-1 and HSV-2) and varicella-zoster virus (VZV) are considered neurotropic on the ground that they can infect nerve endings and be transported by retrograde axonal flow to neuronal nuclei where they can establish latency.[3,4] These viruses avoid recognition by the immune system because of the extremely limited gene expression—in fact no protein products have been detected in cells latently infected with any of the three viruses—and also because the neuronal cells express very little or no MHC class I or class II antigens, thus limiting the possibility of immune recognition. Although all three viruses share these properties, they differ significantly in their biology of latency, reflected in the frequency of reactivation from the latent state.

The human members of the *beta herpesvirinae* subfamily, human cytomegalorvirus (HCMV), human herpesviruses 6 and 7 (HHV-6 and HHV-7), are thought to establish latent infections at least in monocytes and/or T lymphocytes. Of the two human members of the *gamma herpesvirinae,* Epstein-Barr virus (EBV) and human herpesvirus 8 (HHV-8), EBV has been studied most extensively. This virus establishes latency in B lymphocytes. In contrast to the alpha herpesviruses, these viruses express a small set of genes during latent infections. Successful evasion of immune recognition in these cells requires more direct interference with host defense mechanisms.

2.1. Latent HSV-1, HSV-2, and VZV Infections

The alpha herpesviruses HSV-1 and HSV-2 infect and replicate in epithelial cells of mucosal membranes and concurrently or subsequently enter neuronal cells of sensory ganglia that innervate the peripheral sites of infection.[3] At these peripheral sites, the primary infection is cleared by the immune system. The events in the neurons populated by viruses are less clear. A fraction of the neurons undergo productive infection and perish. Gene expression conducive to productive infection is suppressed in the majority of the infected neurons. The viral genome circularizes to form an episome. Although the mechanisms by which the virus establishes latency are unknown, it is likely that viral and neuron cell-specific factors are necessary for this process.

During the latent phase, the virus is quiescent and exhibits only very limited gene expression. The only viral gene products detected in latently infected neurons to date are a set of RNA transcripts known as the latency-associated transcripts or LATs. The largest but least abundant LAT is an 8.5-kb RNA. The most abundant LATs are 1.5 kb and 2 Kb in size, respectively. They are stable introns that accumulate in nuclei and do not encode proteins.[3] Reactivation of both HSV-1 and 2 occurs periodically as a result of a variety of stimuli including stress, exposure to ultraviolet light, hormonal fluctuations, etc. at which time latency is abrogated in a fraction of latently infected cells, a productive infection ensues, the progeny of viral replication are transported by anterograde flow to a site at or near the site of the neuron's infection, and ultimately, the neuron itself is most likely destroyed as much by the viral gene products as by the immune system.[3]

It is conceivable that the progeny of reactivated virus itself populate new neuronal cells and thereby insure frequent reactivation of the virus. However, the difficulty encountered in deliberate superinfection of sensory glanglia with a second, marked HSV, suggests that the HSV-1 and HSV-2 do not make "round trips" to and from sensory neurons. Rather, the diminution of the frequency of reactivations with time suggests the possibility that the pool of reactivable virus becomes exhausted.[3]

It has been postulated that the immune system plays several significant roles in HSV latency. Thus, the inability to superinfect ganglia and the observation that not all "reactivation events" result in clinically apparent lesions has been reasonably ascribed to the development of effective antiviral immunity. Less credence may be placed on the postulate that the immune system enables the establishment of latency or that viral reactivation from latency reflects an immune failure correctable, by therapeutic immunization. That the immune system plays some role in curtailing viral spread in the nervous system during acute, primary infection and establishment of latency, however, is supported by studies that demonstrated a requirement that $CD8^+$ cells clear acutely infected cells in the peripheral nervous system. Mice that were depleted of $CD8^+$ cells showed extensive neuronal destruction and could not clear an acute infection.[5] $CD8^+$ T cell activity was required to limit viral spread and to prevent excessive destruction in the sensory ganglia by the virus. This could be achieved via lysis of infected neuronal cells that harbor replicating virus or by the actions of cytokines released from the $CD8^+$ cells. For example, by protecting uninfected neurons or by directly inhibiting viral replication. Interestingly, acute infection with HSV in mouse sensory ganglia resulted in up-regulation of MHC class I mRNA in sensory neurons, satellite cells, and Schwann cells. Class I antigens, however, could be detected only in satellite cells and Schwann cells, not in the neuronal cells.[6] In rat capsular cells (counterpart to CNS microglia), MHC class I antigens are expressed during acute infection, but these cells are not productively infected.[7] Taken together, there is a complex interplay between virus and host response in which the $CD8^+$ cells are both beneficial and detrimental to the virus. Presentation of viral antigens by MHC class I molecules is likely to result in destruction of the infected cells, but preventing excessive spread in the nervous system ensures that the host will live. That neuronal destruction is limited in the presence of $CD8^+$ CTL suggests that these effector cells do not contain viral spread via lysis of infected cells, but possibly through the actions of cytokines. Studies designed to document the duration of IFN-gamma expression in the nervous system showed that both $CD4^+$ and $CD8^+$ T cells were present in trigeminal ganglia during the acute phase of infection and also well into the latent phase of infection.[8] IFN-gamma secretion also persisted in latently infected ganglia, but it is not known whether this cytokine plays any role during HSV latent infection. Analysis of the role of T cells in sensory ganglia must include consideration of the role of cytokines released by the T cells in the vicinity of latently infected neurons.

Latent infection of varicella-zoster virus (VZV) differs in some details from that of HSV. *In situ* hybridization and PCR analyses suggest that neurons of sensory ganglia are also the home of VZV latent infection. Viral DNA, however, has also been detected in satellite cells.[9] Interestingly, there is no clear homologue of the HSV LAT region in the VZV genome, and *in situ* hybridization with a range of DNA probes and direct analyses of RNA suggest there may be limited transcription from various regions throughout the VZV genome.[4] Unlike HSV-1 and HSV-2, clinical reactivation of VZV occurs only once to cause shingles.

2.2. Latent HCMV, HHV-6, and HHV-7 Infections

The biology of latent infection of the beta herpesviruses is quite different from that of the alpha herpesviruses. As noted previously, these viruses establish latent infections in cells belonging to the immune system, but many of the features surrounding latent infections with these viruses remain unclear. Some of the confusion stems from certain biological characteristics unique to these viruses. For example, it is difficult to distinguish between reactivation of HCMV from a latent state and a low level of persistent replicating virus. Also, the populations of cells harboring virus may differ depending on the immune status of the host. So, for example, although a number of cell types, including polymorphonuclear leukocytes, endothelial cells, and monocytes, are associated with primary or recurring HCMV infection in immunocompromised patients, polymerase chain reaction (PCR) analysis of peripheral blood from healthy carriers has led to the identification of a population of monocytes as the site of latency or viral persistence of HCMV with no apparent involvement of PMNs or endothelial cells.[10] The sensitivity of the PCR has also refuted the previous conventional wisdom that T lymphocytes are a reservoir of latent virus. Although primary infection with HCMV may result in an illness resembling mononucleosis, most of the infected population exhibit subclinical infections, intermittently shedding virus, particularly from salivary glands, but exhibiting no apparent disease or pathology,[11] such as seen in the lesions that result from recrudescence of HSV-1 or of HSV-2 or the lesions associated with reactivation of latent VZV. Gene expression is reported to be limited to very early genes[12–14] but even this notion has been challenged.[15] Only following differentiation of monocytes to macrophages could viral gene transcription be detected, and this was restricted to immediate early genes.[15] It is thought that both cellular and humoral factors play a role in maintaining the HCMV latent state, but the mechanisms governing this process are unknown. The biology of latent infection of HHV-6 and 7 is even less clear. Two variants of HHV-6 have been isolated, variant A which has not been associated with any disease and variant B which is the etiological agent of exanthem subitum.[16] The two variants can be distinguished on the basis of differences in epidemiology, their respective host ranges *in vitro,* reactivity to monoclonal antibodies, restriction endonuclease profiles, and nucleotide sequences. HHV-6 and 7 share many similarities in their infective biology. Both viruses are acquired early in childhood, but only one, HHV-6, has been associated with several diseases including, exanthem subitum (roseola). Both HHV-6 and 7 predominantly infect $CD4^+$ T lymphocytes. Although HHV-6 can infect both $CD4^+$ and $CD8^+$ T cells,[17] however, HHV-7 is more restricted in its tropism. Of T lymphocytes, HHV-7 is restricted to $CD4^+$ cells, and the CD4 molecule is itself an essential component of the HHV-7 receptor.[18] HHV-7 was first isolated from peripheral blood lymphocytes in culture but is readily isolated from the saliva of most adults.[9–22]

HHV-6 and 7 do not establish latent infections as defined by alpha herpesvirus biology. A great deal of circumstantial evidence, however, supports the view that these viruses establish a long-term persistent infection. Most studies have been done with HHV-6, and the most direct evidence comes from analysis of HHV-6 DNA isolated from the blood of an immunosuppressed bone marrow transplant patient. The restriction endonuclease profiles of viral DNA isolated before and after transplant are identical, indicating that the same virus was isolated at both times and suggesting that the virus is maintained within the human host.[23] HHV-6 viral DNA was isolated from a variety of cell types and tissues, including salivary glands, lymph node tissue, and neurons and glial cells in the brain, and can be detected in as many as 90% of peripheral blood specimens obtained from healthy patients.[16] The primary reservoir of HHV-6 *in vivo,* however, remains unknown.

A number of *in vitro* studies have been carried out to shed light on this question. *In vitro,* HHV-6 can be induced to replicate in macrophages from an apparent latent state following treatment with phorbol esters.[24] The rationale for this study comes from the finding that during the acute phase of HHV-6, induced illness, virus could be isolated from nonadherent cells and during the recovery phase, HHV-6 viral DNA can be detected by PCR primarily in adherent cell populations of PBMCs.[24] These investigators found that monocytes isolated from healthy patients and cultured *in vitro* could not be readily infected with HHV-6, but if cultured for 7 days prior to viral infection, viral antigen-positive cells could be detected three days following infection. These cells ceased to express viral antigens after 21 days in culture, but, viral DNA could be detected until 45 days after infection. At this point, the virus could be induced to replicate from PBMCs by treating the cells with TPA (12-*O*-tetradecanoylphorbol 13-acetate).[24] HHV-6 was also induced to replicate following HHV-7 superinfection of peripheral blood lymphocytes that harbored HHV-6 DNA.[19] Other *in vitro* studies showed abortive infection by HHV-6 in glioblastoma cells but the virus could replicate in primary astrocytes.[25]

The presence of neutralizing IgM antibody in HHV-6 immune patients following bone marrow transplant and in a small but stable percentage of the population suggested recent reactivation.[26] Taken together, it seems likely that HHV-6 and possibly also HHV-7 can persist within the host for very long periods of time, probably similarly to HCMV, although none of these results preclude the possibility of reinfection. An identifying characteristic of latent herpesvirus infections is the expression of a limited and unique set of viral genes as distinguished from that which occurs during lytic infection. It is presently unknown whether any viral genes are expressed in those cell populations in which HHV-6 viral DNA is detected. It remains a difficult but important task to identify the main viral reservoir.

2.3. Latent EBV and HHV-8 Infections

The most complex and intriguing mechanism of latent infection is that of B lymphocytes with EBV, but how this virus persists in a rapidly turned over population of cells remains largely a mystery. What is clear is that EBV has evolved mechanisms to persist in cells that are continuously receiving physiological signals to drive the infected cell into a differentiation pathway or lead to their elimination via programmed cell death.[27,28]

EBV is associated with at least three human malignancies, and EBV establishes three different programs of latency in these transformed B cells. These three types of latent infection have been designated Latency I, II and III and are differentiated on the basis of the viral genes expressed and the behavior of the host cell.[27,29,30] Latency I characterizes latently infected Burkitt's lymphoma cells in which only the EBNA I gene and the Epstein-Barr virus encoded RNA (EBERs) 1 and 2 are expressed. Latency II is defined by expression of the EBERs, EBNA1, LMP1, and LMP2 and is characteristic of Hodgkin's lymphoma and nasopharyngeal carcinoma. Latency III is carried out in immunoblastic lymphomas and in B lymphocytes transformed to immortality (LCLs) *in vitro*. In patients with mononucleosis, EBV-carrying B lymphocytes can be detected in the blood, and LCLs can readily be established *in vitro* directly from the blood of previously infected persons.[27] The viral genes expressed in these cells include all six EBNA genes, LMP1, LMP2, and the EBERS.

Another view of EBV latency is based on the assumption that EBV integrates its life cycle into the normal biology of the host cell, so that the environment of the infected B cell determines the characteristics of the latent state.[28] These authors have proposed an alternative nomenclature and suggest that the site of EBV persistence is the resting B cell in which virus is not replicating and expresses LMP2 and possibly EBNA 1. Viral DNA can be detected as an episome in resting B cells indicating that the cells must have been activated to allow circularization of the DNA before their return to a resting state.[31] The common theme, however, is that cells infected with EBV express viral antigens that can be recognized by cytotoxic T cells, yet the latently infected cells are not eliminated by the immune system.

EBV has evolved several strategies that aid in avoiding recognition by antigen-specific T cells. First, this virus inhibits the processing of viral proteins that results in generating peptide antigens that can be presented by MHC class I molecules to effector T cells. It is reported that the EBNA 1 protein contains a *cis*-acting signal consisting of a glycine-alanine repetitive sequence which inhibits antigen processing. Insertion of the sequence encoding this inhibitory signal within the coding domain of the EBNA 4 gene downstream of a region that encodes a known CTL-recognized epitope resulted in decreased lysis of cells expressing this gene.[32]

Second, EBV down-regulates expression of certain cell surface adhesion molecules that are necessary for interactions between effector cells and their targets. CTL-target cell interactions involve specific interactions between peptide-MHC complexes and the T cell receptor and also conjugate formation which involves adhesion of the effector and target cells and is independent of specific antigen recognition.[33] A number of adhesion molecules (adhesins) have been described that are necessary for immune cell interactions. LFA-1 or lymphocyte-associated antigen is expressed by a variety of leukocytes. Its principal ligand, intracellular adhesion molecule-1 or ICAM-1, is expressed by a wide variety of differentiated cells. Another adhesion molecule, LFA-3, is widely distributed and interacts with the T cell specific CD2 antigen, also known as T11 or LFA-2. Both the LFA-1/ICAM1 and LFA-3/CD2 interactions are thought to be important for optimal effector-target conjugation and therefore are important accessory molecules for CTL. EBV infects epithelial cells which express MHC class I molecules and B lymphocytes which express both class I and class II molecules. These molecules are not down-regulated in EBV-infected cells. Rather EBV encodes a function that down-regulates the cell surface expression of the adhesion molecules LFA-3 and ICAM 1.[34]

A detailed analysis of Burkitt's lymphoma-derived cell lines revealed that these cell lines express different levels of adhesins LFA-3 and ICAM 1 at different *in vitro* passage stages. The early passage Burkitt lymphoma cell lines exhibit limited viral gene expression characteristic of the Latency I/II pattern of EBV-latent proteins and also express low levels of LFA-1, ICAM-1, and LFA-3. BL lines at a late passage begin to express the Latency III set of EBV-latent proteins and in addition express higher levels of these adhesion molecules. The differences in cell surface expression of these adhesins is reflected in the morphology of the different cell lines in culture. Those cells that express low levels of adhesins retain a single cell growth morphology, whereas those in which the adhesins were up-regulated, aggregate into clumps. Furthermore, functional analysis of these cell lines suggest that the level of expression of the adhesins is likely to be physiologically relevant because conjugate formation between EBV-specific CTL and BL target cells that express low levels of the adhesins is much less efficient than with BL target cells that express higher levels of the adhesins. The conjugates between targets with low levels of adhesins and CTL form via an LFA-1 pathway. Interestingly, certain EBV+ Burkitt lymphoma cell lines that exhibit low levels of surface adhesins and begin to express Latency III associated proteins remain insensitive to lysis by HLA class I matched EBV-specific CTL. Only when ICAM-1 and LFA-3 were up-regulated in these cell lines did specific lysis occur. Expression of the LFA-3 molecule is particularly important for CTL lysis of target cells. For example, one BL cell line remains resistant to lysis even through late passages, and in this line LFA-3 expression remains down-regulated. Although LFA-1 was expressed, this pathway alone was not sufficient to achieve lysis even though the appropriate viral antigen and HLA-class I complex was available for T cell recognition.[34]

Analyses of Burkitt's lymphoma and nasopharyngeal carcinoma biopsies showed alterations in the expression levels of adhesion molecules in these tumors. Burkitt's lymphoma cells expressed low levels of LFA-1 and undetectable levels of ICAM-1 and LFA-3, whereas the nasopharyngeal carcinoma biopsies showed almost undetectable levels of LFA-3 and elevated levels of ICAM-1, a complete inversion of the expression pattern of these molecules on normal epithelium.[35] Thus, a possible mechanism of tumor escape from T cell-mediated immune surveillance could occur via down-regulation of the cell surface adhesion accessory molecules.

Third, the tropism of EBV for resting B cells precludes T cell recognition due to the absence of certain cell surface molecules required for activating and differentiating effector T cells. Activation of naive T cells requires both antigen recognition via the TCR/MHC molecules and also a secondary signal mediated by ligation of the CD28 antigen on T cells with the costimulatory molecule B7 that is expressed on the B cell. This interaction signals resting T cells to produce IL-2, to proliferate, and to differentiate into effector cells.[36] Resting B cells do not express B7, and although expression of viral antigens may result in recognition by T lymphocytes, these T lymphocytes do not receive the second signal mediated through the B7/CD28 pathway and therefore remain hidden from the host immune system. Although antigen recognition by T cells induces transcription of the IL-2 gene, this alone is insufficient for T cell activation. IL-2 messenger RNA, like other cytokine mRNAs, is very unstable due to an "instability sequence" in its 3′ ends. The signal mediated by ligation of B7 with CD28 results in increased transcription of the IL-2 gene and stabilization of the IL-2 mRNA, resulting in a 100-fold increase in IL-2 production.[36] Furthermore, without the B7 costimulatory signal, T cells that recognize EBV-infected B-lymphocytes may be triggered to undergo apoptosis or become anergic, a type of immune suppression. Suppression of the host immune system could allow the genesis of EBV-induced lymphoproliferative disease characteristic of EBV-infected immunosuppressed patients.

Finally, EBV can block signal transduction that could activate a switch from latent to productive infection.[37] The B cell encounters numerous signals as it circulates from the peripheral blood through the lymphoid organs. To remain in the latent phase, the virus must block signal transduction mediated through the cell surface and triggered by factors in the changing microenvironment of the B cell. LMP2A has been implicated in this process and may be critical in maintaining the latent state by blocking signal transduction mediated by the B cell receptor that could lead to differentiating the B cell and subsequently activating the lytic phase of viral replication. This scenario is further complicated by the fact that LMP2B may down-regulate the action of LMP2A and thus the establishment of latency in resting B lymphocytes could reflect a complex mechanism involving the interplay of several viral and host factors.[37]

HHV-8 is a recently discovered virus whose DNA has been found in 70–100% of both classic and HIV-related Kaposi's sarcoma (reviewed in Offermann).[38] The genomic sequence of HHV-8 indicates that it is a member of the rhadinovirus genus of the *gamma herpesvirinae* subfamily comprising herpesvirus saimiri and herpesvirus ateles[39,40] (and references therein). HHV-8 has also been found in specimens from certain skin tumors, Castleman's disease, and a rare body cavity lymphoma associated with AIDS but has not been reproducibly found in tissues of healthy individuals. Despite the high degree of association between KS and HHV-8 DNA, it is unclear whether HHV-8 is a causative agent in the etiology of these tumors or simply preferentially establishes latency in spindle cells. The body-cavity lymphoma cells may also contain EBV in addition to HHV-8. In one well-studied BC-1 cell line, each virus was selectively activated to enter the lytic cycle by treating the cells with TPA (for EBV) or with n-butyric acid (for HHV-8).[41] In this study, it was found that the majority of the BC-1 cells contain both viral genomes, and in more than 98% of the cells, both viruses are latent. In the other 2%, one or the other of the viruses spontaneously undergoes reactivation.

HHV-8 latent infection has not been extensively characterized, but recent reports suggest that at least one RNA species and possibly more may be present in latently infected cells.[42] The immunology of HHV-8 infection is complicated by the array of cytokines released from Kaposi's sarcoma.

3. IMMUNE EVASION DURING LYTIC INFECTION

3.1. Interference with Antigenic Presentation

Current studies suggest that each of the human herpesviruses has evolved mechanisms that interfere with antigenic presentation to the host immune system. T lymphocytes are a critical component of the host defense system for eliminating viruses. The functions performed by T cells include cytolytic activity (both from $CD8^+$ and $CD4^+$ cells), release of cytokines by $CD4^+$ cells necessary for activating the cytotoxic subset of T cells, activation of B lymphocytes and subsequent production of specific antiviral antibody, and activation of nonspecific inflammatory cells, such as natural killer (NK) cells and macrophages. Specific mechanisms for blocking antigenic presentation have been elucidated for three of the human herpesviruses.

Numerous studies have shown that T cells are required for clearing HSV-1 and HSV-2 viral infections, and $CD8^+$ cells are the critical component for clearing most viruses. It has been reported that patients with T cell defects can develop severe HSV disease, but those with immunoglobulin deficiencies can control HSV infections,[43] although in some models, the antibody response may be important in control of neuroinvasiveness.[44] Early studies in mice designed to determine which subset of T cells confer protection against HSV infection yielded conflicting results. It was later realized that differences in the way the HSV antigen is administered, for primary immunization of the mice and for *in vitro* restimulation of T cells, influences the development of the T cell phenotype.[43] Later studies in mice and in humans have shown that $CD4^+$ cells are required to clear HSV infection in peripheral sites, whereas $CD8^+$ cells are required to limit spread of HSV in sensory ganglia.[5,45–47a,b] Conflicting results remain, however, as it has been reported that $CD8^+$ cells make acute herpes infection worse.[48] The effector activity of the T cells in both cases may be mediated by cytokines. Infiltrating T cells isolated from HSV-2 lesions in human patients were almost entirely $CD4^+$ and mostly devoid of cytotoxic activity, although other investigators have reported isolation of $CD4^+$ cells against HSV that were not cytotoxic.[43,49] Given the limited expression of MHC class II antigens, it appears unlikely that $CD4^+$ T cells are the primary effector cell for clearing HSV infections. Expression of both MHC class I and class II antigens, however, can be induced by IFN-γ, and also stimulation of $CD4^+$ cells could occur via class II positive antigen-presenting cells, such as macrophages or dendritic cells. It is proposed that the role of $CD8^+$ cells in restricting HSV spread in sensory ganglia rests on secretion of cytokines, namely IFN-γ.[8] Local secretion of IFN-γ up-regulated MHC class I expression in cells surrounding the latently infected neurons, but neurons themselves were refractory in that class I mRNA was detected, but not class I antigens.[6] Thus, the neuronal cells may be protected even if surrounding cells are destroyed by the host, but whether the virus plays any role in preventing up-regulation of MHC expression is unknown.

Some *in vitro* studies have shown that human fibroblasts and keratinocytes infected with HSV are resistant to lysis by $CD8^+$ cytotoxic T lymphocytes (CTL).[50,51] These T cells recognize endogenously synthesized antigenic peptides presented on the surface of infected cells by MHC class I molecules. Recent studies *in vitro* have shown that expression of MHC class I molecules at the surface of HSV-infected cells is probably down-regulated through two mechanisms. At early times post infection, a virion-associated protein (vhs) destabilizes host mRNAs resulting in lack of expression of most host proteins. At later time following synthesis of viral proteins, direct interference by a viral protein in the transport of MHC class I molecules has been reported.[52] Recent studies have shown that the infected cell protein No. 47 (ICP47) of HSV-1 interacts directly with proteins that mediate transport of endogenously synthesized peptide antigens into the ER, i.e., the transporters associated with antigen processing, or TAP proteins 1 and 2.[52–54] In cells infected with wild-type virus, cell surface expression of MHC class I molecules containing peptide antigen is precluded thereby avoiding recognition by $CD8^+$ T cells. Whether this interaction of viral protein with class I molecules occurs *in vivo* is unknown. No virally encoded function has been identified that results in interfering with recognition by $CD4^+$ T cells of HSV-infected cells, although it has been reported that MHC class II antigens, induced in nonneuronal cells of mouse brains infected with HSV-1, were retained intracellularly depending on the strain of virus.[55] An additional HSV-encoded activity conferring resistance to $CD8^+$ T cell lysis has been described, but neither the viral gene products involved in this activity nor the mechanism by which they act are known.[56]

HCMV carries several genes that function to interfere with antigenic presentation by MHC class I molecules. Early studies focused on the UL18 gene. The HCMV UL18 gene encodes a homologue to the human MHC class I heavy chain which binds to human β2-microglobulin.[57,58] Because infection of certain cell types with HCMV results in almost a complete loss of cell surface expression by MHC class I proteins, it was postulated that the HCMV class I homologue competes for binding to the light chain of the heterodimer, β-2 microglobulin.[58] However, although the UL18 protein can bind human β2-microglobulin, cellular class I protein can also be found complexed with β2-microglobulin in infected cells,[59] and therefore the HCMV homologue does not completely sequester the host β2-microglobulin protein. Furthermore, deletion of the UL18 gene did not restore surface expression of MHC class I molecules.[60] The picture is further complicated by conflicting results reported from different laboratories. The degree to which MHC class I surface antigenic expression decreases in CMV-infected cells varies among cell lines, possibly because of the multiplicity of viruses used, and in addition, infection with either live or UV-irradiated HCMV can result in an enhancement of cell-surface MHC class I expression.[59,61–63] In any event, the UL18 gene plays no role in modulating MHC class I expression. However, results from recent studies suggest that the viral MHC class I homologue may serve a different purpose for viral escape from immune attack.[64a,b]

In recent years, a number of studies have shed considerable light on both the mechanisms involved in down-regulating MHC class I expression in HCMV-infected cells and on identifying the viral genes involved in these processes. The U_s region of HCMV contains several genes whose products interfere either with transport of antigenic peptides into the ER[65] or with maturation or stability of newly synthesized class I molecules.[66–69,71,72] The U_s2 and U_s11 proteins act by reversing the normal process of translocation of MHC class I heavy chains. U_s2 binds directly to the class I heavy chain and along with U_s11 promotes retrograde transport of newly synthesized MHC class I heavy-chain molecules from the ER into the cytosol via the Sec 61 complex, where they are subsequently deglycosylated and degraded by the cytosolic proteasome.[67,70–72] U_s3 acts in concert with U_s2 and U_s11 by interfering with transport and maturation out of the ER of MHC class I molecules bound to β2-microglobulin.[68,69]

Gene expression during the HCMV nonreplicative phase is not as restricted as that of HSV, and although lack of MHC class I antigenic expression presumably diminishes the host's ability to recognize and kill HCMV-infected cells, peripheral blood cells of healthy seropositive patients contain cytotoxic T cells that recognize HCMV-infected targets.[73] Certain HCMV proteins are exempt from T cell recognition. For example, attempts to generate cytotoxic T cells specific for the HCMV IE protein by stimulating peripheral blood cells *in vitro* were unsuccessful, and the precursor frequency of T cell clones that recognized this antigen were 50–100 times lower than those that recognized other HCMV antigens. These authors suggested that HCMV can selectively inhibit presentation of certain viral antigens.[74] Pretreatment of infected target cells with IFN-γ rendered the infected cells susceptible to lysis by T cells. Because loss of T cell cytotoxicity in immunocompromised hosts is associated with severe HCMV disease, the virus may succeed in partially evading the host T cell cytotoxic response, but certainly these cells are instrumental in controlling HCMV infection. Results obtained *in vitro* cannot be easily extrapolated to those that occur during a natural infection. However, inasmuch as HCMV 1) carries several genes that perform specific functions which result in interference of presentation of viral antigens to cytotoxic T cells, 2) persists in its host for very long periods and, 3) apparently does not establish latency in any immune privileged site, it seems likely that the virus has evolved specific mechanisms for escape from T cell immunity. Then the virus can attenuate host responses so that it can persist, yet the host is able to surmount HCMV disease except in the immunocompromised host.

As mentioned earlier, the EBV EBNA 1 gene encodes a function that is reported to interfere with antigenic presentation at an even earlier step by inhibiting proteolytic processing of viral proteins. Although EBNA 1 is generally expressed in association with latent EBV infection, it is the only EBNA gene that continues to be expressed during lytic infection. Thus this property of the EBNA 1 protein may be even more critical for immune escape in infected B cells that have begun to undergo differentiation.

3.2. Interference with Innate Defense Mechanisms

At the initial stage of infection, the host responds to invading pathogens with a nonadaptive response followed by activation of specific effector cells. The nonspecific innate response occurs within minutes of the infection and is mediated by inflammatory cells, macrophages, natural killer (NK) cells, and the complement pathway. The development of the adaptive, specific response to infection requires several days. Several studies have addressed the role of nonspecific effector cells, such as macrophages and NK cells, in herpesvirus infections. It is reported that herpesvirus infection activates macrophages via induction of interferon and TNFα and that TNFα plays a role in recruiting monocytes into the inflammatory site[75] (and references therein). *In vitro* studies have shown that HSV inhibits the production of factors from monocytes essential for stimulating T cells,[76] but again it is not known whether herpesviruses inhibit the activity of monocytes or macrophages *in vivo*. Although it is reported that NK cell activity in experimental animal systems is important in controlling herpes virus infections,[43,64a,b,77] it is difficult to ascertain the relative importance of NK activity in humans because selective deficiency in NK cells is rare. However, both NK and IFN-γ activity are elevated in HHV-6 infected patients during the acute phase of illness,[78] and there are two documented reports of patients lacking NK cells who were extremely susceptible to herpesvirus infections. One patient who was susceptible to HSV, CMV, and VZV, showed no killer-cell activity, nor could it be induced in her peripheral blood mononuclear cells.[79] The other report describes three siblings, one of whom died of an overwhelming EBV infection. The other two were also susceptible to EBV and showed markedly deficient NK activity and no induction of NK activity with IFN-γ.[80] The most direct evidence to date for viral escape from NK cell killing stems from recent studies of HCMV. The viral MHC class I homologue encoded by UL18 delivers an inhibitory signal to NK cells, so that HCMV-infected cells are resistant to NK-mediated lysis *in vitro*.[64a] In animal models, deletion of the murine CMV class I homologue attenuates the virus, and the host resistance is mediated by NK cells.[64b]

The role of γ/δ T cells in antiviral immunity is poorly understood. However, it has been found that this subset of T cells expands in patients with certain viral infections.[81–85] Most studies involving the role of T cells in herpesvirus immunity have focused entirely on T cells expressing the classical α/β T cell receptor, but in recent years, more attention has been given to γ/δ T cells.

Analyses of peripheral blood lymphocytes from HSV-infected patients showed that T cells with cytotoxic activity expressing either α/β or γ/δ T cell receptors can be isolated but the prevalence of the γ/δ T cells varied widely from patient to patient.[86] Other studies of human γ/δ cells showed that these cells lyse HSV-infected targets and also targets infected with other viruses. These authors proposed that the target antigen recognized by the T cells is of cellular origin, either induced or modified by viral infection.[87] One virally encoded protein has been identified as the antigen recognized by a murine T cell clone expressing the γ/δ T cell receptor.[88] This clone recognizes an epitope encoded in the glycoprotein I of HSV-1 in an MHC class I or II unrestricted manner and the recognition is independent of the known antigen-processing pathways.[89] Recent *in vivo* studies showed that in two mouse models, mice that genetically lacked γ/δ T cell receptors and α/β T cell receptors and $TCR\alpha^{-/-}$ mice treated with anti-$TCR\gamma/\delta$ monoclonal antibodies, infected with HSV-1 by foot pad or ocular routes of infection, $TCR\gamma\delta$ cells reduced morbidity and prevented the development of lethal viral encephalitis.[90] Although in lymphoid organs the $\gamma\delta$ T cells comprise a small percentage of the total T cells, they are quite prominent in certain epithelial tissues, particularly epidermis, and could conceivably play a role in the immune response to HSV, particularly at the earliest stages of infection.

3.3. Infection of Cells of the Immune System

The beta and gamma herpesviruses infect cells of the immune system as an integral part of their life cycle. The alpha herpesviruses do not normally replicate in cells of the immune system. However, numerous studies have shown that under certain conditions *in vitro,* T cells, macrophages, and NK cells can be infected with HSV and this infection generally results in inhibiting effector cell activity.

HSV-infected endothelial cells are resistant to lysis by NK or lymphokine-activated killer (KAK) cells *in vitro,* but following exposure to infected target cells, these killer cells are no longer able to lyse normally susceptible targets.[91] The mechanism of NK cell resistance in HSV-infected cells is not understood. Inhibition of NK and lymphokine-activated killer (LAK) activity requires contact with the infected target cells and may depend on the presence of an HSV-encoded glycoprotein.[92] These results raised the hypothesis that the inhibition of NK activity results from infection of the effector cells by the virus. The proximity of the membranes of NK or LAK cells to those of the target cells may allow entry of the virus into these cells through tight cell junctions. Resistance to NK activity is a phenomenon that occurs late in the HSV life cycle. At an earlier stage of infection, cells are susceptible to NK-mediated lysis.[91] It is not known which viral gene products are responsible for inhibition of these effector cells. However, it does not appear to be related to the general host shut-off phenomena mediated by the virion host shutoff (*vhs*) gene.[92] Similarly, infection and inhibition of T cell activity by HSV does not depend on activity of the virion host shut-off protein but involves cell–cell spread of a virus.[93,94] Although it is not known whether inhibition of immune cells plays any role in HSV immune escape *in vivo,* lytic infection of effector cells is a potential strategy by which viral replication may be enhanced at early stages of primary infection and may result in establishing latency in greater numbers of cells.

Although HHV-6 and 7 infect primarily CD4$^+$ T lymphocytes, CD8$^+$ T cells, γ/δ T cells, and NK cells are permissive for HHV-6 infection, particularly the variant-A strain.[95,96] By infecting immune cells of both innate and adaptive immune responses, these two viruses have the potential to impair the host defense system seriously.[97] HHV-7 selectively infects CD4$^+$ cells, and the CD4 molecule is required for virus entry[18] whereas HHV-6 can infect CD4$^+$, CD8$^+$, γ/δ T cells, NK cells, *in vitro* immortalized B cells, and mononuclear phagocytes, albeit unproductively, although HHV-6 has been detected in circulating monocytes of convalescent children[24] and in tissue macrophages derived from patients with HHV-6 associated pneumonitis.[17,24,95,96,98–100] The cytolytic activity of CD4$^+$ cells infected with HHV-6 or 7 and signal transduction mediated through the CD3 molecule are impaired possibly as a consequence of the changes in cell surface expression of certain molecules.[101] These two viruses exert differential effects in CD4$^+$ cells. Infection with HHV-6 resulted in a decrease in the T cell receptor-associated molecule CD3 which was pronounced in cells infected with HHV-6 variant-A strain and slight in cells infected with HHV-6 variant-B strain, whereas HHV-7 infection caused a marked decrease in expression of the CD4 molecule.[101]

Infection of NK cells also inhibits their effector activity because NK clones susceptible to HHV-6 cannot kill autologous HHV-6 infected targets, although these NK cells can kill other target cells. Conversely, NK cells capable of lysing HHV-6 infected targets could not be infected with HHV-6, suggesting perhaps that interaction of the NK cells with the infected target cell induces resistance to HHV-6.[95,102,103] Exposure of PBMCs to HHV-6 resulted in producing IL-15 and subsequent increased cytotoxicity by NK cells in this population of cells.[103] It is possible that a subpopulation of NK cells that cannot be stimulated with IL-15 are permissive for HHV-6 infection or conversely, that stimulation of NK cells with IL-15 induces resistance to infection with HHV-6.

In vitro studies showed that γ/δ cells isolated from human peripheral blood cells could be infected with HHV-6 but not HHV-7.[96] HHV-6 infection induces expression of the CD4 molecule in γ/δ cells and in α/β CD8$^+$ cells and NK cells and makes these cells susceptible to infection with HIV.[17,95,96] Induction of the CD4 molecule in additional populations of cells may be of critical importance in persons infected with HIV. Furthermore, HHV-6 can coinfect CD4$^+$ cells along with HIV, and the presence of HHV-6 appears to accelerate the kinetics of HIV replication.[104,105]

3.4. Role of the Complement Pathway

Several of the herpesviruses encode molecules that interfere directly or indirectly with the complement pathway. The glycoprotein C of both HSV-1 and HSV-2 binds to the C3b component *in vitro*.[106–109] However, although the HSV-1 infected cells bind C3b, HSV-2 infected cells do not.[109] One of the domains within glycoprotein C has some homology to the C3b receptor CR1.[110] However, several domains are necessary for binding to C3b.[11] gC-1 but not gC-2 can destabilize C3 convertase by inhibiting the binding of properdin to C3b.[112] Recent studies have shown that both gC-1 and gC-2 bind to native C3, and that in contrast to what was previously thought, gC-1 does not act as an analog to cellular CR1 because the region of C3 important for binding of C3b to CR1 is not important for the gC–C3 interaction. Results from this study suggested that gC-1 inhibits complement activation by inhibition of properdin and C5 with C3.[113]

A second mechanism thought to block the classical complement pathway is utilized by HSV, CMV, and VZV. These viruses express Fc receptors, a low-affinity receptor encoded by the viral glycoprotein E and a high-affinity receptor formed as a complex of two viral glycoproteins, gE and gI.[3] Some *in vitro* studies suggest that the low-affinity receptor binds polymeric or aggregated IgG, whereas the high-affinity receptor binds monomeric IgG.[3] One hypothesis suggests that Fc receptors may function in immune escape *in vivo* by forming bipolar bridging if a virus-specific antibody binds to a viral antigen via the Fab region while the Fc protein of the antibody binds the virally encoded Fc receptor.[114] This would potentially interfere with both antibody and complement-mediated destruction of the infected cell, of the virus particle, and with antibody-dependent cellular cytotoxicity.[114,115]

Thus HSV encodes two functions that may interfere with complement-mediated lysis. Sequestering of either nonimmune or virus-specific antibody via the viral Fc receptor could prevent interaction with the C1q component of the classical pathway and block the complement cascade. The most important function of the complement component C3b is opsonization of invading pathogens to facilitate their uptake and subsequent destruction by phagocytic cells. Binding C3b is a critical event in complement activation because it activates the alternative complement pathway, and at this point the main effector activities of both complement pathways are generated. Other possible mechanisms for virus neutralization by complement include coating virus particles to reduce infectivity and direct viral lysis. Avoidance of complement-mediated destruction may enable the virus to initiate replication in enough cells such that it can then establish latent infection. It is not yet clear how important blockage of the complement pathway or of ADCC are for viral clearance *in vivo*. Natural viral isolates generally contain abundant amounts of gC. However some gC-mutants have been isolated. In one case, virus could be inactivated by complement, and infected cells were susceptible to complement-mediated lysis whereas a gC+ revertant virus was resistant.[116] However, recent studies indicate that nonimmune serum from HSV-naive individuals can neutralize a gC-negative virus and that HSV therefore can activate the classical complement pathway in the absence of antibody which may be important in the early phase of infection.[117] A complicating issue in assessing the relevance of these phenomena *in vivo* is that these viral proteins have important functions in the viral life cycle (gC for attachment to some cell types and gE and gI in cell–cell spread), and therefore it would be necessary to create mutations in these proteins that do not interfere with these functions before assessing their role in immune escape. EBV utilizes the receptor for the C3d component of the complement pathway [the CD21 molecule (formerly named CR2)], but whether this plays any role in immune escape is unknown.[118] EBV also encodes a protein that interferes with the alternative complement pathway.[119]

3.5. Role of Cytokines

The role of antiviral cytokines in resolving herpesvirus infection is not well understood. However, results obtained from animal studies suggest that γ-interferon is a critical component of the host defense against HSV infection. It has been reported that IFN-γ secretion from T cells is critical for viral clearance from the skin[120] and is also important for limiting the virus spread in the nervous system, probably by focusing antiviral cytokines at sites of reactivation. Clearance of HSV in ganglia and brain stem was impaired in mice that lacked the gene for the IFN-γ receptor and in mice treated with antibody to IFN-γ.[8] Both *in vitro* and *in vivo* studies have shown that exogenous IFN-γ induces expression of MHC class I antigens in HSV-infected cells.

Type I interferons α and β can exhibit direct antiviral effects on infected cells. The HSV-1 encoded $\gamma_1 34.5$ protein counteracts the effect of β interferon induced by the presence of viral double-stranded RNA molecules. In the presence of double-stranded RNA, the cellular protein kinase PKR phosphorylates the α subunit of the eukaryotic translation initiation factor eIF2 resulting in a block in translation. $\gamma_1 34.5$ binds to the cellular phosphatase I protein which counteracts the phosphorylation of eIF2-α and restores protein synthesis in infected cells.[121]

In recent years the discovery that viruses encode homologues of cytokines that can modulate host immune responses and receptors that can block the action of host cytokines has elicited considerable interest. Among the human beta and gamma herpesviruses, there are several examples of molecular mimicry. The EBV BCRF1 gene encodes a homologue of IL-10 that inhibits IFN-γ production by T helper cells and NK cells.[122,123] This activity mimics that of cellular IL-10 which is thought to suppress at least certain T cell and NK cell responses.[124]

IL-10 can also act as a growth factor for B lymphocytes and as such has complex immunomodulatory effects. In addition, EBV-transformed B lymphocytes produce IL-12 and a protein encoded by the EBI3 locus that is related to the p40 chain of IL-12.[125] This molecule interacts with the p35 chain of IL-12 to produce a secreted IL-12-like protein. This novel protein does not inhibit binding of natural IL-12 to its receptor and its function is still unknown. IL-2 exhibits a variety of immunomodulatory functions including stimulating the growth of T and NK cells, promoting secretion of IFN-γ, and differentiating uncommitted T cells (TH-0) the TH-1 phenotype. HHV-6 inhibits IL-12 synthesis and cellular proliferation. Conversely, HSV infection in the mouse eye induces expression of IL-12 and culminates in a pathogenic response due to the activity of activated TH-1 cells.[126]

HHV-8 encodes a homolog of IL-6.[40] Because IL-6 helps to activate both T and B cells thereby enhancing the host's adaptive immune response, it appears to be deleterious to a virus to encode a homologue of IL-6. However, IL-6 is also thought to be a growth factor for spindle cells which are the predominant and characteristic cell type of Kaposi's sarcoma.[38] Whether the IL-6 homologue plays any role in immune destruction of HHV-8 or in evasion of host immunity is unknown.

Another area in which the herpesviruses practice molecular mimicry is in carrying genes that encode homologues of chemokine receptors.[127] The chemokines are a family of closely related, small polypeptides that generally act as chemoattractants for phagocytic cells and attract macrophages and neutrophils from the blood to infection sites. These proteins are synthesized in a variety of cells, including macrophages, endothelial cells, skin keratinocytes, fibroblasts, and smooth muscle cells of connective tissue.[128] The HHV-6 U83 gene exhibits some features of a cellular chemokine. It is predicted that it is a 10-KDa basic protein with a dicysteine "C–C" motif and is secreted.[128,129] HHV-8 carries two genes, K4 and K6 that, it is predicted, encode homologues of the cellular chemokines vMIPI and vMIPII, respectively.[130,131] Chemokines activate G proteins, and the chemokine receptors are similar to G protein-coupled receptors (GCRs) in that both types of receptors possess seven membrane-spanning domains. HHV-8 encodes a GCR that is homologous to the cellular IL-8 receptor. IL-8, a member of the CXC family of cytokines, binds the HHV-8 GCR.[132] This receptor binds other chemokines of both the CXC and CC families. It is of interest that the HHV-6 U12 and U51 genes are similar to the HCMV-encoded GCRs UL33, U_s27, and U_s28 and to the lymphocyte-specific G protein-coupled peptide receptor induced by EBV, EBI 1.[129,133,134] The U_s28 encoded GCR is a functional receptor.[135] HHV-7 carries two genes that, it is predicted, encode GCRs, the U12 and U51 genes.[136] HHV-6 and 7 also induce the EBI 1 gene in $CD4^+$ cells.[137]

3.6. Interference with Host Suicide Mechanisms

Viral infections frequently activate intrinsic pathways for programmed cell death. Considering that the herpesviruses carry genes whose products shut off host protein synthesis, thus diverting host functions and resources to support viral replication, it is not surprising that these viruses also carry genes that function to prevent the cell from dying prematurely, that is, before completing viral replication, assembly, and egress. HSV encodes at least two functions to preclude programmed cell death. The first of these deals with a problem occurring in cells infected with nearly all viruses: synthesis of antisense transcripts which stimulate the double-stranded RNA-activated protein kinase (PKR).[138] Activated PKR phosphorylates the α subunit of the translation initiation factor eIF-2. In consequence, all protein synthesis ceases.[139] As a rule viruses express a function to block the activation of PKR. The strategy of HSV differs from known mechanisms in that PKR does become phosphorylated, but a protein encoded by HSV-$\gamma_1$34.5 binds to protein phosphatase 1α and redirects its activity to the dephosphorylation of eIF-2α.[121]

The second common response to viral infection is classical apoptosis resulting in changes in the morphology of the cells and degradation of chromosomal DNA. HSV, EBV, and HCMV have all been reported to express functions that block apoptosis. In recent studies HSV-1 mutants lacking the gene encoding ICP4, the major regulatory protein, have been shown to induce apoptosis[140] and additional studies suggest that ICP4 is required but may not be sufficient to block it (Leopardi and Roizman, unpublished results). Of interest are the observations that HSV also blocks apoptosis induced by osmotic or thermal shock in at least some cell lines in culture (Galvan-Girado, Leopardi, and Roizman, unpublished results). These results suggest that apoptosis induced by infection with HSV shares pathways induced by other factors.

HCMV encodes two proteins IE1 and IE2 which prevent apoptosis, the mechanism of which is presently unknown.[141] VZV induces apoptosis in cells in which the virus can establish lytic infection. Nerve cells infected with VZV do not undergo apoptosis, but no virally encoded protective function has been described.[142] EBV expresses at least two functions that protect against apoptosis. The EBV BHRF1 gene encodes a homologue of the antiapoptosis protein bcl-2, and the virus also up-regulates the expression of cellular bcl-2 via the EBV-encoded protein LMP1.[143–145] BHRF1 is present in all virus isolates but is not essential *in vitro* for transformation or replication. BHRF1 is not likely to play a role during latency because it is never expressed, but it is abundantly expressed as an early protein during the EBV lytic cycle, so that although ultimately EBV infection is cytocidal, prevention of apoptosis may prolong cell life and thereby maximize viral replication. LMP1 has also been implicated in mediating protecting against apoptosis. EBV-positive Burkitt's lymphoma cell clones expressing

EBNA-1 could be protected against apoptosis induced by Ca^{2+} ionophores or serum starvation if the Latency III pattern of gene expression were activated. The protective function has been attributed to three latency-associated proteins, LMP1, EBNA 2, and EBNA 4. All three of these proteins up-regulate bcl-2 expression,[143,145–147] raising the possibility that during the normal course of infection, all three of these proteins may act synergistically to avoid programmed cell death and allow the virus to persist in the memory B cell population. Paradoxically, it has also been reported that overexpression of LMP1 induces apoptosis in cell culture. This finding led to the suggestion that cells undergoing transformation require a signal to replicate continuously and also a signal to inhibit apoptosis.[148] HHV-8 also carries a gene homologous to bcl-2 (ORF16), but its role in protecting against apoptosis is unknown.[131]

4. SUMMARY AND CONCLUSIONS

Human herpesviruses have evolved a wide variety of strategies to evade or obstruct host immune defenses. The key features highlighted in this chapter about herpesvirus interactions with the host immune system are summarized in Table I. Although some similarities exist among the different viruses, the mechanisms employed by each virus clearly reflect the unique biological properties of the individual viruses. Common to all the herpesviruses is the ability to establish a lifetime latent infection. The limited viral gene expression and the type of cell in which latency is established create a privileged site in which the virus can reside for very long periods of time. Current knowledge suggests that all of the human herpesviruses establish latency either in cells which are immune-privileged, such as HSV infection of neuronal cells that lack expression of MHC class I or class II molecules, or in cells which are themselves components of the immune system and in which the virus can directly modulate host responses (e.g., infection of B cells by EBV and T cells by HHV-6 and 7).

Interference with host immune responses during the lytic phase of replication is multifaceted and encompasses both innate and adaptive immune responses. Several of the herpesviruses encode molecules that interfere with the complement pathway. At least two viral homologues of cellular immunomodulatory cytokines and a number of G protein-coupled chemokine receptors have been identified. Other examples are predicted on the basis of genomic sequence data. HSV, CMV, and EBV encode functions that enable escape from T lymphocyte-mediated killing. These mechanisms involve interference with antigenic processing, presentation to effector cells, and modulation of effector cell activity. At least HSV, CMV, and EBV have in addition evolved means to prevent defense mechanisms that may be indirectly immune-mediated, namely, apoptosis or protein synthesis shutoff. Finally, all of the human herpesviruses are associated with

TABLE I
Immune Evasion Strategies Utilized by Human Herpes Viruses

Evasion strategy	HSV	HCMV	EBV	HHV-6,7	HHV-8
Latency	Sensory neurons	Monocytes	B-cells	Unknown	Unknown
Block antigen presentation	α47, U_L41 (vhs)	U_S2, U_S3, U_S11	EBNA 1		
Modulation of immune cells accessory molecules			LFA 3 ↓[a] ICAM 1 ↓[a]	HHV-6:CD3 ↓ CD3 ↓, CD4 ↑ HHV-7:CD4 ↓	
Block humoral response	U_S7, U_S8 (Fc recept.) U_L44 (gC)	Fc receptor	[a]		
Block nonspecific killer cells	NK-cells monocytes[a]	UL18: NK-cells		NK, monocytes[a]	
Infection of immune cells		Monocytes	B-cells	HHV-6: CD4+, CD8+, γ/δ, NK,	

				B-cells, monocytes HHV-7: CD4+	
Molecular mimicry		UL18: MHC class I U_S27, (GCR) U_S28, (GCR) UL33: (GCR)	BCRF1 (IL-10) BHRF1 (bcl-2) EBI-3 (IL-12 p40 homolog)	HHV-6: U83 HHV-7 U12, U51	K2 (vIL-6) K4 (vMIPI) K6 (vMIPII) ORF 74 GPCR (IL-8R homolog ORF16 (bcl-2)
Modulation of host immune response proteins			EBI-1 ↑ (GCR) IL-12 ↑	HHV-6: IL-15 ↓ HHV-6,7: EBI-1 ↓	
Apoptosis	α4, US3	IE1, IE2	BHRF-1 LMP1 EBNA 2, 4		
Block protein synthesis shut off by activated PKR	$\gamma_1 34.5$				

[a]Viral genes involved have not been identified.

some measure of immunosuppression, either indirectly or by direct infection of cells that are part of the immune system.

The combination of immediate innate defense mechanisms and adaptive, specific immune responses presents a formidable barrier to invading viruses. Although animal models generally are employed to identify a critical component in the immune response against the virus, current knowledge supports the idea that virtually every aspect of the host defense system plays some role, even if minor, in contributing to the balance between viral infection and host defense. These viruses usually do not kill their hosts, and in immunocompetent individuals, infection with any of the herpesviruses may have extremely discomforting albeit not life-threatening consequences. The severe consequences that occur in immunocompromised individuals, however, clearly illustrate the critical role of the complex host immune response. The capture of certain cellular genes by the viruses to be evolutionarily modified for optimal immune escape provides a fascinating glimpse of the adaptability of viruses to coexist within a quite hostile environment.

ACKNOWLEDGMENTS. When not employed in writing chapters, our research is supported by grants from the National Cancer Institute (CA47451), the United States Public Health Service.

REFERENCES

1. Banks, T. A., and Rouse, B. T., 1992, Herpesviruses-immune escape artists?, *Clin. Inf. Dis.* **14:**933–941.
2. Roizman, B., 1996, Herpesviridae, in: *Fields Virology,* 3rd ed., Volume 2 (B. N. Fields, D. M. Knipe, P. Howley, R. M. Chanock, M. S. Hirsch, J. L. Melnick, T. P. Monath, and B. Roizman, eds.), Lippincott-Raven, Philadelphia, pp. 2221–2230.
3. Roizman, B., and Sears, A., 1996, Herpes simplex viruses and their replication, in: *Fields Virology,* 3rd ed., Volume 2 (B. N. Fields, D. M. Knipe, P. Howley, R. M. Chanock, M. S. Hirsch, J. L. Melnick, T. P. Monath, and B. Roizman, eds.), Lippincott-Raven, Philadelphia, pp. 2231–2295.
4. Cohen, J. I., and Straus, S. E., 1996, Varicella-zoster virus and its replication, in: *Fields Virology,* 3rd ed., Volume 2 (B. N. Fields, D. M. Knipe, P. Howley, R. M. Chanock, M. S. Hirsch, J. L. Melnick, T. P. Monath, and B. Roizman, eds.), Lippincott-Raven, Philadelphia, pp. 2525–2546.
5. Simmons, A., and Tscharke, D. C., 1992, Anti-CD8 impairs clearance of herpes simplex virus from the nervous system: Implications for the fate of virally infected neurons, *J. Exp. Med.* **175:**1337–1344.
6. Pereira, R. A., Tscharke, D. C., Simmons, A., 1994, Upregulation of class I major histocompatibility complex gene expression in primary sensory neurons, satellite cells, and Schwann cells of mice in response to acute but not latent herpes simplex virus infection in vivo, *J. Exp. Med.* **180:**841–850.

7. Weinstein, D. L., Walker, D. G., Akiyama, H., and McGeer, P. L., 1990, Herpes simplex virus type I infection of the CNS induces major histocompatibility complex antigen expression on rat microglia, *J. Neurosci. Res.* **26:**55–65.
8. Cantin, E. M., Hinton, D. R., Chen, J., and Openshaw, H., 1995, Gamma interferon expression during acute and latent nervous system infection by herpes simplex virus type 1, *J. Virol.* **69:**4898–4905.
9. Croen, K. D., Ostrove, J. M., Dragovic, L. J., and Straus, S. E., 1988, Patterns of gene expression and sites of latency in human nerve ganglia are different for varicella-zoster and herpes simplex viruses, *Proc. Natl. Acad. Sci. USA* **85:**9773–9777.
10. Taylor-Wiedeman, J., Sissons, J. G. P., Borysiewicz, L. K., and Sinclair, J. H., 1991, Monocytes are a major site of persistence of human cytomegalovirus in peripheral blood mononuclear cells, *J. Gen. Virol.* **72:**2059–2064.
11. Britt, W. J., and Alford, C. A., 1996, Cytomegalovirus in: *Fields Virology,* 3rd ed., Volume 2 (B. N. Fields, D. M. Knipe, P. Howley, R. M. Chanock, M. S. Hirsch, J. L. Melnick, T. P. Monath, and B. Roizman, eds.), Lippincott-Raven, Philadelphia, pp. 2493–2523.
12. Jordan, M. C., 1983, Latent infection and the elusive cytomegalovirus, *Rev. Infect. Dis.* **5:**205–215.
13. Schrier, R. D., Nelson, J. A., Oldstone, M. B. A., 1985, Detection of human cytomegalovirus in peripheral blood lymphocytes in a natural infection, *Science* **230:**1048–1051.
14. Saltzman, R. L., Quirk, M. R., Jordan, M. C., 1988, Disseminated cytomegalovirus infection. Molecular analysis of virus and leukocyte interactions in viremia, *J. Clin. Invest.* **81:**75–81.
15. Taylor-Wiedeman, J., Sissons, P., and Sinclair, J., 1994, Induction of endogenous human cytomegalovirus gene expression after differentiation of monocytes form healthy carriers, *J. Virol.* **68:**1597–1604.
16. Pellett, P. E., and Black, J. B., 1996, Human herpesvirus 6, in: *Fields Virology,* 3rd ed., Volume 2 (B. N. Fields, D. M. Knipe, P. Howley, R. M. Chanock, M. S. Hirsch, J. L. Melnick, T. P. Monath, and B. Roizman, eds.), Lippincott-Raven, Philadelphia, pp. 2587–2608.
17. Lusso, P., Malnati, M., De Maria, A., De Rocco, S., Markham, P. D., and Gallo, R. C., 1991, Productive infection of $CD4^+$ and $CD8^+$ mature human T cell populations and clones by human herpesvirus 6. Transcriptional down-regulation of CD3, *J. Immunol.* **147:**685–691.
18. Lusso, P., Secchier, P., Crowley, R. W., Garzino-Demo, A., Berneman, Z. N., and Gallo, R. C., 1994, CD4 is a critical component of the receptor for human herpesvirus 7: Interference with human immunodeficiency virus, *Proc. Natl. Acad. Sci. USA* **91:**3872–3876.
19. Frenkel, N., Schirmer, E. C., Wyatt, L. S., Katsafanas, G., Roffman, E., Danovich, R. M., and June C. H., 1990, Isolation of a new herpesvirus from human $CD4^+$ T cells, *Proc. Natl. Acad. Sci. USA* **87:**748–752.
20. Wyatt, L. S., and Frenkel, N., 1992, Human herpesvirus 7 is a constitutive inhabitant of adult human saliva, *J. Virol.* **66:**3206–3209.
21. Berneman, Z. N., Ablashi, D. V., Li, G., Eger-Fletcher, M., Reitz, M. S., Hung, C. L., Brus, I., Komaroff, A. L., and Gallo, R. C., 1992, Human herpesvirus 7 is a T-lymphotropic virus and is related to, but significantly different from human herpesvirus 6 and human cytomegalovirus, *Proc. Natl. Acad. Sci. USA* **89:**10552–10556.
22. Frenkel, N., and Roffman, E., 1996, Human herpesvirus 7, in: *Fields Virology,* 3rd ed., Volume 2 (B. N. Fields, D. M. Knipe, P. Howley, R. M. Chanock, M. S. Hirsch, J. L. Melnick, T. P. Monath, and B. Roizman, eds.), Lippincott-Raven, Philadelphia, pp. 2609–2635.
23. Yoshikawa, T., Nakashima, T., Asano, Y., Suga, S., Yazaki, T., Kojima, S., Mukai, T., and Yamanishi, K., 1992, Endonuclease analyses of DNA of human herpesvirus-6 isolated from blood before and after bone marrow transplantation, *J. Med. Virol.* **37:**228–231.
24. Kondo, K., Kondo, T., Okuno, T., Takahashi, M., and Yamanishi, K., 1991, Latent human herpesvirus 6 infection of human monocytes/macrophages, *J. Gen. Virol.* **72:**1401–1408.

25. He, J., McCarthy, M., Zhou, Y., Chandran, B., and Wood, C., 1996, Infection of primary human fetal astrocytes by human herpesvirus 6, *J. Virol.* **70:**1296–1300.
26. Suga, S., Yoshikawa, T., Asano, Y., Nakashima, T., Yazaki, T., Fukuda, M., Kojima, S., Matsuyama, T., Ono, Y., and Oshima, S., 1992, IgM neutralizing antibody responses to human herpesvirus-6 in patients with exanthem subitum or organ transplantation, *Microbiol. Immunol.* **36:**495–506.
27. Masucci, M. G., and Ernberg, I., 1994, Epstein–Barr virus: Adaptation to a life within the immune system, *Trends Microbiol.* **2:**125–130.
28. Thorley-Lawson, D. A., Miyashita, E. M., and Khan, G., 1996, Epstein–Barr virus and the B cell: That's all it takes, *Trends Microbiol.* **4:**204–208.
29. Kieff, E., 1996, Epstein–Barr virus and its replication, in: *Fields Virology,* 3rd ed., Volume 2 (B. N. Fields, D. M. Knipe, P. Howley, R. M. Chanock, M. S. Hirsch, J. L. Melnick, T. P. Monath, and B. Roizman, eds.), Lippincott-Raven, Philadelphia, pp. 2343–2396.
30. Rickinson, A. B., and Kieff, E., 1996, Epstein–Barr virus, in: *Fields Virology,* 3rd ed., Volume 2 (B. N. Fields, D. M. Knipe, P. Howley, R. M. Chanock, M. S. Hirsch, J. L. Melnick, T. P. Monath, and B. Roizman, eds.), Lippincott-Raven, Philadelphia, pp. 2397–2446.
31. Decker, L. L., Klaman, L. D., and Thorley-Lawson, D. A., 1996, Detection of the latent form of Epstein–Barr virus DNA in the peripheral blood of healthy individuals, *J. Virol.* **70:**3286–3289.
32. Levitskaya, J., Coram, M., Levitsky, V., Imreh, S., Steigerwald-Mullen, P. M., Klein, G., Kurilla, M. G., and Masucci, M. G., 1995, Inhibition of antigen processing by the internal repeat region of the Epstein–Barr virus nuclear antigen-1, *Nature* **375:**685–688.
33. Springer, T. A., 1990, Adhesion receptors of the immune system, *Nature* **346:**425–434.
34. Gregory, C. D., Murray, R. J., Edwards, C. F., and Rickinson, A. B., 1988, Down-regulation of cell adhesion molecules LFA-3 and ICAM-1 in Epstein–Barr virus-positive Burkitt's lymphoma underlies tumor cell escape from virus-specific T cell surveillance, *J. Exp. Med.* **167:**1811–1824.
35. Busson, P., Zhang, Q., Fuillon, M.-M., Gregory, C. D., Young, L. S., Clausse, B., Lipinski, M., Rickinson, A. B., and Tursz, T., 1992, Elevated expression of ICAM1 (CD54) and minimal expression of LFA3 (CD58) in Epstein–Barr virus-positive nasopharyngeal carcinaoma cells, *Int. J. Cancer* **50:**863–867.
36. Janeway, C. A. Jr., and Travers, P., 1994, T cell-mediated immunity in: *Immunobiology; The Immune System in Health and Disease* (M. Robertson, ed.), Current Biology/Garland Publishing, New York and London, pp. 1–49.
37. Longnecker, R., and Miller, C. L., 1996, Regulation of Epstein–Barr virus latency by latent membrane protein 2, *Trends Microbiol.* **4:**38–42.
38. Offermann, M. K., 1996, HHV-8: a new herpesvirus associated with Kaposi's sarcoma, *Trends in Microbiol.* **4:**383–385.
39. Moore, P. S., Gao, S. J., Dominguez, G., Cesarman, E., Lungu, O., Knowles, D. M., Garber, R., Pellett, P. E., McGeoch, J. J., and Chang, Y., 1996, Primary characterization of a herpesvirus agent associated with Kaposi's sarcoma, *J. Virol.* **70:**549–558.
40. Neipel, F., Albrecht, J.-C., Ensser, A., Huang, Y.-Q., Li, J., Friedman-Kien, A. E., and Fleckenstein, B., 1997, Human herpesvirus 8 encodes a homolog of interleukin-6, *J. Virol.* **71:**839–842.
41. Miller, G., Heston, L., Grogan, E., Gradoville, L., Rigsby, M., Sun, R., Shedd, D., Kishnaryov, V. M., Grossberg, S., and Chang, Y., 1997, Selective switch between latency and lytic replication of Kaposi's sarcoma herpesvirus and Epstein–Barr virus in dually infected body cavity lymphoma cells, *J. Virol.* **71:**314–324.
42. Staskus, K. A., Zhong, W., Gebhard, K., Herndier, B., Wang, H., Renne, R., Beneke, J., Pudney, J., Anderson, D. J., Ganem, D., and Haase, A. T., 1997, Kaposi's sarcoma-associated herpesvirus gene expression in endothelial (spindle) tumor cells, *J. Virol.* **71:**715–719.

43. Schmid, D. S., and Rouse, B. T., 1992, The role of T cell immunity in control of herpes simplex virus, *Curr. Top. Microbiol.* **179:**57–74.
44. Mitchell, B. M., and Stevens, J. G., 1996, Neuroinvasive properties of herpes simplex virus type 1 glycoprotein variants are controlled by the immune response, *J. Immunol.* **156:**246–255.
45. Nash, A. A., Jayasuriya, A., Phelan, J., Cobbold, S. P., Waldmann, H., and Prospero, T., 1987, Different roles for L3T4$^+$ and Lyt 2$^+$ T cell subsets in the control of an acute herpes simplex virus infection of the skin and nervous system, *J. Gen. Virol.* **68:**825–833.
46. Nash, A. A., Phelan, J., and Wildy, P., 1981, Cell-mediated immunity in herpes simplex virus-infected mice: H-2 mapping of the delayed-type hypersensitivity response and the antiviral T-cell response, *J. Immunol.* **126:**1260–1262.
47a. Manickan, E., and Rouse, B. T., 1995, Roles of different T-cell subsets in control of herpes simplex virus infection determined by using T cell-deficient mouse models, *J. Virol.* **69:**8178–8179.
47b. Manickan, E., Rouse, R. J. D., Yu, Z., Wire, W. S., and Rouse, B. T., 1995, Genetic immunization against herpes simplex virus: Protection is mediated by CD4$^+$ T lymphocytes, *J. Immunol.* **155:**259–265.
48. Ikemoto, K., Pollare, R. B., Fukumoto, T., Morimatsu, M., and Suzuki, F., 1995, CD8$^+$ type-2 T cells enhance the severity of acute herpes virus infection in mice, *Immunol. Lett.* **47:**63–72.
49. Koelle, D. M., Corey, L., Burke, R., Eisenberg, R. J., Cohen, G. H., Pichyangkura, R., and Triezenberg, S. J., 1994, Antigenic specificities of human CD4$^+$ T cell clones recovered from recurrent genital herpes simplex virus type 2 lesions, *J. Virol.* **68:**2803–2810.
50. Koelle, D. M., Tigges, M. A., Burke, R. L., Symington, F. M., Riddell, S. R., Abbo, H., and Corey, L., 1993, Herpes simplex virus infection of human fibroblasts and keratinocytes inhibits recognition by cloned CD8$^+$ cytotoxic T lymphocytes, *J. Clin. Invest.* **91:**961–968.
51. Tigges, M. A., Leng, S., Johnson, D. C., and Burke, R., 1996, Human herpes simplex virus (HSV)-specific CD8$^+$ CTL clones recognize HSV-2-infected fibroblasts after treatment with IFN-γ or when virion host shutoff functions are disabled, *J. Immunol.* **156:**3901–3910.
52. York, I. A., Roop, C., Andrews, D. W., Riddell, S. R., Graham, F. L., and Johnson, D. C., 1994, A cytosolic herpes simplex virus protein inhibits antigen presentation to CD8$^+$ T-lymphocytes, *Cell* **77:**525–535.
53. Fruh, K., Ahn, K., Djaballah, H., Sempe, P., Van Endert, P. M., Tampe, R., Peterson, P. A., and Yang, Y., 1995, A viral inhibitor of peptide transporters for antigen presentation, *Nature* **375:**415–418.
54. Hill, A., Jugovic, P., York, I., Russ, G., Bennink, J., Yewdell, J., Pleogh, H., and Johnson, D., 1995, Herpes simplex virus turns off the TAP to evade host immunity, *Nature* **375:**411–415.
55. Lewandowski, G. A., Lo, D., and Bloom, F. E., 1993, Interference with major histocompatibility complex class II-restricted antigen presentation in the brain by herpes simplex virus type 1: A possible mechanism of evasion of the immune response, *Proc. Natl. Acad. Sci. USA* **90:**2005–2009.
56. Horohov, D. W., Wyckoff, H. H. III, Moore, R. N., and Rouse, B. T., 1986, Regulation of herpes simplex virus-specific cell-mediated immunity by a specific suppressor factor, *J. Virol.* **58:**331–338.
57. Beck, S., and Barrell, G. B., 1988, Human cytomegalovirus encodes a glycoprotein homologous to MHC class-I antigens, *Nature* **331:**269–272.
58. Browne, H., Smith, G., Beck, S., and Minson, T., 1990, A complex between the MHC class I homologue encoded by human cytomegalovirus and β2 microglobulin, *Nature* **347:**770–772.
59. Yamashita, Y., Shimokata, K., Mizuno, S., Yamaguchi, H., and Nishiyama, Y., 1993, Down-regulation of the surface expression of class I MHC antigens by human cytomegalovirus, *Virology* **193:**727–736.
60. Browne, H., Churcher, M., and Minson, T., 1992, Construction and characterization of a

human cytomegalovirus mutant with the UL18 (class I homolog) gene deleted, *J. Virol.* **66:** 6784–6787.

61. Grundy, J. E., Ayles, H. M., McKeating, J. A., Butcher, R. G., Griffiths, P. D., and Poulter, L. W., 1988, Enhancement of class I HLA antigen expression by cytomegalovirus: Role in amplification of virus infection, *J. Med. Virol.* **25:**483–495.
62. van Dorp, W. T., Jonges, E., Bruggeman, C. A., Daha, M. R., van Es, L. A., and van der Woude, F. J., 1989, Direct induction of MHC class I but not class II expression on endothelial cells by cytomegalovirus infection, *Transplantation* **48:**469–472.
63. Hosenpud, J. D., Chou, S., and Wagner, C. R., 1991, Cytomegalovirus-induced regulation of major histocompatibility complex class I antigen expression in human aortic smooth muscle cells, *Transplantation* **52:**896–903.
64a. Reyburn, H. T., Mandelboim, O., Vales-Gomez, M., Davis, D. M., Pazmany, L., and Strominger, J. L., 1997, The class I MHC homologue of human cytomegalovirus inhibits attack by natural killer cells, *Nature* **386:**514–517.
64b. Farrell, H. E., Vally, H., Lynch, D. M., Fleming, P., Shellam, G. R., Scalzo, A. A., and Davis-Poynter, N. J., 1997, Inhibition of natural killer cells by a cytomegalovirus MHC class I homologue in vivo, *Nature* **386:**510–516.
65. Hengel, H., Flohr, T., Hammerling, G. J., Koszinowski, U. H., and Momburg, F., 1996, Human cytomegalovirus inhibits peptide translocation into the endoplasmic reticulum for MHC class I assembly, *J. Gen. Virol.* **77:**2287–2296.
66. Jones, T. R., Hanson, L. K., Sun, L., Slater, J. S., Stenberg, R. M., and Campbell, A. E., 1995, Multiple independent loci within the human cytomegalovirus unique short region down-regulate expression of major histocompatibility complex class I heavy chains, *J. Virol.* **69:**4830–4831.
67. Wiertz, E. J. H. J., Jones, T. R., Sun, L., Bogyo, M., Geuze, H. J., and Ploegh, H. L., 1996, The human cytomegalovirus U_s11 gene product dislocates MHC class I heavy chains from the endoplasmic reticulum to the cytosol, *Cell* **84:**769–779.
68. Ahn, K., Angulo, A., Ghazal, P., Peterson, P. A., Yang, Y., and Fruh, K., 1996, Human cytomegalovirus inhibits antigen presentation by a sequential multistep process, *Proc. Natl. Acad. Sci. USA* **93:**10990–10005.
69. Jones, T. R., Wiertz, E. J. H. J., Sun, L., Fish, K. N., Nelson, J. A., and Ploegh, H. L., 1996, Human cytomegalovirus U_s3 impairs transport and maturation of major histocompatibility complex class I heavy chains, *Proc. Natl. Acad. Sci. USA* **93:**11327–11333.
70. Beersma, M. F. C., Vijlmakers, M. J. E., and Ploegh, H. L., 1993, Human cytomegalovirus down-regulates HLA class I expression by reducing the stability of class I H chains, *J. Immunol.* **151:**4455–4464.
71. Wiertz, E. J. H. J., Tortorella, D., Bogyo, M., Yu, J., Mothes, W., Jones, T. R., Rapoport, T. A., Nd Ploegh, H. L., 1996, Sec61-mediated transfer of a membrane protein from the endoplasmic reticulum to the proteasome for destruction, *Nature* **384:**432–438.
72. Jones, T. R., and Sun, L., 1997, Human cytomegalovirus U_s2 destabilizes major histocompatibility complex class I heavy chains, *J. Virol.* **71:**2970–2979.
73. Mocarski, E. S., Jr., 1996, Cytomegaloviruses and their replication, in: *Fields Virology*, 3rd ed., Volume 2 (B. N. Fields, D. M. Knipe, P. Howley, R. M. Chanock, M. S. Hirsch, J. L. Melnick, T. P. Monath, and B. Roizman, eds.), Lippincott-Raven, Philadelphia, pp. 2447–2492.
74. Gilbert, M. J., Riddell, S. R., Li, C.-R., and Greenberg, P. D., 1993, Selective interference with class I major histocompatibility complex presentation of the major immediate-early protein following infection with human cytomegalovirus, *J. Virol.* **67:**3461–3469.
75. Heise, M. T., and Virgin, H. W., IV, 1995, The T cell-independent role of gamma interferon and tumor necrosis factor alpha in macrophage activation during murine cytomegalovirus and herpes simplex virus infections, *J. Virol.* **69:**904–909.

76. Hayward, A. R., Read, G. S., and Cosyns, M., 1993, Herpes simplex virus interferes with monocyte accessory cell function, *J. Immunol.* **150:**190–196.
77. Kohl, S., 1990, Protection against murine neonatal herpes simplex virus infection by lymphokine-treated human leukocytes, *J. Immunol.* **144:**307–312.
78. Takahashi, K. E., Segal, T., Kondo, T., Mukai, M., Moriyama, M., Takahashi, M., and Yamanishi, K., 1992, Interferon and natural killer cell activity in patients with exanthem subitum, *J. Pediatr. Inf. Dis.* **11:**369–373.
79. Biron, C. A., Byron, K. S., and Sullivan, J. L., 1989, Severe herpesvirus infections in an adolescent without natural killer cells, *N. Eng. J. Med.* **320:**1731–1735.
80. Fleisher, G., Star, S., Koven, N., Kamiya, H., Douglas, S. D., and Henle, W., 1982, A non-X-linked syndrome with susceptibility to severe Epstein–Barr virus infections, *J. Pediatr.* **100:**727–730.
81. Carding, S. R., Allan, W., Kyes, S., Hayday, A., Bottomly, K., and Doherty, P. C., 1990, Late dominance of the inflammatory process in murine influenza by γδ T cells, *J. Exp. Med.* **177:**1225–1231.
82. De Paoli, P., Gennari, D., Martelli, P., Cavarzerani, V., Comoretto, R., and Santini, G., 1990, γδ T cell receptor-bearing lymphocytes during Epstein–Barr virus infection, *J. Infect. Dis.* **161:**1013–1016.
83. De Paoli, P., Gennari, D., Martelli, P., Basaglia, G., Crovatto, M., Battistin, S., and Santini, G., 1991, A subset of γδ lymphocytes is increased during HIV-1 infection, *Clin. Exp. Immunol.* **83:**187–191.
84. Hou, S., Katz, J. M., Doherty, P. C., and Carding, S. R., 1992, Extent of γδ T cell involvement in the pneumonia caused by Sendai virus, *Cell Immunol.* **143:**183–193.
85. Maccario, R., Comoli, P., Percivalle, E., Montagna, D., Locatelli, F., and Gerna, G., 1995, Herpes simplex virus-specific human cytotoxic T cell colonies expressing either γδ or αβ T cell receptor: Role of accessory molecules on HLA-unrestricted killing of virus-infected targets, *Immunology* **85:**49–56.
86. Maccario, R., Revello, M., Comoli, P., Montagna, D., Locatelli, F., and Gerna, G., 1993, HLA-unrestricted killing of HSV-1-infected mononuclear cells, *J. Immunol.* **150:**1437–1445.
87. Bukowski, J. F., Morita, C. T., and Brenner, M. B., 1994, Recognition and destruction of virus-infected cells by human γδ CTL, *J. Immunol.* **153:**5133–5140.
88. Johnson, R. M., Lancki, D. W., Sperling, A. I., Dick, R. F., Spear, P. G., Fitch, F. W., and Bluestone, J. A., 1992, A murine $CD4^-$, $CD8^-$ T cell receptor-γδ T lymphocyte clone specific for herpes simplex virus glycoprotein I, *J. Immunol.* **148:**983–988.
89. Sciammas, R., Johnson, R. M., Sperling, A. I., Brady, W., Linsley, P. S., Spear, P. G., Fitch, F. K., and Bluestone, J. A., 1994, Unique antigen recognition by a Herpesvirus-specific TCR-γδ cell, *J. Immunol.* **152:**5392–5397.
90. Sciammas, R., Kodukula, P., Tang, Q., Smith, L., Hendricks, R. L., Bluestone, J. A., 1997, T cell receptor-gamma/delta cells protect mice from herpes simplex type 1-induced lethal encephalitis, *J. Exp. Med.* **185:**1969–1975.
91. Confer, D. L., Vercellotti, G. M., Kotasek, S., Goodman, J. L., Ochoa, A., and Jacob, H. S., 1990, Herpes simplex virus-infected cells disarm killer lymphocytes, *Proc. Natl. Acad. Sci. USA* **87:**3609–3613.
92. York, I. A., and Johnson, D. C., 1993, Direct contact with herpes simplex virus-infected cells results in inhibition of lymphokine-activated killer cells because of cell-to-cell spread of virus, *J. Inf. Dis.* **168:**1127–1132.
93. Posavad, C. M., Newton, J. J., and Rosenthal, K. L., 1993, Inhibition of human CTL-mediated lysis by fibroblasts infected with herpes simplex virus, *J. Immunol.* **151:**4865–4873.
94. Posavad, C. M., Newton, J. J., and Rosenthal, K. L., 1994, Infection and inhibition of human cytotoxic T lymphocytes by herpes simplex virus, *J. Virol.* **68:**4072–4074.

95. Lusso, P., Malnati, M. S., Garzino-Demo, A., Crowley, R. W., Long, E. O., and Gallo, R. C., 1993, Infection of natural killer cells by human herpesvirus 6, *Nature* **362:**458–462.
96. Lusso, P., Garzino-Demo, A., Crowley, R. W., and Malnati, M. S., 1995, Infection of γ/δ T lymphocytes by human herpesvirus 6: Transcriptional induction of CD4 and susceptibility to HIV infection, *J. Exp. Med.* **181:**1303–1310.
97. Lusso, P., and Gallo, R. C., 1995, Human herpesvirus 6 in AIDS, *Immunol. Today* **16:**67 – 71.
98. Carrigan, D. R., Drobyski, W. R., Russler, S. K., Tapper, M. A., Knox, K. K., and Ash, R. C., 1991, Interstitial pneumonitis associated with human herpesvirus-6 infection after marrow transplantation, *Lancet* **338:**147–149.
99. Lusso, P., Markham, P. D., Tschachler, E., Di Marzo-Veronese, F., Salahuddin, S. Z., Ablashi, D. V., Pahwa, S., Krohn, K., and Gallo, R. C., 1988, In vitro cellular tropism of human B-lymphotropic virus (human herpesvirus-6), *J. Exp. Med.* **167:**1659–1670.
100. Ablashi, D. V., Lusso, P., Hung, C., Salahuddin, S. Z., Josephs, S. F., Llana, T., Kramarsky, B., Biberfeld, P., Markham, P. D., and Gallo, R. C., 1988, Utilization of human hematopoetic cell lines for the propagation, detection, and characterization of HBLV (human herpesvirus 6) *Int. J. Cancer* **42:**787–791.
101. Furukawa, M., Yasukawa, M., Yakushijin, Y., and Fujita, S., 1994, Distinct effects of human herpesvirus 6 and human herpesvirus 7 on surface molecule expression and function of $CD4^+$ T-cells, *J. Immunol.* **152:**5768–5775.
102. Malnati, M. S., Lusso, P., Ciccone, E., Moretta, L., and Long, E. O., 1993, Recognition of virus-infected cells by natural killer cell clones is controlled by polymorphic target cell elements, *J. Exp. Med.* **178:**961–969.
103. Flamand, L., Stefanescu, I., and Menezes, J., 1996, Human herpesvirus-6 enhances natural killer cell cytotoxicity via IL-15, *J. Clin. Invest.* **97:**1373–1381.
104. Lusso, P., Ensoli, B., Markham, P. D., Ablashi, D. V., Salahuddin, S. Z., Tschachler, E., Wong-Staal, F., and Gallo, R. C., 1989, Productive dual infection of human $CD4^+$ T lymphocytes by HIV-1 and HHV-6, *Nature* **337:**370–373.
105. Ensoli, B., Lusso, P., Schachter, F., Joseph, S. F., Rappaport, J., Negro, F., Gallo, R. C., Wong-Staal, F., 1989, Human herpes virus-6 increases HIV-1 expression in co-infected T cells via nuclear factors binding to the HIV-1 enhancer, *EMBO J.* **8:**3019–3027.
106. Friedman, H. M., Cohen, G. H., Eisenberg, R. J., Seidel, C. A., Cines, D. B., 1984, Glycoprotein C of herpes simplex virus 1 acts as a receptor for the C3b complement component on infected cells, *Nature* **309:**633–635.
107. McNearney, T. A., Odell, C., Holers, V. M., Spear, P. G., and Atkinson, J. P., 1987, Herpes simplex virus glycoproteins gC-1 and gC-2 bind to the third component of complement and provide protection against complement-mediated neutralization of viral infectivity, *J. Exp. Med.* **166:**1525–1535.
108. Seidel-Dugan, C., Ponce de Leon, M., Friedman, H. M., Fries, L. F., Frank, M. M., Cohen, G. H., and Eisenberg, R. J. 1988, C3b receptor activity on transfected cells expressing glycoprotein C of herpes simplex virus types 1 and 2, *J. Virol.* **62:**4027–4036.
109. Eisenberg, R. J., Ponce de Leon, M., Friedman, H. M., Fries, L. F., Frank, M. M., Hastings, J. C., and Cohen, G. H., 1987, Complement component C3b binds directly to purified glycoprotein C of herpes simplex virus type 1 and 2, *Microb. Pathog.* **3:**423–435.
110. Kubota, Y., Gaither, T. A., Cason, J., O'Shea, J. J., and Lawley, T. J., 1987, Characterization of the C3 receptor induced by herpes simplex virus type 1 infection of human epidermal, endothelial, and A431 cells, *J. Immunol.* **138:**1137–1142.
111. Hung, S.-L., Srinivasan, S., Friedman, H. M., Eisenberg, R. J., and Cohen, G. H., 1992, Structural basis of C3b binding by glycoprotein C of herpes simplex virus, *J. Virol.* **66:**4013–4027.
112. Hung, S.-L., Peng, C., Kostavasili, I., Friedman, H. M., Lambris, J. D., Eisenberg, R. J., and

Cohen, G. H., 1994, The interaction of glycoprotein C of herpes simplex virus types 1 and 2 with the alternative complement pathway, *Virology* **203:**299–312.
113. Kostavasili, J., Sahu, A., Friedman, H. M., Eisenberg, R. J., Cohen, G. H., and Lambris, J. D., 1997, Mechanism of complement inactivation by glycoprotein C of herpes simplex virus, *J. Immunol.* **258:**1763–1771.
114. Frank, I., and Friedman, H. M., 1989, A novel function of the herpes simplex virus type 1 Fc receptor: Participation in bipolar bridging of antiviral immunoglobulin G, *J. Virol.* **63:**4479–4488.
115. Dubin, G., Socolof, E., Frank, I., and Friedman, H. M., 1991, Herpes simplex virus type 1 Fc receptor protects infected cells from antibody-dependent cellular cytotoxicity, *J. Virol.* **65:**7046–7050.
116. Hidaka, Y., Sakai, Y., Toh, Y., and Mori, R., 1991, Glycoprotein C of herpes simplex virus inactivation and lysis of virus-infected cells, *J. Gen. Virol.* **72:**915–921.
117. Friedman, H. M., Wang, L., Fishman, N. O., Lambris, J. D., Eisenberg, R. J., Cohen, G. H., and Lubinski, J., 1996, Immune evasion properties of herpes simplex virus type 1 glycoprotein gC, *J. Virol.* **70:**4253–4260.
118. Nemerow, G. R., Houghton, R. A., Moore, M. D., and Cooper, N. R., 1989, Identification of an epitope in the major envelope protein of Epstein–Barr virus that mediates viral binding to the B lymphocyte EBV receptor (CR2), *Cell* **56:**369–377.
119. Mold, C., Bradt, B. M., Nemerow, G. R., and Cooper, N. R., 1989, Epstein–Barr virus regulates activation and processing of the third component of complement, *J. Exp. Med.* **168:**949–969.
120. Smith, P. M., Wolcott, R. M., Chervenak, R., and Jennings, S. R., 1994, Control of acute cutaneous herpes simplex virus infection: T cell-mediated viral clearance is dependent upon interferon-γ (IFN-γ), *Virology* **202:**76–88.
121. He, B., Gross, M., and Roizman, B., 1997, The $\gamma_1$34.5 protein of herpes simplex virus 1 complexes with protein phosphatase 1α to dephosphorylate the α subunit of the eukaryotic translation initiation factor 2 and preclude the shutoff of protein synthesis by double-stranded RNA-activated protein kinase, *Proc. Nat. Acad. Sci. USA* **94:**843–848.
122. Hsu, D. H., DeWaal-Malefyt, R., Fiorentino, D. F., Dang, M.-N., Vieira, P., DeVries, J., Spits, H., Mosmann, T. R., and Moore, K. W., 1990, Expression of interleukin-10 activity by Epstein–Barr virus protein BCRF1, *Science* **250:**830–832.
123. Vieira, P., DeWaal-Malefyt, R., Dang, M. N., Johnson, K. E., Kastelein, R., Fiorentino, D. F., DeVries, J. E., Roncarolo, M.-G., Mosmann, T. R., and Moore, K. W., 1991, Isolation and expression of human cytokine synthesis inhibitory factor cDNA clones: Homology to Epstein–Barr virus open reading frame BCRF1, *Proc. Natl. Acad. Sci. USA* **88:**1172–1776.
124. Fiorentino, D. F., Zlotnik, A., Vieira, P., Mosmann, T. R., Howard, M., Moore, K. W., and O'Garra, A., 1991, IL-10 acts on the antigen presenting cell to inhibit cytokine production by Th1 cells, *J. Immunol.* **146:**3444–2451.
125. Devergne, O., Humme, M., Koeppen, H., Le Beau, M. M., Nathanson, E. C., Kieff, E., and Birkenbach, M., 1996, A novel interleukin-12 p40-related protein induced by latent Epstein–Barr virus infection B lymphocytes, *J. Virol.* **70:**1143–1153.
126. Kanangat, S., Thomas, J., Gangappa, S., Babu, J. S., and Rouse, B. T., 1996, Herpes simplex virus type 1-mediated up-regulation of IL-12 (p40) mRNA expression, *J. Imunol.* **156:**1110–1116.
127. Murphy, P. M., 1994, Molecular piracy of chemokine receptors by herpesvirus, *Inf. Agents Dis.* **3:**137–154.
128. Oppenheim, J. J., Zachariae, C. O. C., Mukaida, N., and Matsushima, K., 1991, Properties of the novel proinflammatory supergene "intercrine" cytokine family, *Ann. Rev. Immunol.* **9:**617–648.
129. Gompels, U. A., Nicholas, J., Lawrence, G., Jones, M., Thomson, B. J., Martin, M. E. D., Efstathiou, S., Craxton, M. and Macaulay, H. A., 1995, The DNA sequence of human herpesvirus-6: Structure, coding content, and genome evolution, *Virology* **209:**29–51.

130. Moore, P. S., Boshoff, C., Weiss, R. A., and Chang, Y., 1996, Molecular mimicry of human cytokine and cytokine response pathway genes by KSHV, *Science* **274:**1739–1744.
131. Russo, J. J., Bohenzky, R. A., Chien, M.-C., Chen, J., Yan, M., Maddalena, D., Parry, J. P., Peruzzi, D., Edelman, I. S., Chang, Y., and Moore, P. S., 1996, Nucleotide sequence of the Kaposi sarcoma-associated herpesvirus (HHV8), *Proc. Natl. Acad. Sci. USA* **93:**14862–14867.
132. Arvanitakis, L., Geras-Raaka, E., Varma, A., Gershengorn, M. C., and Cesarman, E., 1997, Human herpesvirus KSHV encodes a constitutively active G-protein-coupled receptor linked to cell proliferation, *Nature* **385:**347–350.
133. Chee, M. S., Bankier, A. T., Beck, S., Bohni, R., Brown, C. M., Cerney, R., Horsnell, T., Hutchinson, C. A., Kouzarides, T., Martignetti, J. A., Preddie, E., Satchwell, S. C., Tomlinson, P., Weston, K. M., and Barrell, B. G., 1990, Analysis of the protein coding content of the sequence of human cytomegalovirus strain AD169, *Curr. Top. Microbiol. Immunol.* **154:**125–169.
134. Birkenbach, M., Josefsen, K., Yalamanchili, R., Lenoir, G., and Kieff, E., 1993, Epstein–Barr virus-induced genes: First lymphocyte-specific G protein-coupled peptide receptors, *J. Virol.* **67:**2209–2220.
135. Gao, J. L., and Murphy, P. M., 1994, Human cytomegalovirus open reading frame U_s28 encodes a functional beta chemokine receptor, *J. Biol. Chem.* **269:**28539–28542.
136. Nicholas, J., 1996, Determination and analysis of the complete nucleotide sequence of human herpesvirus 7, *J. Virol.* **70:**5975–5989.
137. Hasegawa, H., Utsunomiya, U., Yasukawa, M., Yanagisawa, K., and Fujita, S., 1994, Induction of G protein-coupled receptor EBI 1 by human herpesvirus 6 and 7 infection in $CD4^+$ T-cells, *J. Virol.* **68:**5326–5329.
138. Proud, C. G., 1995, PKR: A new name and new roles, *Trends Biochem. Sci.* **20:**241–246.
139. Sarre, T. F., 1989, The phosphorylation of eukaryotic initiation factor 2: A principle of translational control in mammalian cells, *BioSystems* **22:**311–325.
140. Leopardi, R., and Roizman, B., 1996, The herpes simplex virus major regulatory protein ICP4 blocks apoptosis induced by the virus or by hyperthermia, *Proc. Natl. Acad. Sci. USA* **93:**9583–9587.
141. Zhu, H., Shen, Y., and Shenk, T., 1995, Human cytomegalovirus IE1 and IE2 proteins block apoptosis, *J. Virol.* **69:**7960–7970.
142. Sadzot-Delvaux, C., Thonard, P., Schoonbroodt, S., Piette, J., and Rentier, B., 1995, Varicella-zoster virus induces apoptosis in cell culture, *J. Gen. Virol.* **76:**2875–2879.
143. Henderson, S., Rowe, M., Gregory, C., Croom-Carter, D., Wang, F., Longnecker, R., Kieff, E., and Rickinson, A., 1991, Induction of bcl-2 expression by Epstein–Barr virus latent membrane protein 1 protects infected B cells from programmed cell death, *Cell* **65:**1107–1115.
144. Henderson, S., Huen, D., Rowe, M., Dawson, C., Johnson, G., and Rickinson, A., 1993, Epstein–Barr virus-coded BHRF1 protein, a viral homologue of bcl-2, protects human B cells from programmed cell death, *Proc. Natl. Acad. Sci. USA* **90:**8479–8483.
145. Rowe, M., Peng-Pilon, M., Huen, D. S., Hardy, R., Croom-Carter, D., Lundgren, E., and Rickinison, A. B., 1994, Up-regulation of bcl-2 by the Epstein–Barr virus latent membrane protein LMP1: a B-cell-specific response that is delayed relative to NF-κB activation and to induction of cell surface markers, *J. Virol.* **68:**5602–5612.
146. Finke, J., Fritzen, R., Ternes, P., Trivedi, P., Bross, K. J., Lange, W., Mertelsmann, R., and Dolken, G., 1992, Expression of bcl-2 in Burkitt's lymphoma cell lines: Induction by latent Epstein–Barr virus genes, *Blood* **80:**459–469.
147. Silins, S. L., and Sculley, T. B., 1995, Burkitt's lymphoma cells are resistant to programmed cell death in the presence of the Epstein–Barr virus latent antigen EBNA-4, *Int. J. Cancer* **60:**65–72.
148. Lu, J.J.-Y., Chen, J.-Y., Hsu, T.-Y., Ye, W. C. Y., Su, I.-J., and Yang, C.-S., 1996, Induction of apoptosis in epithelial cells by Epstein–Barr virus latent membrane protein 1, *J. Gen. Virol.* **77:**1883–1892.

2

Immunopathology of Herpesvirus Infections

BARRY T. ROUSE and SALLY S. ATHERTON

1. INTRODUCTION

Herpesviruses are among the most prevalent human viral infections. Many of the nine known members infect the majority of the population yet they fail to cause epidemics which periodically devastate the population. By several measures herpesviruses must be considered highly successful pathogens, and even the most remote human tribes are infected. Characteristically the production of progeny virus is irreversibly destructive to the infected cell, but such episodes of cell destruction are usually short-lived and are successfully controlled by a normal immune system.[1] The key to the success of herpesviruses is that the cytodestructive phase is indefinitely in the host. Persistence takes the form of limited gene expression in target cells and this special relationship is usually called latency.[1] Latency is best understood in members of the alpha herpesvirus and gamma herpesvirus subfamilies. The former includes the herpes simplex virus (HSV) types and varicella-zoster virus (VZV). These agents establish latency mainly in nervous tissue, and the prolonged relationship has no functional effects on latently infected cells. The majority of disease expression with alpha viruses occurs during primary or recurrent episodes of productive infection. Lesions, especially in the early phase, result largely from viral[14] induced cell destruction. In later phases, however, the

BARRY T. ROUSE • Department of Microbiology, University of Tennessee, Knoxville, Tennessee 37996-0845. SALLY S. ATHERTON • Department of Cellular and Structural Biology, University of Texas Health Sciences Center at San Antonio, San Antonio, Texas 78284-7762.

Herpesviruses and Immunity, edited by Medveczky *et al.* Plenum Press, New York, 1998.

inflammatory response involves immune-mediated mechanisms, and these may cause significant functional damage in certain organs (see section 3).

With Epstein–Barr viruses (EBV), the gamma herpesviruses, latency is maintained mainly in B lymphocytes, and this can take a number of different type of virus–cell relationships.[3] In one form, usually called latency type III, limited gene expression causes cell proliferation, and B cells more or less take on a behavior pattern that occurs physiologically in their response to antigenic stimulation.[3] Latency type III can involve a large number of B cells following primary infection, and these signal the induction of a reactive $CD8^+$ T cell response to them. The affected patient expresses the signs of infectious mononucleosis, a disease whose pathogenesis involves an immunopathological mechanism. Latency types I and II, which result from the induction of viral genes from the usage of primary promoter different than occurs during latency III gene expression, does not cause spontaneously proliferating B cells or CDB^+ T Cell recognition.[3] Long-term persistence by EBV is maintained by memory B cells expressing type I and type II latency transcripts of, and evidence for lesions, immunopathological or otherwise, is lacking (see section 5).

The beta herpesvirus subfamily includes human cytomegalovirus along with some newly recognized lymphotropic herpesviruses, such as HHV 6 and 7 which colonize T-lymphocytes.[4,5] Of the beta viruses, only HHV 6 produces disease in immunocompetent adults. The other viruses, best exemplified by HCMV, act as pathogens in individuals who have immature or compromised immune systems, particularly those affecting T lymphocyte function.[6] A completed replication cycle by HCMV results in stimulation of host cell DNA, RENA, and protein synthesis unlike the inhibitory effects characteristic of most other herpesviruses. Our discussion does not further consider beta herpesviruses because the pathogenesis of such agents, at least in humans, is unlikely to include elements of immunopathology.

2. IMMUNOPATHOLOGICAL MECHANISMS AGAINST VIRUSES

All exogenous viruses which systemically infect their animal hosts induce an immune response of some kind. At one time it was thought that infections by some agents during immunological immaturity could result in immunological tolerance.[7] The role model for such a situation was lymphocytic choriomeningitis virus (LCMV). Certainly, infection of the fetus with LCMV results in the absence of detectable $CD8^+$ T-cell immunity, but as shown in the seminal paper by Oldstone and Dixon,[8] mice develop antibodies to the persistent viral infection. Moreover complexes of viral antigens and antibodies are found in the circulation and these may become entrapped in various tissues, activate complement, and cause inflammatory reactions. The commonest sites for this so-called immune complex disease are the glomerulus and blood vessels of the joints, skin, and the choroid plexus.[9] The LCMV model stands as the first example where it was shown that the pathogenesis of a viral infection has an immunopathological mechanism.[10]

Immune complex disease occurs only where persistence of large amounts of viral antigens occurs in the presence of an antibody response. This is an uncommon scenario with human virus infections, but it can occur with chronic hepatitis B viral infection.[11] In this case, some have fluted complexes from kidney tissue and shown that a viral antigen is present.[11] Immune complex disease is not suspect during herpesvirus infections presumably because the form of persistence which such viruses adopt does not produce an abundance of cell-free viral antigens to form pathologic complexes with antibodies. Some older reports advocate that some rare manifestations of HSV infections, such as erythema multiforme, represent an immune complex disease,[12] but this notion awaits confirmation.

An additional mechanism by which antibodies interact with viruses to provide immunopathological tissue damage is illustrated by dengue hemorrhagic fever/dengue shock syndrome (DHF/DSS) in humans and a disease in cats called feline infectious peritonitis. In both situations, antibodies facilitate the entry of a virus into target cells which in turn are floridly activated to produce an abundance of cytokines.[13,14] This so-called cytokine storm damages blood vessels and other structures. The pathogenesis of the severe, but fortunately rare disease, DHF/DES is incompletely understood, but it occurs only when individuals are infected who already have heterologous nonprotecting antibodies resulting either from infections with other viral strains or received passively in utero.[13] These antibodies enhance infection of macrophages and these cells generate an abundance of cytokines including tissue-damaging TNFα and activators of other cell types, such as $CD4^+$ and $CD8^+$ T cells which in turn contribute to the cytokine storm. An excellent summary of the likely pathogenesis of DHF/DSS was written by Kurane and Ennis.[13]

Various investigators have advocated that viruses might cause immunopathological disease by acting as molecular mimics for self-proteins.[15,16] The theory is that the antibodies induced act as autoantibodies which somehow participate in autoinflammatory responses. Despite some determined searching, no acceptable candidate virus has emerged to demonstrate such a mechanism. Clearly this is the case with the known human herpesviruses.

The more common immunopathological reactions against viruses involve T cells. The usual condition is persistence of a mildly or noncytopathic virus and an enduring T cell-mediated response that is slow or ineffective in removing the virally infected cells. Once again the role model and still best studied example is provided by LCMV. With this infection $CD8^+$ T cells play the principal role in both protection and immunopathology.[17] Their function becomes immunopathological where there is a high frequency of CD8+ cytotoxic T lymphocytes (CAL) in the presence of a large number of infected cells, particularly in delicate tissues, such as the choriomeninges. Rapid cell destruction by CD8+ T cells occurs, perhaps exploiting the mechanism of lysis as commonly measured *in vivo* by the ^{51}Cr release assay.[18] However, exactly how CAL functions *in viva* to subserve either protective or immunopathological effects is an unresolved issue. The alternatives include lysis, apoptosis, and the release of protective or proinflammatory cytokines or other molecules by antigen-stimulated CD8+ T cell-mediated immunopathology is infectious mononucleosis caused by EBV infection (discussed in section 5).

A viral infection model that has helped reveal possible cellular and molecular mechanisms in play during the pathogenesis of a CD8+ T cell-mediated immunopathology has been the study of mice expressing hepatitis B virus as a transgene.[19] Such mice remain normal unless given adoptive transfers of CD8+ T cells after which they develop liver lesions in stepwise involving several mechanisms.[19] The earliest detectable step involves direct attachment of CTL to HBV antigen-positive hepatocytes and the latter are killed by apoptosis. Given the nature of apoptosis, the death of such hepatocytes is unlikely to release agonists which drive the inflammatory response. However, after the initial apoptotic phase, antigen-nonspecific inflammatory cells, such as neutrophils and monocytes, are recruited. These cause far more hepatic cell damage than CTL, and the damage zone extends way beyond the sites of CTL-mediated apoptosis.[20] Such events are presumed to be mediated by ctyokines, particularly IFNγ, probably released by CTL, which can be directly cytotoxic to hepatocytes that express abundant levels of HBV surface Ag.[21] In such mice, the administration of Ab to IFNγ or depletion of macrophages protects against the hepatitis. Consequently, in the HBV transgene model the immunopathology is initiated and likely orchestrated mainly by CD8+ T cells, but the principal immunopathological effects are mediated nonspecifically by cytokines and recruited cells, such as macrophages. This pattern of events is more commonly found in CD4+ T cell-mediated immunopathology, as discussed subsequently.

Of the two principal types of αβTCR T cells, CD4+ T cells are most likely to be involved in an inflammatory response. Thus CD4+ T cells, although capable at least *in vivo* of expressing cytotoxicity, more commonly subserve their effector function by producing several cytokines and other signalling molecules.[22] The cytokines and chemokines produced recruit and activate other cell types, such as macrophages and granulocytes, which then may release their own proinflammatory mediators and reactive oxyradicals. The components of the resulting inflammatory reaction may contribute to rapid viral removal and hence recovery or it may be prolonged because of persistent viral replication or perhaps sustained induction by agonists, such as viral or host antigens. The latter circumstances are considered immunopathological.

As first noted by Mossman and Coffman[23] with mouse T cells, the $CD4^+$ population can be divided into two distinct subsets based on the pattern of their cytokine production. Cells adopting the Th1 pattern of reactivity produce mainly IFNγ, IL-2, and LTα, and these cells are usually involved in conducting inflammatory reactions against viruses and bacteria which usually end up being protective. However, several persistent viral infections, especially those infecting the nervous system induce prolonged reactions considered immunopathological.[24] One of the best studied examples is Theiler's murine encephalomyelitis virus (TMEV) in SJL mice.[25] The model may mimic the pathogenesis of the human disease multiple sclerosis although incriminating a known virus in this disease has proven to be a difficult task. The lesions in TMEV infection result in white matter demyelination, likely performed by macrophages[26] following activation by cytokines produced by $CD4^+$ Th1 T cells. The actual demyelination may result from the action of cytokines produced by macrophages, such as IL-6 and TNFα, or perhaps because of oxyradical or nitric oxide production.[27] Interestingly, in the TMEV model, procedures, which neutralize the cytokines produced by Th1 cells, or immunization approaches which limit Th1 induction minimize the immunopathological lesions.[28] As discussed in a following section, immunopathological effects orchestrated by $CD4^+$ Th1 cells are the mechanisms in play during expression of a disease caused by HSV, called herpetic stromal keratitis.[29]

The other major subset of $CD4^+$ T cells is called Th2 cells. Their signature cytokines are IL-4, IL-5, and IL-10. These cells function as the principal effectors of immunity or immunopathology against parasitic infections. Indeed still the best understood model to show the relevance of TH2 T cell function is persistent infection with Leishmania.[30] There it was shown that only strains of mice that respond to the protozoan by generating predominantly a Th2-mediated response suffer immunopathology and disease susceptibility.[30] Respiratory syncytial virus (RSV) infection[31] is a human viral disease where Th2 cells appear as the main effectors and in so doing mediate an immunopathological disease.[31] In RSV, the immunopathological disease occurs as the virus is being eliminated. In a mouse model for RSV, lesions can be attained only by adoptive transfers of $CD4^+$ T cells which express a type 2 cytokine profile.[32] Surprisingly, one of the proteins of RSV, the G protein, induces the pathologic Th2 response whereas the F protein induces Th1 and $CD8^+$ T cells which play a protective role. To the authors' knowledge, none of the lesions induced by herpesviruses involve Th2-controlled immunopathological reactions.

Finally, viruses may be considered involved in immunopathological reactions by somehow inducing an autoimmune inflammatory response. This topic excites much speculation but sparse concrete evidence, particularly with human viral infections. The subject has received several recent reviews, and those should be consulted for a full discussion of the issues.[33,34] Hypotheses to account for the role of viruses in autoimmune disease abound. The most popular are molecular mimicry and the revelation of cryptic self-epitopes. The latter is the most widely accepted and there are several suggestions to explain how cryptic epitopes might be made accessible to the immune system. These include neoantigens generated as a consequence of virally imposed novel antigen processing and up-regulation of crucial host proteins involved in antigen recognition.[35] It is also suggested that the latter viruses act as polyclonal activators of the immune system[36] or even encode recombinase molecules that could reinitiate T cell repertoire expansion.[37] It is suggested that circumstance occurs in EBV infection, and it is discussed in section 5. In the following section, evidence is discussed that ocular infection with HSV may initiate an autoinflammatory response in the cornea.

3. IMMUNOPATHOGENESIS OF HSV INFECTION OF THE CORNEA

Herpes simplex virus causes death by necrosis in all cells in which it completes a replication cycle.[38] Productive replication is usually confined to mucocutaneous sites, and these foci are soon limited by the innate and later T cell-mediated, adaptive immune response. Even primary infections may be limited in size and often go unnoticed. Aspects of innate immunity responsible for early containment remain ill defined, but cellular destruction by NK cells and cellular protection by interferons are favored explanations.[39] The earliest adaptive aspect of immunity to occur after primary infection is a T cell response, and T cell function, rather than antibody, plays the predominant role in controlling infection.[40] Following reactivation from latency and lesion reexpression, a characteristic of HSV infection in humans, the T cell response is rapidly recalled from memory and then is the principal vehicle of lesion resolution. Of the two major types of αβTCR T cells, $CD4^+$ cells are likely those primarily, if not exclusively, responsible for effecting immunity.[41] This is true in the mouse model and in humans, the natural host.[41,42] The mechanism of immunity involves the generation of an inflammatory response following recognition of viral antigen-expressing MHC class II positive cells, most probably Langerhans cells of the dendritic cell series. Of the two principal functional subsets of $CD4^+$ T cells, those of the type 1 cytokine producing group appear as the predominant cells involved[41] and their secretory product IFNγ in both humans and the mouse model is instrumental in mediating immunity.[42,43] In humans, levels of IFNγ secretion have been correlated with the severity of lesions, [44] and in the mouse suppression of IFNγ exacerbates lesions.[45] Indeed HSV infections in IFNγ knockout mice are far more severe than in normal control littermates.[46] Nevertheless, IFNγ is not the only cytokine involved in defense, because recovery can occur in the absence of the cytokine, although the process is less efficient.[47]

The inflammatory response involved during resolution of HSV infections persists beyond the time when replicating virus is present and in the later stages can be considered an immune driven event to viral or perhaps host antigens. Any discomfort or bystander tissue damage accruing from the immunological reaction is a small price to pay for the ultimate outcome of recovery. This is true in most mucocutaneous locations but it is not acceptable in the eye. In this location, any inflammatory reaction along the visual axis can impair vision. Indeed, the eye has several strategies which minimize inflammatory reactions to antigen. This notion of ocular immune privilege was initially popularized by Kaplan and Strelein[48] and was extended subsequently by several workers and shown to result from the activity of multiple factors. The topic recently received an excellent review by Ferguson.[49] In addition Ferguson and colleagues have shown that whereas HSV infection of the ocular anterior chamber is of minimal consequence, in animals genetically lacking Fas^+ expression and so unable to trigger apoptosis in T cells, assumed to be one mechanism of ocular privilege, lesions were profound.[50] In large part such lesions in Gld mice are likely to be immunopathological.

Ocular privilege does not operate successfully in the cornea, at least following infection with HSV. Accordingly, upon infection of the cornea, the natural route of infection for HSV, the virus causes epithelial cell destruction followed usually by an inflammatory response in the underlying stroma. The latter response occurs to some degree in about one fifth of patients infected with HSV.[51] In certain mouse strains infected with the RE strain of HSV-1, the odds of developing HSK are 90% or greater.[52] The mouse model is currently the most valuable for understanding the pathogenesis of HSV infection. However, the model does have an almost fatal flaw. In humans HSK occurs at least threefold more commonly as a sequel to recurrent infection, and humans can suffer several recurrences in their lifetime.[57] Rarely if ever are these associated with reinfection. Instead recurrences represent reactivation from latency. In contrast, mice seldom if ever suffer clinically evident recurrences. Indeed the failure to achieve clinical reexpression or even silent viral shedding, spontaneously or following the use of various types of artificial stress, continues to be a source of disappointment to experimentalists. Two groups, however, claim to have a reproducible model of recurrent disease in the mouse eye.[53,54] This model involves infection with a highly virulent strain in the presence of antiviral monoclonal Ab followed by exposure to UV light to cause disease expression. It is claimed that the model has closer similarity to human HSK, but it is very inconvenient to use, and few insights have so far emerged from the limited results published to date.

Most of our recent knowledge of HSV pathogenesis comes from studies on primary infection of mice with HSV-1. Results clearly depend on the genetics of both the host and the virus and on other factors, such as dose and whether or not corneal abrasion was used at the time of infection. The mouse model best mimics the form of stromal keratitis observed in man, which is immunologically mediated and is more frequently evident following recurrent infection. The evidence for immune mediation in the human disease is indirect and based largely on favorable responses to treatment with anti-inflammatory drugs.[51]

Infection of the corneas of immunocompetent BALB/c, CB17 scid mice or BALB/c nu/nu mice has similar initial consequences. Virus replicates in the epithelium for a few days, and virus is readily recoverable from ocular swabs. Epithelial lesions are evident, but these soon resolve, and by 4–5 days postinfection nothing in the eye appears amiss. This situation usually persists in scid and nu/nu mice, but starting 12–14 days after infection animals show signs of encephalitis which are ultimately lethal.[55] At death, their eyes appear grossly and histologically normal. The scenario in BALB/c mice is quite different. Here, starting 8–10 days p. i., signs of inflammatory infiltrates appear in the cornea stroma, neovascularization becomes evident in the cornea, and eyes show signs of corneal edema and opacity.[56] Ocular lesions progress in severity, and some may show signs of necrosis by 16–20 days p. i. Mice, however, survive but only in few instances do lesions regress, and this never occurs in animals that advance to severe necrotizing lesions. Such observations, some of which were made first by Metcalf and colleagues in 1979,[57] firmly established that HSK in the mouse represents an immunopathological disease. The observations were subsequently confirmed by Russell *et al.*,[29] who showed that disease susceptibility in nude mice could be reconstituted with T but not other cell types. Initially, $CD8^+$ T cells were suspected of involvement in mediating the ocular pathology, and some investigators still adhere to that viewpoint.[58] However, most recent observations are consistent with the hypothesis that HSK is primarily mediated by $CD4^+$ T cells.[59] This first became evident from the result of experiments showing disease prevention by suppressing $CD4^+$ but not $CD8^+$ T cells with MAb treatment.[60] The notion was confirmed by showing that the transfer of susceptibility is achieved with $CD4^+$ T cells, but not with $CD8^+$ T cells.[55] Indeed some observation indicate that $CD8^+$ T cells may even serve a down-modulating activity in HSK.[60]

Regarding the role of different functional subsets of $CD4^+$ T cells in mediating HSK, the bulk of evidence favors a principal role for Th1 cytokine-producing cells. Accordingly, the pathogenesis of HSK resembles the state of affairs described in section 2 for TMEV infection of SJL mice. There is, however, one major difference. Whereas with TMEV, the $CD4^+$ Th1 T cell response is directed against a persisting largely noncytodestructive virus, HSV is a notoriously cytolytic virus. Moreover, a search for replicating virus when lesions become clinically noticeable find none.[61] In fact, replicating virus appears to be cleared from the cornea long before HSK becomes evident. Thomas *et al.*[62] observe virus present for only 3–5 days, and both viral mRNA and viral antigen are usually undetectable by 7 days, some 2 days or more before HSK starts. Such observations beg the question whether the immunopathological event in the stroma is driven by CD4+ T cell recognition of viral antigens or whether some other mechanisms are in play. Accordingly, it is conceivable that during the initial phase of HSV-1, virus replication in the cornea may up-regulate some host-derived component in the cornea which in turn drives the inflammatory response. Looked at this way, HSK would represent an autoinflammatory response set off by HSV infection in a tissue, the cornea, that is normally an immunologically privileged site and, in lacking vascularization, not subject to T cell surveillance.

Other lines of evidence support the possibility that HSK, at least in the mouse, represents an autoinflammatory response. This possibility was first raised by the observation of Mercadal *et al.*[55] that adoptive transfer of HSK is achieved in HSV infected SCID mice by HSV immune CD4+ T cells and also by naive CD4+ T cells. Moreover, HSK becomes evident before animals develop detectable HSK-specific immune responses. Even more provocative responses. Even more provocative evidence for the possible role of autoreactivity was provided in the observations of Avery *et al.*[63] They employed two strains of mice congenic to each other and markedly different in susceptibility to HSK. Resistance was associated with the possession of the Igh^b allele, responsible for encoding the $Ig2a^b$ allotype of immunoglobulin. By rendering susceptible (Igh^a) mice tolerant to $IG2a^b$, they became resistant to HSK expression. In addition, disease could be conferred on HSV-infected nude mice by the transfer of CD4+ T cell clones specific for a peptide derived from the $IgG2a^b$ molecule. Their data may add up to the possibility that HSV infection causes the up-regulation of a corneal antigen which cross-reacts with the $Ig2a^b$ molecule.

Other lines of evidence supporting autoreactivity include the observation that HSK can also be adoptively transferred to HSV-infected SCID mice with cell lines that are autoreactive and which lack HSV-specific reactivity[64] and that rats with HSK develop T cell reactivity with peptides, such as PLP, derived from a myelin protein.[65]

It remains to be seen if HSK in humans is also an autoreactive response. Many have suspected that this is the case, but the issue is difficult to evaluate. In humans, HSK is always treated once diagnosed and anti-inflammatory agents form a major part of the drug arsenal. The disease usually responds favorably, just as do most other autoimmune diseases. This is of course, insufficient evidence to label HSK as an autoimmune disease in humans.

4. IMMUNOPATHOGENESIS OF HERPETIC RETINAL DISEASE

In humans, acute retinal necrosis, which occurs in immunocompetent individuals, may be the consequence of infection by several members of the alpha herpesviruses (HSV-1, HSV-2, VZV)[66–68] and perhaps even by HCMV[69] (see below). Acute retinal necrosis occurs in euthymic BALB/c mice after primary uniocular anterior chamber infection with HSV-1.[70] The model includes both immunopathogenic and immunoprotective components, and the outcome of infection depends on the timing of viral infection, viral virulence, and the site being studied.[71-73] When a relatively nonneuroinvasive, nonneurovirulent strain of HSV-1 (for example, KOS) is inoculated into one anterior chamber, a sequence of events involving the injected eye, the brain, and the uninjected contralateral eye occurs (See Fig. 1 for an illustration of pathways of virus spread). Virus replicates abundantly within the first 24–48 hours in the anterior segment (iris, ciliary body, and corneal endothelium) of the injected eye, and this is accompanied by massive inflammation in the anterior segment.[74] Within 1–2 days following infection, virus leaves the injected eye via the postganglionic parasympathetic neurons which supply the anterior segment. Once in this pathway, virus spreads to neurons which are synaptically connected to the ipsilateral Edinger–Westphal nucleus. From this site, virus spreads to the ipsilateral suprachiasmatic nucleus of the hypothalamus and thence to the optic nerve and retina of the contralateral uninjected eye.[75] Virus is first detected in the contralateral retina by day 7 p. i., and subsequent destruction of the retina begins 9–10 days p. i. (See Fig. 1).[76-77]

Paradoxically, although the retina and optic nerve of the uninjected eye are infected with virus, the optic nerve and retina of the injected eye are spared from infection in euthymic mice.[70] How the protection occurs is not yet clear, but various results indicate that the mechanism is a T cell-dependent process.[78-80] Thus, uniocular anterior chamber inoculations of HSV-1 in athymic or $CD4^+$- and $CD8^+$-depleted mice result in infection of the optic nerve and retina of both eyes.[81] Whereas levels of virus in the contralateral eye remained the same regardless of the subset of T cells depleted, the extent of retinitis and retinal necrosis between groups was significantly different. When $CD4^+$ cells were present, retinitis proceeded more quickly to retinal necrosis, and the extent of inflammation in the retina increased. In the absence of $CD4^+$ or both $CD4^+$ and $CD8^+$ cells, retinitis in the uninoculated eye was milder.[82] Such results incriminate $CD4^+$ T cells as subserving an immunopathological role. However, the mechanism by which $CD4^+$ T cells cause acute retinal necrosis remains unclear. The issue is further confounded by the fact that *in vivo* normal retinal cells do not usually express high levels of necessary class II MHC molecules necessary for recognition by $CD4^+$ T cells.[83,84] Additional experiments indicate that during acute retinitis $CD4^+$ cells in the contralateral eye have both Th1 and Th2 cytokine profiles, suggesting that neither Th1 nor Th2 predominates in the mouse model of acute retinal necrosis.[85] It could be that atypical immunoinflammatory events are in play because, as discussed previously, the eye employs mechanisms that usually succeed in limiting inflammatory responses.

Although a $CD4^+$-mediated response is immunodestructive in the uninoculated eye, the T cell immune system is also immunoprotective. If $CD4^+$ and

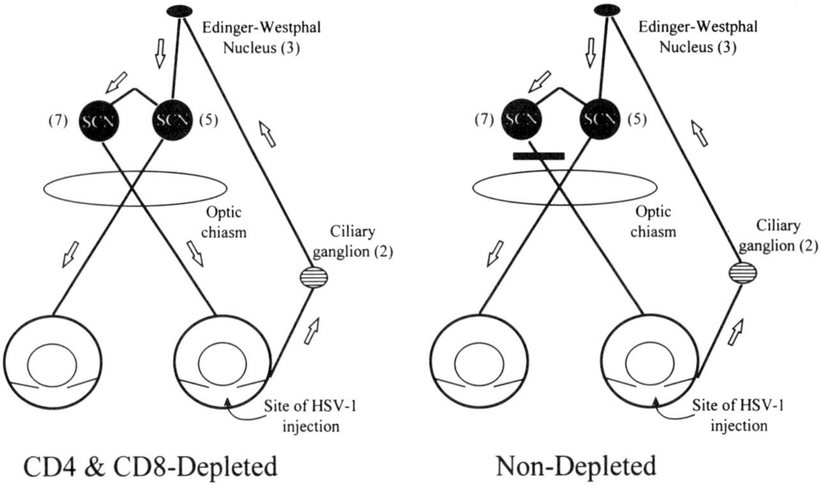

FIGURE 1. Timing and sites of virus spread in T cell-depleted and euthymic mice following anterior chamber inoculation of HSV-1. (Right) Following uniocular anterior chamber inoculation of the KOS strain of ISV-1 in euthymic BALB/c mice, the virus replicates in the anterior segment of the injected eye, leaves the injected eye, and infects the ipsilateral ciliary ganglion by day 2 p. i. The virus then spreads sequentially via synaptically connected neurons from the ipsilateral ciliary ganglion to the ipsilateral Edinger–Westphal nucleus (day 3 p. i.), to the ipsilateral suprachiasmatic nucleus (SCN) by day 5 p. i. and from the ipsilateral SCN to the contralateral optic nerve and retina by day 7 p. i. Although the virus spreads to the contralateral SCN by day 7 p. i., the ipsilateral optic nerve and retina do not become virus-positive in euthymic mice. (Left) Following uniocular anterior chamber inoculation of the KOS strain of HSV-1 in BALB/c mice depleted of both $CD4^+$ and $CD8^+$ T cells, the route and timing of the virus spread to the ipsilateral ciliary ganglion and to the ipsilateral Edinger–Westphal nucleus do not differ from that in euthymic BALB/c mice. However, in T cell-depleted mice, both the ipsilateral and the contralateral SCN become virus-positive at day 5 p. i., and the optic nerve and retinal of the injected eye and the uninjected eye are infected with virus by day 6 or day 7. p. i.

$CD8^+$ T cells are lacking, the ipsilateral eye remains unprotected from viral invasion.[86] Surprisingly, IFNγ, a Th1-associated cytokine, is not observed in euthymic mice in the area of the suprachiasmatic nuclei until day 9 p. i. At this time, virus has already disseminated to the optic nerve and retina of the uninoculated eye but is prevented from spreading from the contralateral suprachiasmatic nucleus to the ipsilateral optic nerve and retina. In the absence of T cells, virus spreads to infect both optic nerves and retinas, but the retinitis observed in this instance is slowly progressive without the component of acute retinal necrosis observed only in euthymic mice or in mice with a full complement of $CD4^+$ T cells.

Overall, experiments on the retinitis model indicate that $CD4^+$ T cells are immunopathological in that they contribute to destruction of the retina of the uninoculated eye and are immunoprotective in that they are needed within the CNS to protect the retina of the injected eye from posterior to anterior spread of the virus. The role of $CD8^+$ T cells in a destructive and protective function is less well defined, but preliminary observations indicate that such cells participate in protective events in the retinitis model. Taken together, the results of studies to examine the virologic and immunologic components of the mouse model of acute retinal necrosis have shown that the sites of virus infection in the brain, the timing of virus spread, and the T cell-mediated immune response conspire together to destroy the retina of the uninjected eye while preventing infection and damage to the optic nerve and retina of the injected eye.

Results from studies using the mouse model of acute retinal necrosis provide several lessons for understanding the pathogenesis of the human disease. Because acute retinal necrosis occurs in the setting of immunocompetence, it is likely that multiple viral and immune events occur. Many patients with either primary or reactivated herpesvirus keratitis additionally experience inflammation of the anterior uvea resulting from viral access to the anterior chamber. It is likely that in most patients the virus then follows the trigeminal nerves supplying the anterior segment of the eye and becomes latent in the corresponding trigeminal ganglion. However, it is postulated that in a small subset of patients, the virus enters the parasympathetic nerve pathway which has connections to the suprachiasmatic nuclei. Once in one or both suprachiasmatic nuclei, the virus could theoretically spread to one or both retinas resulting in retinitis followed by acute retinal necrosis.[87,88] However, additional studies using human patients with acute retinal necrosis are neeeded to verify pathways of viral spread and to correlate these with the status of T cell immunity.

The role of the immune system in the pathogenesis of retinitis by other members of the herpesvirus family is less well defined and depends, at least in part, on differences in the host's ability to mount an immune response to these viruses. For example, a functional immune system is generally considered to be beneficial in controlling HCMV infections.[89–91] Commonly, symptomatic local or systemic HCMV infections are usually observed only in immunosuppressed individuals, such as AIDS patients, or in immunosuppressed patients following transplantation, especially of bone marrow. However, acute retinal necrosis due to HCMV has occasionally been described in these patients. Laby *et al.*,[69] reported a case of progressive outer retinal necrosis caused by HCMV in an HIV-infected patient. The authors suggested that because the patient's immune system was still relatively intact at the time of disease development, the ability to mount an immune response may have resulted in an acute retinal necrosis episode rather than the slowly progressive retinal infection usually observed in more profoundly immunosuppressed patients with cytomegalovirus retinitis.

5. PATHOGENESIS OF INFECTIOUS MONONUCLEOSIS

Many of the features and complications of infectious mononucleosis (IM) associated with Epstein–Barr virus (EBV) infection represent an immunopathological process.[92] Evidence from a variety of studies suggests that there may be two types of immunopathological responses to this virus: (1) a "classic" CTL response[93] and (2) an autoimmune response.[96] The relationship of the first response to the second is not clear, but because only some patients develop autoimmune complications, it is likely that the second type of response occurs less frequently than the first. The first response is a CTL reaction against EBV-infected B cells. In a normal, immunocompetent patient with infectious mononucleosis, $CD8^+$ cytotoxic T cells (atypical lymphocytes) which recognize EBV-infected B cells are probably responsible for many of the signs and symptoms which define EBV infectious mononucleosis (fever, lymphadenopathy, splenomegaly). However, EBV-specific $CD8^+$ T cells are also beneficial in that they eliminate large numbers of virus-infected B cells from the body.[95] The importance of this latter function becomes evident in immunocompromised patients following transplantation of T cell-depleted bone marrow when lack of CTL specific for EBV-infected B cells allows uncontrolled proliferation of B cells and increases the risk of EBV-induced lymphoproliferative disease in these patients.[96] The exaggerated T cell response in infectious mononucleosis likely reflects cytokine secretion by EBV-infected B cells which, in turn, results in a considerably amplified T cell reaction. Although generally this amplified CTL response is beneficial to the host, it is possible that some of the occasional long-term complications of EBV infection might be caused by other T cell-related activities, such as an anomalous NK-like function which has been reported for EBV-infected T cells or through lymphokine release inducing massive infiltration of non-virus-specific cells into a variety of tissues.

The second type of immunopathogenic response is less well defined but may result in autoimmunity. Results of recent studies suggest that EBV-induced T cell proliferation and altered T cell receptor repertoire contribute to such a response.[94] Because EBV infects T cells and B cells, it has been suggested that expression of EBV lytic genes by T cells alters T cell proliferation and also affects genes involved in developing the T cell receptor repertoire. It is postulated that EBV infection of immature thymocytes disrupts thymocyte maturation. One interesting recent idea is that altered T cell selection and development of the T cell repertoire result from RAG protein activation and interaction between immunoglobulin RAG proteins and the EBV protein BALF-Z, which may be viral homologue of the RAG protein. Once in possession of an altered repertoire which may be a viral homologue of the cellular RAG proteins, these T cells would not be subject to normal selection and by persisting and replicating, might be able to inappropriately target and attack self-antigens. EBV also encodes proteins, such as CD21 binding protein, which are expressed on EBV-infected thymocytes. It is possible that expression of such viral proteins on thymocytes also plays a role in altering the development of the T cell receptor repertoire.

Consequently as with herpes simplex viruses, immune responses to EBV also have both beneficial and detrimental components. The classic EBV-specific $CD8^+$ CTL response is probably responsible for many of the pathological features of acute infectious mononucleosis. However, these cells are also beneficial because in most patients, they are responsible for controlling a virus-infected B cell population with the potential for unrestrained growth and proliferation. The consequences of EBV infection of thymocytes and other T cells are just beginning to be elucidated, but it is likely that complications from this latter type of infection will be longer lasting, more deleterious, and more difficult to treat than an EBV infection resolved without apparent difficulty. Given the extremely high rate of EBV infection and seropositivity among the general population, the idea that a usually benign infection with EBV may result in autoimmunity should be of concern to both immunologists and physicians. Studies are needed to define which EBV-infected individuals are at highest risk of developing one or more of the complications observed after EBV infection, including autoimmune disorders resulting from altered T cell reactivity.

REFERENCES

1. Roizman, B., 1996, *Herpesviridae,* in: *Fields Virology,* 3rd ed. (B. N. Fields, D. M. Knipe, P. M. Howley et al., eds.), Lippencott-Raven, Philadelphia, pp. 2221–2230.
2. Kieff, E. 1996, Epstein–Barr virus and its replication, in: *Fields Virology,* 3rd ed. (B. N. Fields, D. M. Knipe, P. M. Howley *et al.,* eds.), Lippencott-Raven, Philadelphia, pp. 2343–2396.
3. Rickinson, A. B., and Kieff, E., 1996, Epstein–Barr virus, in: *Fields Virology,* 3rd ed. (B. N. Fields, D. M. Knipe, P. M. Howley *et al.,* eds.), Lippencott-Raven, Philadelphia, pp. 2397–2446.
4. Pellet, P. E., and Black, J. B., 1996, Human herpesvirus 6, in: *Fields Virology,* 3rd ed. (B. N. Fields, D. M. Knipe, P. M. Howley *et al.,* eds.), Lippencott-Raven, Philadelphia, pp. 2587–2608.
5. Frenkel, N., and Roffman, E., 1996, Human herpesvirus 7, in: *Fields Virology,* 3rd ed. (B. N. Fields, D. M. Knipe, P. M. Howley *et al.,* eds.), Lippencott-Raven, Philadelphia, pp. 2605–2622.
6. Britt, W. J., and Alford, C. A., 1996, Cytomegalovirus, in: *Fields Virology,* 3rd ed. (B. N. Fields, D. M. Knipe, P. M. Howley *et al.,* eds.), Lippencott-Raven, Philadelphia, pp. 2493–2523.
7. Burnet, F. M., and Fenner, F., 1949, Production of antibody, Monographs of Walter and Eliza Hall Institute, McMillan, Melbourne.
8. Oldstone, M. B. A., and Dixon, F. J., 1969, Pathogenesis of chronic disease associated with persistent lymphocytic choriomeningitis viral infection. I. Relationship of antibody production to disease in neonatally infected mice, *J. Exp. Med.* **129:**483–505.
9. Cochrane, C. A., and Dixon, F. J., 1976, Antigen-antibody complex induced disease, in: *Textbook of Immunology and Pathology* (P. A. Miescher and H. J. Muller-Eberhard, eds.), Arune-Stratton, New York, pp. 137–156.
10. Buchmeier, M. J., Welsh, R. M., Dutko, F. J., and Oldstone, M. B. A., 1980, The virology and immunobiology of lymphocytic choriomeningitis virus infection, *Adv. Immunol.* **30:**275–331.

11. Chisari, F. B., Ferrari, C., 1995, Hepatitis B virus pathogenesis, *Ann. Rev. Immunol.* **13:**29–60.
12. Kazmierowski, J. A., Piezner, D. S., and Wuepper, K. D., 1982, Herpes Simplex antigen in immune complexes of patients with erythema multiforme: Presence following recurrent herpes simplex infections, *JAMA* **247:**2547–2551.
13. Kurane, I., and Ennis, F. A., 1994, Cytotoxic T lymphocytes in dengue virus infection, *Curr. Top. Microbiol. Immunol.* **189:**93–108.
14. Trautwein, G., 1992, Immune mechanisms in the pathogenesis of viral disease: A review, *Vet. Microbiol.* **33:**19–34.
15. Oldstone, M. B. A., 1989, Molecular mimicry as a mechanism for the cause and as a probe uncovering etiologic agents of molecular mimicry autoimmune disease, *Curr. Top. Microbiol. Immunol.* **145:**127–135.
16. Wucherpfennig, K. W., and Strominger, J. L., 1995, Molecular mimicry in T cell-mediated autoimmunity: Viral peptides activate human T-cell clones specific for myelin basic protein. *Cell* **80:**695–705.
17. Kagi, D., Ledermann, B., Burki K., Zinkernagel, R. M., and Hengartner, H., 1995, Lymphocyte-mediated cytotoxicity *in vitro* and *in vivo:* Mechanisms and significance, *Immunol. Rev.* **146:**95–115.
18. Franco, M. A., Tin, C., Rott, L. S., Van Cott, J. L., McGhee, J. R., and Greenberg, H. B., 1997, Evidence for $CD8^+$ T cell immunity to murine rotavirus in the absence of perforin, Fas and gamma interferon, *J. Virol.* **71:**479–486.
19. Guidotti, L. G., Matze, B., Schaller, H., and Chisari, F. V., 1995, High level hepatitis B virus replication in transgenic mice, *J. Virol.* **69:**6158–6169.
20. Ando, K., Moriyama, T., Guidotti, L. G., Wirth, S., Schreiber, R. D., Schlicht, H. J., Huang, S. N., and Chisari, F. V., 1993, Mechanisms of class I restricted immunopathology. A transgenic mouse model of fulminant hepatitis, *J. Exp. Med.* **178:**1541–1554.
21. Guidotti, L. G., Ishikawa, T., Hobbs, M. V., Matzke, B., Schreiber, R., and Chisari, F. V., 1995, Intracellular inactivation of the hepatitis B virus by cytotoxic T lymphocytes, *Immunity* **4:**25–36.
22. Meltzer, M. S., and Nacy, C. A., 1989, Delayed hypersensitivity and the induction of activated cytotoxic macrophages, in:*Fundamental Immunology,* 2nd ed. (W. E. Paul, ed.), Raven Press, New York, pp. 765–780.
23. Mosmann, T. R., and Coffman, R. L., 1989, Th1 and Th2 cells: Different patterns of lymphokine secretion lead to different functional properties, *Annu. Rev. Immunol.* **7:**145–173.
24. Fazakerley, J. K., and Buchmeier, M., 1993, Pathogenesis of induced demyelination, *Adv. Virus Res.* **42:**249–324.
25. Miller, S. D., McRae, B. L., Vnaderlugt, C. L., Nikcevich, K. M., Pope, J. G., Pope, L., and Karpus, W. J., 1995, Evolution of T cell repertoire during the course of experimental immune mediated demyelinating disease, *Immunol. Rev.* **144:**225–244.
26. Dal-Canto, M. C., and Lipton, H. L., 1975, Primary demyelination in Theiler's virus infection. An ultrastructural study, *Lab. Invest.* **33:**626–637.
27. Campbell, I. L., Abraham, C. R., Mashliah, E., Kemper, P., Inglis, J. D., Oldstone, M. D. A., and Murke, L., 1993, Neurologic disease induced in transgenic mice by cerebral overexpression of interleukin 6, *Proc. Natl. Acad. Sci. USA* **90:**10061– 10065.
28. Peterson, J. D., Karpus, W. J., Clatch, R. J., and Miller, S. D., 1993, Split tolerance of Th1 and Th2 cells in tolerance to Theiler's murine encephalomyelitis virus, *Eur. J. Immunol.* **23:**46–55.
29. Russell, R. G., Naisse, M. P., Larsen, H. S., and Rouse, B. T., 1984, Role of T lymphocytes in the pathogenesis of herpetic stromal keratitis, *Invest. Ophthalmol. Vis. Sci.* **25:**938–944.
30. Reiner, S. L. and Locksley, R. M., 1995, The regulation of immunity to Leishmania major, *Ann. Rev. Immunol.* **13:**151–177.
31. Openshaw, P. J. M., 1995, Immunopathological mechanisms in respirtory syncytial virus disease, *Springer Semin. Immunopathol.* **17:**187–201.

32. Alwan, W. H., Kozlowska, W. J., and Openshaw, P. J. M., 1994, Distinct types of lung disease caused by functional subsets of antiviral T cells, *J. Exp. Med.* **179:**81–89.
33. Aichele, P., Bachmann, M. F., Hengartner, H., and Zinkernagel, R. M., 1996, Immunopathology or organ specific autoimmunity as a consequence of virus infection, *Immunol. Rev.* **152:**21–45.
34. von Herrath, M. G., and Oldstone, M. B. A., 1996, Virus induced autoimmune disease, *Curr. Opinion Immunol.* **8:**878–885.
35. Sercarz, E. E., Lehmann, P. V., Ametani, A., Benichou, G., Miller, A., and Moudgil, K., 1993, Dominance and crypticity of T cell antigenic determinants, *Ann. Rev. Immunol.* **11:**729–766.
36. Schere, M. T., Ignatowicz, L., Winslow, G. M., Kappler, J. W., and Marrack, P., 1993, Superantigens: Bacterial and viral proteins that manipulate the immune system, *Ann. Rev. Cell Biol.* **9:**101–128.
37. Dreyfus, D. H., Kelleher, C. A., Jones, J. F., and Gelfand, E. W., 1996, Epstein–Barr virus infection of T-cells: Implications for altered T lymphocyte activation, repertoire development and autoimmunity. *Immunol. Rev.* **152:**89–110.
38. Roizman, B., and Sears, A. E., 1996, Herpes simplex virus and their replication, in: *Fields Virology*, 3rd ed. (B. N. Fields, D. M. Knipe, P. M. Howley *et al.*, eds.), Lippencott-Raven, Philadelphia, pp. 2231–2396.
39. Whitley, R. J., 1996, Herpes simplex viruses, in: *Fields Virology*, 3rd ed. (B. N. Fields, D. M. Knipe, P. M. Howley *et al.*, eds.), Lippencott-Raven, Philadelphia, pp. 2297–2342.
40. Schmidt, D. S., and Rouse, B. T., 1992, The role of cytotoxic T lymphocytes in control of HSV, *Curr. Top. Microbiol. Immunol.* **179:**57–74.
41. Manickan, E., Francotte, M., Kuklin, N., Dewerchin, M., Molitor, C., Gheysen, D., Slaoui, M., and Rouse, B. T., 1995, Vaccination with recombinant vaccinia viruses expressing ICP27 induces protective immunity against herpes simplex virus through $CD4^+$ T cells, *J. Virol.* **69:**4711–4716.
42. Cunningham, A. L., Turner, R. R., Miller, A. C., Para, M. F., and Merigan, T. C., 1985, Evolution of recurrent herpes simplex lesions: An immunohistologic study, *J. Clin. Invest.* **75:**226–233.
43. Smith, P. M. Wolcott, R. M., Chervenak, R., and Jenning, S., 1994, Control of acute cutaneous herpes simplex virus infection. T cell-mediated viral clearance is dependent on interferon γ IIFNγ), *Virology* **202:**76–88.
44. Cunningham, A. L., and Merigan, T. C., 1983, γ interferon production appears to predict time of recurrence of herpes labialis, *J. Immunol.* **133:**422–427.
45. Hendricks, R. L., Tumpey, T. M., and Finnegan, A., 1992, IFNγ and IL-2 are protective in skin but pathologic in the corneas of HSV-1 infected mice, *J. Immunol.* **149:**3023–3028.
46. Bouley, D. M., Kanangat, S., Wire, W., and Rouse, B. T., 1995, Characterization of herpes simplex virus type 1 infection and herpetic stromal keratitis development in IFNγ knockout mice, *J. Immunol.* **155:**3964–3971.
47. Yu, Z., Manickan, E., and Rouse, B. T., 1996, Role of interferon gamma in immunity to herpes simplex virus, *J. Leukocyte Biol.* **60:**528–532.
48. Kaplan, H. J., and Strelein, J. W., 1977, Immune response to immunization via the anterior chamber of the eye. I. F Lymphocyte-induced immune deviation, *J. Immunol.* **118:**809–814.
49. Ferguson, T. A., and Griffith, T. S., 1997, Cell death and the immune response: Lessons from the privileged, *J. Clin. Immunol.*, in press.
50. Griffith, T. S., Brunner, T., Fletcher, S. M., Greer, D. R., and Ferguson, T. A., 1995, Fas ligand-induced apoptosis as a mechanism of immune privilege, *Science* **270:**1189–1192.
51. Pepose, J., Leib, D. A., Stuart, P. M., and Easty, D. L., 1996, in: *Ocular Infection and Immunity* (J. S. Pepose, G. N. Holland, and K. R. Wilhelmus, eds.), Mosby, St. Louis, pp. 905–932.
52. Doymaz, M. Z., and Rouse, B. T., 1992, Herpetic stromal keratitis: An immunopathological disease mediated by $CD4^+$ T lymphocytes, *Invest. Ophthalmol. Vis. Sci.* **33:**2165–2173.

53. Laycock, K. A., Lee, S. F., Brady, R. H., and Pepose, J. S., 1991, Characterization of a murine model of recurrent herpetic keratitis induced by ultraviolet B radiation, *Invest. Ophthalmol. Vis. Sci.* **32:**2741–2746.
54. Shimeld, C., Hill, T. J., Blyth, W. A., and Easty, D. L., 1990, Reactivation of latent infection and induction of recurrent herpetic eye disease in mice, *J. Gen. Virol.* **71:**397–404.
55. Mercadal, C., Bouley, D., DeStephano, D., and Roiuse, B. T., 1993, Herpetic stromal keratitis in the reconstituted SCID mouse model, *J. Virol.* **67:**3404–3408.
56. Babu, J. S., Kanangat, S., and Rouse, B. T., 1995, T cell cytokine mRNA expression during the course of the immunopathologic ocular disease herpetic stromal keratitis, *J. Immunol.* **154:**4822–4829.
57. Metcalf, J. G., Hamilton, D. S., and Reichert, R. W., 1979, Herpetic keratitis in athymic (nude) mice, *Infect. Immunol.* **26:**1164–1171.
58. Hendricks, R. L., and Tumpey, T., 1990, Contribution of virus and immune factors to herpes simplex virus type 1 induced corneal pathology. *Invest. Ophthalmol. Vis. Sci.* **31:**1929–1939.
59. Rouse, B. T., 1996, Virus-induced immunopathology, *Adv. Virus Res.* **47:**353–376.
60. Newell, C. K., Martin, S., Sendele, D., Mercadal, C. H., and Rouse, B. T., 1989, Herpes simplex virus stromal keratitis: Role of T lymphocyte subsets in the immunopathology, *J. Virol.* **63:**769–775.
61. Babu, J., Thomas, J., Kanangat, S., Morrison, L. A., Knipe D. M., and Rouse, B. T., 1996, Viral replication is required for induction of ocular immunopathology by herpes simplex virus, *J. Virol.* **70:**101–107.
62. Kuklin, N., Daheshia, M., Karem, K., Manickan, E., and Rouse, B. T., 1997, Induction of mucosal immunity against herpes simplex virus by plasmid DNA immunization, *J. Virol.*, in press.
63. Avery, A. C., Zhao, Z. S., Rodriquez, A., Bikoff, E. K., Soheilian, M., Foster, C. S., and Cantor, H., 1995, Resistance to herpes stromal keratitis conferred by an IgG2a-derived peptide, *Nature* **376:**431–434.
64. Daheshia, M., Kuklin, N., Kanangat, S., Manickan, E., and Rouse, B. T., 1997, Suppression of ongoing ocular inflammatory disease by topical administration of plasmid DNA encoding IL-10, submitted.
65. Clark, L., Fareed, M., Miller, S. D., Merryman, C., and Heber-Katz, E., 1996, Corneal infection with herpes simplex virus type 1 leads to autoimmune responses in rats, *J. Neuroscience Res.* **45:**770–775.
66. Lewis, M. L., Culbertson, W. W., Post, J. D., MIller, D., Kokame, G. T., and Dix, R. D., 1989, Herpes simplex virus type 1: A cause of the acute retinal necrosis syndrome, *Ophthalmology* **96:**875–878.
67. Thompson, W. S. Culbertson, W. W., Smiddy, W. E. Robertson, J. E., and Rosenbaum, J. T., 1994, Acute retinal necrosis caused by reactivation of herpes simplex virus type 2, *Am. J. Ophthalmol* **118:**205–211.
68. Culbertson, W. W., Blumenkranz, M. S., Pepose, J. S., Stewart, J. A., and Curtin, V. T., 1986, Varicella-zoster virus is a cause of the acute retinal necrosis syndrome, *Ophthalmology* **93:**559–569.
69. Laby, D. M., Nasrallah, F. P., Butrus, S. I., and Whitmore, P. V., 1993, Treatment of outer retinal necrosis in AIDS patients, *Graefes Arch. Clin. Exp. Ophthalmol.* **231:**271–273.
70. Whittum, J. A., McCulley, J. P., Neiderkorn, J. Y., and Streilein. J. W., 1984, Ocular disease induced in mice by anterior chamber inoculation of herpes simplex virus, *Invest. Ophthalmol. Vis. Sci.* **25:**1065–1073.
71. Whittum-Hudson, J. A., and Pepose, J. S., 1988, Herpes simplex virus type 1 induced anterior chamber-associated immune deviation (ACAID) in mouse strains resistant to intraocular infection, *Curr. Eye. Res.* **7:**125–130.
72. Kielty, D., Cousins, S. W., and Atherton, S. S., 1987, HSV-1 Retinitis and delayed hypersensitivity in DBA/2 and C57BL/6 mice, *Invest. Ophthalmol. Vis. Sci.* **28:**1994–1999.

73. Atherton, S. S., Kanter, M., and Streilein, J. W., 1991, ACAID requires early replication of HSV-1 in the injected eye, *Curr. Eye Res.* **10S:**75–80.
74. Margolis, T. P., LaVail, J. H., Setzer, P. Y., and Dawson, C. R., 1989, Selective spread of herpes simplex virus in the central nervous system after ocular inoculation, *J. Virol.* **63:**4756–4761.
75. Vann, V. R., and Atherton, S. S., 1991. Neural spread of herpes simplex virus after anterior chamber inoculation, *Invest. Ophthalmol. Vis. Sci.* **32:**2462–2472.
76. Atherton, S. S., and Streilein, J. W., 1987. Two waves of virus following anterior chamber inoculation of HSV-1, *Invest. Ophthalmol. Vis. Sci.* **28:**571–578.
77. Cousins, S. W., Gonzalez, A., and Atherton, S. S., 1989, Herpes simplex retinitis in the mouse: Clinicopathologic correlations, *Invest. Ophthalmol. Vis. Sci.* **30:**1485–1494.
78. Whittum-Hudson, J. A., and Pepose, J. S., 1987, Immunologic modulation of virus-induced pathology in a murine model of acute herpetic retinal necrosis, *Invest. Ophthalmol. Vis. Sci.* **28:**1541–1548.
79. Atherton, S. S., Altman, N. H., and Streilein, J. W., 1989, Consequences of intraocular infection with HSV-1 in athymic mice: Evidence for an immunopathogenic process in virus-induced retinitis, *Curr. Eye Res.* **8:**1179–1192.
80. Atherton, S. S., and Vann, V. R., 1993, Immunologic control of neural spread of herpes simplex virus type 1 (HSV-1) following anterior chamber inoculation, in: *Recent Advances in Uveitis,* J. P. Dernouchamps, C. Verougstraete, L. Caspers-Velu, and M. J. Tassignon, eds.), Kugler Amsterdam, pp. 23–28.
81. Azumi, A., and Atherton, S. S., 1994, Sparing of the ipsilateral retina after anterior chamber inoculation of HSV-1: Requirement for either $CD4^+$ or $CD8^+$ T cells, *Invest. Ophthalmol. Vis. Sci.* **35:**3251–3259.
82. Azumi, A., Cousins, S. W., Kanter, M. Y., and Atherton, S. S., 1994, Modulation of murine herpes simplex virus type 1 retinitis in the uninoculated eye by $CD4^+$ T lymphocytes, *Invest. Ophthalmol. Vis. Sci.* **35:**54–63.
83. Drescher, K. M., and Whittum-Hudson, J. A., 1996, Modulation of immune-associated surface markers and cytokines production by murine retinal glial cells, *J. Neuroimmunol.* **64:**71–81.
84. Percopo, C. M., Hooks, J. J., Shinohara, T., Caspi, R., and Detrick, B., 1990, Cytokine-mediated activation of a neuronal retinal resident cell provokes antigen presentation, *J. Immunol.* **145:**4101–4107.
85. Atherton, S. S., Azumi, A., Thomas, C., Babu, J. S., and Rouse, B. T., 1995, Relationship of T-cell cytokines to HSV-1 retinitis, *Invest. Ophthalmol. Vis. Sci.* **36:**S386.
86. Matsubara, S., and Atherton, S. S., T-cell depletion correlates with early spread of HSV-1 to the suprachiasmatic nuclei, submitted.
87. Zhao, M., Azumi, A., and Atherton, S. S., 1995, T lymphocyte infiltration in the brain following anterior chamber inoculation of HSV-1, *J. Neuroimmunol.* **58:**11–19.
88. Matsubara, S., and Atherton, S. S., 1996, TNF-α and spread of HSV-1 to the suprachiasmatic nuclei (SCN), *Invest. Ophthalmol. Vis. Sci.* **37:**S952.
89. Sergott, R. C., Anand, R., Belmont, J. B., Fischer, D. H., Bosley, T. M., and Savino, P. J., 1989, Acute retinal necrosis neuropathy: Clinical profile and surgical therapy, *Arch. Ophthalmol.* **197:**692–696.
90. Duker, J. S.,and Blumenkranz, M. S., 1991, Diagnosis and management of the acute retinal necrosis (ARN) syndrome, *Surv. Ophthalmol.* **35:**327–343.
91. Moeller, M. D., Gutman, R. A., and Hamilton, J. D., 1982, Acquired cytomegalovirus retinitis: Four new cases and a review of the literature with implications for management, *Am. J. Nephrol.* **2:**251–255.
92. Holland, G. N., Pepose, J. S., Pettit, T. H., Gottleib, M. S., Yee, R. D., and Foos, R. Y., 1983, Acquired immune deficiency syndrome: Ocular manifestations, *Ophthalmology* **90:**859–873.
93. Fiala, M., Payne, J. E., Berne, T. V., Moore, T. C., Henle, W., Montgomerie, J. Z., Chatterjee,

S. N., and Guze, L. B., 1975, Epidemiology of cytomegalovirus infection after transplantation and immunosuppression, *J. Infect. Dis.* **132:**421–433.
94. Straus, S. E., Cohen, J. I., Tosato, G., and Meier, J., 1993, Epstein–Barr virus infections: Biology, pathogenesis, and management, *Ann. Int. Med.* **118:**45–58.
95. Khanna, R., Burrows, S. R., and Moss, D. J., 1995, Immune regulation in Epstein–Barr virus-associated diseases, *Microbiol. Rev.* **59:**387–405.
96. Dreyfus, D. H., Kelleher, C. A., Jones, J. F., and Gelfand, E. W., 1996, Epstein–Barr virus infection of T cells:Implications for altered T-lymphocyte activation, repertoire development and autoimmunity, *Immunol. Rev.* **152:**89–110.
97. Rickinson, A. B., and Kieff, E. 1996. Epstein–Barr virus, in: *Fields Virology* (B. N. Fields, D. M. Knipe, P. M. Howley, R. M. Chanock, T. P. Monath, J. L. Melnick, B. Roizman, and S. E. Straus, eds.), Lippincott-Raven, Philadelphia, pp. 2397–2446.
98. Shapiro, R. S., McClain, K., Frizzera, G., Gajl-Peczalska, K. J., Kersey, J. H., Blazar B. R., Arthur, D. C., Patton, D. F., Greenberg, J. S., Burke, B., Ramsay, N. K. C., McGlave, P., and Filipovich, A. H., 1988, Epstein–Barr virus associated B-cell proliferative disorders following bone marrow transplantation, *Blood* **71:**1234–1243.

3

HSV Gene Expression during Latent Infection and Reactivation

EDWARD K. WAGNER and DAVID C. BLOOM

1. INTRODUCTION

The basic replication process of HSV has been a subject of intense experimental investigation and reviews in the last three decades. In particular, we have recently provided a detailed discussion of the experimental investigation of various aspects of gene regulation operating during such productive infection as well as more general outlines.[1-3] It is notable in the context of the present chapter that the bulk of the earliest gene products manifest in productive infection function to stimulate a nominally quiescent cell into a relatively high level of transcriptional activity. As outlined below, latency is a manifestation of the failure of this cascade. The role of HSV gene expression in establishment, maintenance, and reactivation from latency is subtle, and presently our knowledge is often defined by the elimination of obvious and readily testable models.

EDWARD K. WAGNER • Department of Molecular Biology and Biochemistry, University of California, Irvine, Irvine, California 92697-3900. DAVID C. BLOOM • Department of Microbiology, Arizona State University, Tempe, Arizona 85287-2701.

Herpesviruses and Immunity, edited by Medveczky *et al.* Plenum Press, New York, 1998.

2. THE HSV GENOME IN THE ESTABLISHMENT AND MAINTENANCE OF LATENT INFECTIONS IN NEURONS

2.1. Establishment of Latent Infections in Sensory Neuron

The establishment of latent infection following HSV replication in peripheral tissue is a process which can be readily examined only in animal models, and most studies have been carried out using various routes of latent infection in mice. The establishment of latency in neurons is essentially or completely a passive phenomenon. No viral gene product is involved in the process.

Margolis and co-workers utilized an HSV-1 recombinant virus expressing the *lacZ*(β-galactosidase) gene under the control of LAT promoter and compared viral gene expression in various tissues in mice following peripheral infection,[4] They concluded that those neurons in which latent infection was established were infected as rapidly as those undergoing productive virus replication. Further, they determined that the major group of neurons in which latent (i.e., nonproductive) infection was occurring were those with the SSEA-3 surface marker. Their conclusion that latent infection occurs as an early event in infection and, thus, defines a restricted group of neurons is consistent with the conclusions of Speck and Simmons.[5]

Latchman and colleagues carried out an extensive analysis of potential mechanisms to account for the failure of the HSV productive cycle in establishing latency and conclude that expression of specific transcription factors important in the expression of the immediate-early regulatory proteins of the virus are altered in cultures of neuronal cells compared to nonneuronal cells. They suggest that such differential populations of transcription factors could play a major role in restricting productive infection in certain classes of neuronal cells, notably by interfering with the first activation *via* α-TIF.[6–9]

The greatest support for the idea that latent infection results from failure of the productive cascade comes from numerous studies on replication-impaired and replication-deficient HSV mutants. In all cases studied, there is no evidence that *any* gene expression at all is required. Most convincingly, a number of studies have shown that viral mutants lacking a functional α 4 or α-*trans*-inducing factor (α-TIF) which can express little if any productive cycle gene products can readily establish latent infections.[10–15] These latter results also demonstrate that virus replication is not *required* to establish latent infection, and Margolis and co-workers directly established that virus replication does not occur in neurons fated to be latently infected with a replication-competent virus.[4]

There are a myriad of reports of efficient establishment of latent infections in numerous animal models with virus mutants which do not express the latency associated transcript (LAT).[16–21] Where carefully studied, such mutants have no, or only a very limited. effect on the actual number of latently infected neurons in such animals.[17,22–26] Thus, it is clear that LAT expression is not a requisite for establishing latency per se.

2.2. Viral Genomes in Latently Infected Neurons

Viral DNA is readily detected in peripheral and central nervous system tissue of latently infected animals and in human autopsy material. Rock and Fraser demonstrated that viral genomes are either episomal or concatenated in animal neurons,[27,28] and this observation was confirmed and also shown to be the case in human neurons by workers with Wildy.[29] A more complete demonstration of an episomal state for viral DNA in mouse brains was subsequently presented by Mellerick and Fraser.[30]

Calculations show that the number of viral genomes in latently infected neurons is in the range of 10 to 100 copies per latently infected neurons.[27,29,31,32] The lower numbers correspond to neurons isolated from the CNS. This range of values is within the detection limits of standard *in situ* hybridization methods, but with one exception,[33] there has been a general failure to detect viral genomes utilizing them. This may reflect some aspect of the physical state of the viral genome in latently infected cells, and only PCR amplification *in situ* has provided positive results in both animal and human tissue.[34–37]

2.3. Most Latent HSV Genomes in Neurons Are Not Transcriptionally Active

The ability to detect LATs expressed from the HSV genome during latent infections in neurons provided a great impetus to the study of HSV gene function during latency and reactivation. Indeed, the high abundance of the stable intron processed from LAT makes it a reliable marker for assaying latent infections, and the fact that it is expressed at a very low level compared to productive cycle transcripts during productive infection makes its presence a useful marker for assaying latent versus productive or lytic infection in studying the establishment of latent infections outlined above.

Despite its convenience as a marker of latently infected cells, in mice and rabbits the number of neurons latently infected with HSV expressing LATs ranges between 10 to 30% or less of the total neurons containing viral genomes.[32,34,36,38–40] Because this transcription unit is correlated with efficient reactivation (see below), it is important to devise methods of correlating HSV transcription during latency with reactivation per se.

It should be noted in passing that Ramakrishnan and colleagues have found a much higher proportion of latently infected neurons expressing LAT in rat brains latently infected with a ribonucleotide reductase mutant.[41,42] Whether this difference reflects the animal used or the viral mutant is unclear at this time.

2.4. HSV Genomes Are Stably Maintained in Latently Infected Neurons

As is the case with establishing latent infection by HSV, maintaining viral genomes appears to be a passive phenomenon. This is in keeping with the stability of the neuronal population itself. The number of viral genomes and the number of neurons evidencing the presence of this stable intron processed from LAT do not change with time following the establishment of latent infection.[27,43,44] In a careful investigation of any possible role for latency-associated transcription in maintaining viral genomes, Sedarati, Stevens, and colleagues did a very careful quantitative analysis of viral DNA recoverable from 1 to 11 months following establishment of latent infection in murine dorsal root ganglia. Following an initial decline in genomes after productive infection ceased, no changes were seen.[17] Further, these workers showed that a replication-deficient mutant genome is fully stable over time in latently infected neurons.[12] Indeed, there are no reports in the literature to suggest that any replication-impaired or replication-deficient mutants studied for possible viral gene functions required to establish latent infections evidenced a time-correlated loss of latently infected neurons.

3. HSV GENE EXPRESSION DURING LATENT INFECTION IN NEURONS

The DNA encompassing the HSV latent phase transcription unit is situated in the long repeat regions of the viral genome and is thus diploid. It is colinear with a number of productive phase transcripts of both known and unknown functions expressed in the same and in the antisense orientation.

Much of our picture of how latent phase transcripts are expressed and how they function in reactivation is based on the detailed characterization of these transcripts using northern blot, S_1, and RNase resistance analysis of RNA:DNA and RNA:RNA hybrids.[44–59] For this review, all of our mapping data have been correlated with the sequence of the 17*syn*$^+$ strain, and minor differences in sequence between strains noted in the primary references have no impact on the general picture outlined below.

Northern blots of RNA from latently infected ganglia hybridized with defined DNA and short oligonucleotide probes demonstrate the presence of at least two partially colinear, relatively abundant poly(A−) RNA species mapping within the limits of the strong *in situ* hybridization signal noted above. The most abundant RNA species is approximately 2.0 kb in size, and there is a less abundant one of 1.5–1.4 kb.

The evidence that these two RNA species are introns is overwhelming. Most convincingly, Feldman and co-workers have expressed the primary latency-associated transcription unit in transfected cells and have demonstrated processing this unit to produce poly(A−) LAT.[60] Precise mapping of the 5′ and 3′ ends of the ply(A−) LAT species from latently infected tissue was carried out using both S_1 nuclease and RNAse protection analysis, and the 5′ and 3′ ends map to canonical splice signals.[44,48] Further, 2-kb poly(A−) LAT is uncapped,[49] and can be isolated as a lariat form.[61] Feldman's work[60] also demonstrates that this intron is stable in transfected cells and accumulates as it does in neurons, although we have found that the intron is not stable in all cells in which it is expressed, for example, mouse neuroblastoma cells. More recently, several laboratories have demonstrated that the two stable RNA species can be isolated as lariats.[61,62]

The smaller species is related to the larger by the removal of another 600-base intron.[44] This "respliced" LAT intron is seen only in latently infected tissue, and the relative amounts seen depend on the strain of virus used to establish the latent infection. Although the fact that secondary splicing of introns is not a well understood or described phenomenon has led some to speculate that an independently expressed small species is generated from a "second" LAT promoter, the recent demonstration that it is isolatable in lariat form is clearly consistent with a resplicing mechanism for its formation. Finally, although some laboratories have reported that the lower abundance smaller species is actually made up of two partially continguous trnascripts differing by 50 bases,[46–48,53] the report by Wu and colleagues demonstrates that these two species are actually the same RNA but differ in their physical configurations.[61]

Northern blot analysis of poly(A^+) RNA from productively infected cultured cells reveals the presence of an 8.5-kb transcript, expressed from the same DNA strand as that encoding LAT strand and beginning just 3′ of the LAT promoter and extending to a polyadenylation signal in the short repeat region.[49,50] The presence of such a primary transcript can be inferred from *in situ* hybridization data of latently infected neurons (reviewed in [63]), and on laboratory has reported detecting the primary transcript using northern blots of RNA from latently infected neural tissue.[45] Further, transcripts in latently infected tissue initiating at the expected cap site and extending to near the putative polyadenylation signal have been detected using RNAse protection analysis,[49,50] and confirmed by PCR analysis.[22,23]

RNAse protection assays demonstrate that the 5′ end of this poly(A^+) transcript is located approximately 25 bases 3′ of the 'TATA box' element of the LAT promoter (see below) and that RNA from latently infected murine sensory nerve ganglia protects the same fragment as seen with productive cycle ply(A^+) RNA. Although the 3′ end of LAT is not as precisely located as the 5′ end of the primary transcript, it is quite clear that the polyadenylation signal situated in the short repeat region just 3′ and on the opposite strand of DNA from that controlling the polyadenylation of α4, is utilized in lytic infection. RNAse protection probes from that region are protected by poly(A)-containing RNA isolated from productively infected cells, and PCR products from oligo-dT primed cDNA from latently infected rabbit and murine sensory nerve ganglia have been generated with primers located upstream, but not downstream, of this polyadenylation signal. This strongly suggests that a major portion of latent phase transcription terminates here. It should be noted, however, that weak *in situ* hybridization signals have been reported with HSV-2 probes extending into the C-terminal portion of the α4 protein.[53] Therefore, there may be more than one 3′ end to the latent transcript.

Because the 2- and 1.4-kb stable, poly(A^-) latent phase transcripts are introns, it is very clear that processed forms of the exons in the primary poly(A^+) transcript must occur. To date, these have not been extensively characterized, but Feldman and co-workers detected processed RNA, from which the major LAT intron had been spliced out, in their *in vitro* system, thereby demonstrating the stability of the LAT intron.[60] Evidence for such a processed RNA was also seen in PCR analysis of trigeminal neuron-derived RNA following induced reactivation of latently infected rabbits,[23] but we have not been able to isolate such a poly(A^+) containing RNA from productively infected or uninduced latently infected tissue.

A 2-kb poly(A) containing RNA extending from the LAT cap site to a position near the 3′ end of α0 mRNA can be readily isolated from productively infected cells,[49] However, it is not yet clear whether its contiguous 3′ end communicates with the LAT polyadenylation signal some 7 kb downstream via a splice or whether polyadenylation occurs at a noncanonical termination/polyadenylation signal. No role is known for this transcript.

As will be discussed, there are some data to support the idea of an independent promoter or, at least, transcription start site near the 5′ end of the stable 2-kb intron. Although it has been posited that this promoter gives rise to an RNA species partially colinear with the 2-kb and 1.4-kb LAT introns, this mechanism has not been convincingly demonstrated.

Other processed forms, such as one utilizing the splice signals seen in the 1.4-kb LAT intron along with those seen in the primary transcript, can also be posited, but no direct evidence for their occurrence in either lytic or latent infections exists at this time. It should be noted, however, that the expression of the translational reading frame present in the region of the viral genome encompassed by the 2-kb intron during infections of cultured cells[63,64] should involve the expression of some polyadenylated transcript mapping in this region. Yet none has been clearly identified, although Nicosia and colleagues have suggested that it is colinear with the 5′ portion of the latency-associated transcription unit.[65]

One of the problems of working with transcripts of such low abundance is that no biological function related to the expression of LAT maps to the regions identified. Another major problem is that there are a number of productive-phase, low-abundance transcripts mapping within the contiguous limits of the latency-associated tanscription unit whose kinetics of expression and regulation are unusual in that they are inhibited by immediate-early transcription.[66,67] Such transcripts can greatly complicate the analysis of low-abundance transcripts thought to be associated with latent infection because much of the analysis must be done in cultured cells. Finally, transcription patterns from LAT and other transcripts colinear with its 5′ portions are also expressed during productive infection and may or may not be under the control of novel promoters.

4. THE PROMOTER CONTROLLING HSV LATENT-PHASE TRANSCRIPTION

With HSV-1, detectable transcription is abolished in latent infections with a recombinant virus (17Δ*Pst*) containing a deletion of approximately 200 bases containing polII promoter elements located at -1700 bases 3′ of the α0 transcript terminus.[22,23,50] This is also the case with more extensive deletions encompassing this region.[26,50,68,69] That this region indeed contains the HSV-1 LAT promoter was fully confirmed by showing that high levels of a reporter gene mRNA are expressed when recombined in the appropriate position just downstream and RNAse protection experiments have shown that the LAT primary transcript initiates within the 202-base region deleted in the recombinant, just 3′ of a canonical 'TATA'box.[70] The location of this latent-phase promoter coincides exactly with a canonical polII promoter element suggested by sequence analysis. There is a high degree of sequence homology between HSV-1 and HSV-2 in the immediate region of this promoter but not immediately upstream or downstream of it.[70]

4.1. Analysis of Functional Elements of the HSV-1 Latent-Phase Promoter by Transient Expression Assay

The sequence of this latent-phase promoter contains a number of potential control elements which could have a role in transcription during latent infections. These include canonical cyclic AMP response elements (CRE); CAT box; Sp1, USF, YY1, AP-2 sites, an early growth response element and a sequence element, just at the cap site, which is quite similar to the strong α4 DNA binding site involved in that regulatory protein's autoregulation.[47,59,71–76] Although determining the actual importance of such sequences in the latent-phase activity of this promoter requires animal studies, a number of studies have been carried out in cell culture which suggest that some of these and other elements have a potential function. Thus, promoter activity is responsive to cAMP levels in cell culture, and the α4 binding site represses the ability of LAT expression to be induced by α4 in transient assays.[72,74,77–79]

The latent-phase promoter behaves unusually in transient expression assays in cultured cells. It has a rather high basal activity and is not highly activated by HSV superinfection compared with productive cycle promoters.[49,80,81] Further, basal promoter activity is markedly enhanced in cells of neuronal origin, and a region of the LAT promoter corresponding to that which we have implicated in such neuronal activity binds to a possible transcription factor found in neuronal-derived IMR-32 cells.[45]

It is important to note that despite the evidence that sequence elements influence neuronal expression in the latent-phase promoter, this promoter is not profoundly specific for neural cells because its activity is readily detectable in productive infection of cultured cells and in animals.[4,44,46,49,54,58,82] Further, the analysis of the LAT promoter in cultured cells and during productive infection leaves at least one rather obvious conundrum. Because LAT promoter activity appears to be repressible by α4 protein and the LAT promoter has rather high basal levels of activity, it is predicted that LAT should act somewhat like a weak immediate-early transcript during lytic infection. Despite this, there is no evidence for its expression under conditions of infection where only such immediate-early transcription occurs. This suggests that the actual states of the transcription template and the LAT promoter itself are important in LAT expression in latent infection.

4.2. Analysis of Latent-Phase Promoter Elements *in Vivo*

The study of functional elements of the latent-phase promoter is complicated by the fact that it is not clear just how active this promoter is during latent infection. As discussed above, viral genomes in many latently infected neurons are transcriptionally silent, and the high stability of the introns processed from the primary latent-phase transcript in transcriptionally active neurons makes a ready estimate of promoter activity impossible. Further, Margolis and colleagues have shown that reporter gene expression from the latent-phase promoter declines with time after establishment of latency.[83] It is important to note, however, that because the virus utilized had a deletion downstream of the LAT cap site, it is possible that this deletion affected the long-term expression characteristics of this virus. Adding to this argument is the observation that the total number of reporter-positive neurons dropped dramatically with time,[83] a pattern which does not mirror wild-type LAT expression. Still, all taken together, the best interpretation of all data is that the very small amount of RNA corresponding to the primary transcript recoverable suggests extremely low promoter activity during the latent phase of infection in neurons.

Despite the difficulties in study, it is clear that the functional elements of the latent-phase promoter seen in neurons *in vivo* do not fully correspond to those identified *in vitro*. Lokensgard, Feldman, and colleagues suggested that DNA sequence elements downstream of the latent-phase transcript cap have an important role in latent-phase transcription because they found that even extensive portions of upstream promoter elements controlling a reporter gene did not maintain neuronal expression when recombined into the unique long region of the viral genome.[84] Although such an analysis does not preclude that a role for template structure is important instead of the presence of an actual promoter element or elements, more recent studies by Soares, *et al.*[76] provide very strong evidence for the presence of a neuronal "enhancer" element within 60 bases or so immediately downstream of the transcript start site. Whether this element has a role in expression *in vivo* is unclear, however. Lastly, analysis of mutant virus, in which upstream promoter elements have been altered, was used to demonstrate the functionality of the cAMP response element nearest to the latent transcript start site. Further, the importance of the 'TATA' box in maximal expression was established.[77,78]

Clearly, further analysis is necessary, but it is also clear that the relatively laborious approach of generating viruses bearing defined mutations in putative control elements will provide the only reliable means of identifying elements in the latent-phase promoter that demonstrate neuron and latent-phase specificity. The fact that viruses which are lacking or impaired in latent-phase transcription are inefficient reactivators *in vivo* makes surveys of potentially important sites a somewhat more tractable task. For example, recent work by Bloom, Hill, and colleagues has demonstrated a partial role of the proximal cAMP response element in epinephrine-induced reactivation in rabbits.[86]

4.3. Is There a Second Latent-Phase Promoter?

Although deletions of the latent-phase promoter abolish readily measurable transcription during latent infection, as assayed by *in situ* hydridization and PCR analysis,[22,23,49,50,87,88] some workers continue to speculate about the existence of a second latent-phase promoter just upstream of the splice acceptor region of the stable 2-kb intron. This speculation is based on the following observations: (1) Insertion of a β-galactosidase reporter gene upstream of the intron results in low levels of reporter protein expression in latently infected mice even though a transcript initiating at the latent-phase transcript start site would have a very long leader.[89] (2) A β-galactosidase reporter gene construct controlled by the putative 'TATA-less' promoter from this region, when recombined into the long unique region of the viral genome, expressed a small amount of enzyme activity for a very long period following neuronal infection.[90] (3) A 2-kb transcript which is partially colinear with the 2-kb intron is expressed during productive infection by mutant viruses lacking a functional latent-phase promoter.[87,91] (4) Goins, Glorioso, and colleagues have found that a putative 'TATA-less' promoter is active in transient expression assays, although some 10-fold less so than the weak latent-phase promoter itself.[90]

Although the high stability of the functional β-galactosidase enzyme makes it unsuitable as a reporter to measure promoter activity in a given instance,[83] the presence of the enzyme in neuronal tissue and the transient expression data suggest that there is a low-activity promoter functioning upstream of the stable 2-kb intron. However, the great preponderance of the published data indicate that this promoter is active in *productive* not *latent* infection. Therefore it is invalid to define it as a second latent-phase promoter. Despite their nomenclature (latency-associated promoter 2 or LAP-2), Glorioso and colleagues agree with this assessment,[76] and thus the terminology is unfortunate and potentially misleading.

5. THE ROLE OF LATENT-PHASE TRANSCRIPTION IN LATENCY AND REACTIVATION

Because a latent infection can be established in the absence of measurable viral gene expression in all animal models tested, it is clear that latent-phase transcription cannot be a requirement. Further, careful quantitative analysis of the levels of viral genomes within neurons latently infected with virus mutants that do not express such transcripts argues against any required role of latency-associated transcription in maintaining viral genomes. By elimination, then, it is expected that any role for such latent-phase transcription is in the reactivation process, and, indeed, this is the case in rabbit, guinea pig, and some murine models. Despite this, the actual details of the role of such gene expression may differ depending on the model system utilized, and the specific regions of the latency-associated transcription unit critical for the effects seen do not precisely co-map between murine and *in vivo* models.

5.1. Latent-Phase Transcription Is Required for Efficient Reactivation in *in Vivo* Models

The fortuitous finding of a latent transcription negative mutant of strain 17*syn*$^+$ of HSV-1 (x10-13) with a 1200-base deletion of the latency-associated promoter and a 5′ region of the primary transcript derived from stocks of virus generated in the formation of HSV-1xHSV-2 recombinants[21,44] allowed Hill *et al.* to show that the lack of such gene expression correlates with the low frequency of induced viral reactivation *in vivo* in the rabbit eye model.[19] Similarly, a second strain 17*syn*$^+$ mutant with an extensive deletion is a poor spontaneous reactivator.[92]

The extensive nature of the deletions in these viruses required comprehensive confirmation of the role of latent-phase transcription *per se* with a mutant of strain 17*syn*+ containing a defined, engineered deletion of the latent-phase promoter itself.[23] These latter studies demonstrated a measurable effect on both spontaneous and induced viral reactivation as measured by recovery of infectious virus in the tear film. PCR analysis also demonstrated both the essential lack of any observable transcription extending to the polyadenylation signal of the primary transcript and that the levels of viral DNA measured by ratios to cellular DNA were essentially the same in latent infections with both mutant, *wt,* and latent-phase transcription-restored rescue virus. The reduced spontaneous reactivation in a rabbit model was confirmed with an independent mutant virus by Perng and colleagues.[93]

Because HSV-2 is a much more efficient spontaneous reactivator than HSV-1 in the guinea pig vaginal model, a completely independent mutant with a similar deletion of the LAT promoter was tested by Krause and colleagues,[20] and an essentially identical result was found. Latent-phase transcription has a significant role in efficient *in vivo* reactivation. Again, no effect was observed on the amount of viral DNA seen in latently infected tissue.

5.2. The Region of the Latent-Phase Transcript Important in the Efficient Reactivation Phenotype in Rabbits Is Confined to a Region of 480 bp or Less within Its Extreme 5′ End

Bloom *et al.* described a series of mutant of HSV-1 in which a polyadenylation signal was inserted between the cap site of LAT and its 3′ end. Termination of transcripts even 1,500 bases downstream of the cap had no effect on induced reactivation in the rabbit eye model.[94] In this same communication, it was shown that the specific deletion of a 348-bp DNA element situated between −205 and −554 bp 3′ of the cap site, containing three potential translation initiator codons, results in a virus which expresses both normal amounts of LAT and the stable introns, but which is not efficiently reactivated upon epinephrine induction. Agreeing with this observation, Hill *et al.* showed that a strain 17*syn*+ mutant containing a 370-bp deletion encompassed by two *Sty*I restriction sites, which partially overlap the critical 348-bp region, did not efficiently reactivate upon epinephrine induction.[95] Despite this agreement, however, a similar mutant in the McKrae strain of HSV-1 did not evidence reduced spontaneous reactivation.[96] The exact reason for this difference between induced and spontaneous reactivation in this case is not clear now. Despite such differences, all data taken together suggest that the smallest contiguous DNA element required for reactivation is situated within the 138-bp region from −77 to −216 downstream of the cap site, or, alternatively, multiple partially redundant sites exist within the full 478-bp region spanned by the two mutants. Current data using mutants, whose sequences have been mutagenized instead of deleted, confirm the observation that no elements downstream of the point 554 bases 3′ of the start of LAT are involved in efficient induced reactivation. Other deletion and substitution mutants are being currently examined to characterize this region further.

5.3. Evidence that Modulation of Expression of LAT during the Latent Phase or at the Initiation of Reactivation Has a Role in Efficient Induction of Virus in the Rabbit Eye Model

The fact that a new class of reactivation mutants now exist which express normal amounts of LAT during the latent infection[94] reinforces the observation that the amount of LAT produced by naturally occurring HSV strains does not easily correlate with their relative reactivation potentials.[39,97] This suggests that LAT's influence on reactivation may be exerted as a transient change in the quantity of transcription following the reactivation stimulus. Supporting the idea that LAT may be induced during reactivation are results with a recombinant in which one of the CRE elements within the LAT promoter was mutagenized.[86] Although a similar mutant examined in the mouse showed only slight reduction in reactivation efficiencies by cocultivation,[85] the mutant (17CRE) showed a more dramatic reduction in reactivation in the rabbit, particularly with respect to spontaneous reactivation frequencies. This clearly suggests that the CRE elements in the LAT promoter may be important in inducing LAT transcription during the early phase of reactivation. These studies also highlight the apparent differences between the various animal models used for these studies.

5.4. Latent-Phase Transcription Facilitates but Is Not Required for Efficient Recovery of Infectious Virus from Explanted Latently Infected Murine Ganglia

The murine explant model, which is not an *in vivo* reactivation model, exhibits a more varied pattern of dependence on LAT expression. Thus, latent-phase transcription with at least some strains of HSV-1 has an effect on the course of virus production in explanted latently infected murine ganglia. This was first reported by Leib and colleagues[16] who showed that fewer trigeminal ganglia, explanted following latent infection with a mutant of the KOS strain of HSV-1 specifically deleted in the latent-phase promoter, can produce virus than the *wt* control. Both extensive and more defined mutants of the 17*syn*$^+$ strain of HSV-1 showed similar reduced and/or delayed recovery levels from latently infected murine dorsal root and trigeminal ganglia.[18,22,98] Despite this, Izumi *et al.*[26] and Devi-Rao *et al.*[22] did not see such a delay or reduction in virus recovery from sacral ganglia latently infected with several KOS(M) strain-derived, latent-phase transcription negative mutants. The reason for such a difference may be related to the fact that the KOS(M) viral strain has different pathogenic properties in mice.[99]

5.5. The HSV-1 Latent Phase Transcript May Have a Role in the Efficiency in Establishing Latent Infection in Murine Trigeminal Ganglia

Sawtell and Thompson,[24,100] utilizing their hyperthermic reactivation model assaying productive-phase gene expression in the neuron itself, reported a reduction in reactivation frequency with a latent-phase transcription negative mutant of the KOS(M) strain of HSV-1. They also reported that the number of latently infected neurons seen in trigeminal ganglia in mice infected in the eye with the latent-phase transcription negative mutant was measurably lower than those for mice infected with the *wt* virus and suggested that the reduced reactivation frequency and delayed kinetics of reactivation observed in explant models could be explained by this difference.

Although the significance of this finding is not altogether clear, two other labs have also reported a slightly reduced efficiency in establishing latent infections by LAT $(^{-})$ viruses in murine eye models. Devi-Rao *et al.* assayed viral DNA levels using PCR and found a slight but measurably significant reduction in viral DNA recovered from murine trigeminal ganglia latently infected with latency-associated transcription negative mutants of strain 17*syn*$^{+}$ and KOS(M) but observed no difference with KOS(M)-derived mutant virus in sacral ganglia.[22] A measurable reduction in latently infected neurons in trigeminal ganglia following infection of mice with a LAT$(^{-})$ mutant of strain 17*syn*$^{+}$ compared to infections with *wt* virus was also reported by Block and colleagues using PCR amplified *in situ* hybridization.[40]

If such a role for latent-phase transcription in establishing latent infection in murine trigeminal neurons is confirmed, then there is a significant difference in the phenomenology of establishing latency in mice compared to rabbits. This follows from the fact that there is no evidence for delayed recovery of latent-phase promoter-defective virus from latently infected rabbit trigeminal ganglia,[23] and, as noted above, no quantitative difference in viral DNA recovery was found in rabbits latently infected with this or *wt* virus. As discussed below, however, such a difference or differences does not necessarily require that the mechanism of action of latent-phase transcription differs markedly in the different systems.

5.6. Murine Explant Models Do Not Reveal a Region Critical for Virus Recovery Equivalent to That Characterized for Rabbit Reactivation *in Vivo*

It is striking that despite the excellent agreement between different mutants and different strains of HSV-1 in localizing the reactivation function to a specific region in rabbit models, this region is not critical for recovery of virus from explanted latently infected murine trigeminal or dorsal root ganglia. Thus, Maggioncalda and colleagues showed that the deletion of the 370 bp between the *Sty*I sites in strain 17*syn*$^+$, which so profoundly affects induced reactivation in rabbits, has no effect at all on the recovery of virus from explanted latently infected trigeminal ganglia.[101] Bloom *et al.*[94] reported the same for their 348-bp deletion mutant. The possible interpretation of such results in the light of comparing the mode of action of latency-associated transcription in rabbits compared to mice is briefly considered in the following section.

6. THE MECHANISM OF ACTION OF HSV LAT IN REACTIVATION

Because the expression of the 5′ portion of the latent-phase transcription unit is involved in efficient reactivation in several *in vivo* models, it would seem no great trick to determine the anatomical site at which it works and the molecular mechanism of its action. Unfortunately, this trick has not been successfully performed to date. There is a large accumulating body of evidence that the mode of action is not simple. LAT mediates reactivation in highly differentiated tissue and clearly in only a very limited subpopulation of cells, possibly a single neuron. In addition, the basal level of latent-phase LAT expression varies among neurons. All of these factors combine to make the study of what is really going on in HSV reactivation very difficult.

6.1. There Is No Evidence for a Major Antisense-Mediated Repressive Action in Animal Models

The fact that the latency-associated transcription unit and its stable intron are antisense to the α0 transcript that is critical in mediating low multiplicity virus replication suggested that LAT works as an antisense modulator of immediate early gene expression to initiate the productive cascade.[56] Such a modulation might protect certain critical neurons or peripheral cells from productive virus infection. Although such a role is counter to the positive effect of latent-phase transcription on reactivation, one could posit that certain critical cells must have been protected at the very initiation of the latent phase of infection or that the modulation protects the nerve ganglia during the process of reactivation itself.

This model is attractive, and it might explain the significant differences between the course of reactivation seen with HSV and that of the "naturally" latent-phase transcription negative alpha herpesvirus, VZV. Unfortunately, none of the model systems utilized to examine reactivation support such a role. Thus, although several groups reported that high levels of LAT intron in productively infected cells had a small but measurable negative effect on the course of productive infection,[60,102] levels normally seen in productive infection have no effect.[49]

Also, the histology of reactivating rabbit ganglia following latent infection with LAT($^-$) virus shows no difference in cytopathology compared to *wt* virus (J. M. Hill, unpublished). Most critically, of course, is the fact that mutant viruses can be constructed which express the stable LAT intron but which do not reactivate efficiently in the rabbit model, and viruses can be constructed which do not express the latent-phase transcript intron and which react normally.[23,96,103]

6.2. There Is No Evidence for a Latent-Phase-Expressed Viral Protein Involved in Reactivation

The prominent open translational reading frame within the HSV-1 LAT intron prompted speculation and experimental study of the possibility that it or another protein expressed *during* the latent phase of infection mediates the reactivation phenotype.[44] The expression of this open reading frame during productive infection has been detected, [63,64] but elimination or mutation of the open reading frame has no effect on efficient spontaneous or induced reactivation in rabbits.[23,104] Nor does this open reading frame play a role in explant-induced recovery of virus from latently infected mouse neurons.[105]

Expression of other translational reading frames during latent infection can also be eliminated. Thus, Bloom and colleagues specifically set out to eliminate the three potential translation initiators found in the critical 348-bp region in the 5′ portion of the latent-phase transcription unit to asses the possible expression of a protein and clearly showed no effect on efficient recovery of virus from epinephrine-induced latently infected rabbits.[94] Further, the normal reactivation of viruses in which the vast majority of latent phase transcripts were terminated by inserting an efficient polyadenylation signal 1500 bases 3′ of the transcript start site effectively eliminates appreciable expression of any translational reading frames downstream of this.

Of course, only specific mutational deletion of each open translational reading frame found within the whole latent-phase transcription unit will *completely* rule out the expression of a minute amount of some protein having a role, and it will be required to eliminate fully expression of one of these translational reading frames during the early events following sucessful reactivation in a specific cell. This is a laborious task and somewhat difficult to justify as a high priority with the data at hand. Still, sufficient work has been done to rule out the expression of the ICP34.5 protein expressed by a transcript from the same DNA strand that encodes the α0 transcript and has a role in neurovirulence with an influence on HSV recovery from rabbits *in vivo* or mice *in vitro*.[106–111] Similarly, the ORF-P protein that is expressed from one or another low-abundance transcripts which are partially colinear but independently controlled by productive-phase promoters does not have any measurable role in the reactivation process.[67,94,112,113]

6.3. Possible *cis*-Acting Mechanisms for the Influence of LAT on Reactivation

The best interpretation of all of the evidence discussed above is that HSV latent-phase transcription mediates reactivaion *via* some *cis*-acting mechanism. Two possible models for such activity are discussed below.

6.3.1. *Methylation and Transcriptional Availability*

The region at the 5′ end of LAT is critical for epinephrine-induced reactivation. It is possible that this region downstream of the LAT promoter is a transcriptional enhancer and is required to maintain the appropriate structure of the latent DNA that allows the LAT promoter to be accessible to reactivation mediators. A model for such a structural enhancer is based on the observation that this region contains a cluster of CpGs that extend into the LAT promoter.[94] CpG islands have long been known to be implicated in transcriptional regulation of eukaryotic gene expression through methylation. Aside from directly interfering with binding specific transcription factors, methylation of CpGs can change the twist of the DNA and impact nucleosomal placement. Then it is possible that the deletion of a critical region at this region 3′ to the LAT promoter alters the DNA structure so as to alter the accessibility of binding elements within the LAT promoter to transcriptional mediators of reactivation. The idea that methylation could play a role in reactivation is not a new one: It has long been known that 5-azacytidine affects transcription of thymidine kinase and induces reactivation in *in vivo* systems.[114,115] The actual significance of these *in vivo* studies was difficult to interpret, however, because of the potential global and indirect effects of 5-azacytidine. In addition, initial studies of latent HSV DNA demonstrated that the latent genome was not extensively methylated.[116] However, the identification of the CpGs in this reactivation critical region of the LAT makes investigation of the methylation status of this specific region of the latent HSV genome particularly important.

6.3.2. *Could the Critical Region of Latent Phase Transcription Unit Act as a Cell-Specific Origin of Genome Replication?*

The potential methylation patterns defined by CpG islands in the critical *cis*-acting region of the latent-phase transcript are similar to those seen near the origins of replication of the virus. This and the requirement for transcription could be interpreted to imply that the critical region acts as a tissue-specific origin or replication operating only in the cell or cells immediately involved in productive reactivation. Certainly, there is no evidence based on the recovery of plasmids from neuronal cells transfected with this region of the latent-phase transcription unit and then superinfected with virus that this region acts as a readily measurable origin of replication in cultured cells.[49] The only way to really test such a possibility is to identify a tissue or cell type in which it can be demonstrated that this critical region is required for genomic amplification.

6.4. Does Latent-Phase Transcription Supply an Essential Function to Neurons or Peripheral Cells Initiating Reactivation?

One of the major difficulties in developing a full understanding of the role of HSV latent-phase transcription in the reactivation process is that we do not yet have a definitive picture of *where* it works. Although it may be assumed that it must play its role in a subset of neurons, it can also be posited that the critical site for LAT to act is in a peripheral cell or tissue either just before establishing latency or during the early stages following the appearance of infectious virus at the neuronal periphery. With this information in hand, a more critical investigation of the mechanism of latent-phase transcription in facilitating reactivation may be feasible.

It has long been known that very small amounts of HSV and/or viral transcripts and proteins can be detected in some neuronal ganglia in the absence of evidence of reactivation, and recently this has been carefully described using PCR analysis in murine ganglia.[43,117–119] Although this can be envisioned as evidence of very early steps in the reactivation process involving productive virus replication in the neuronal ganglia, these sporadic events could also result from abortive, atypical, or "dead end" virus production with no clear role in the actual process of reactivation itself. This second conclusion follows from the fact that careful PCR analysis of RNA from ganglia latently infected with LAT expressing and LAT negative viruses in the few hours following either induction of reactivation in rabbits with epinephrine or following the insult of explant in mice showed essentially no difference in the levels of timing of the appearance of productive cycle transcripts.[22,23] Although such results were tentatively interpreted to indicate that the differences in reactivation phenotypes seen with the two viruses must, therefore, be either the result of a critical event in a limited population of neurons or in a peripheral tissue, it should be remembered that a full survey of viral transcripts is not yet at hand. It is possible that differences might be seen if transcripts requiring, say, DNA replication were assayed. Unfortunately, PCR primers sensitive enough for such studies have been difficult to characterize.

Because the expression of LAT has no measurable role in virus replication in cultured cells, investigation of possible growth restriction of LATI(−) virus mutants is not straightforward. One promising report of such a role[120] may be invalid because the mutant virus utilized eliminates transcription from upstream of LAT, an area known to mediate some aspects of neuropathology and virus replication in CV-1 cells from the work of Rosen-Wolf and colleagues.[121] Still, one way of rationalizing the differences seen between the lack of any role for latent-phase transcription in the efficiency of establishing latent infections in rabbits, guinea pigs, and murine dorsal root ganglia compared to that seen in the mouse eye model is that LAT($^-$) virus slightly restricts replication in some peripheral tissue. This restriction is generally seen in the critical first step following the appearance of virus at the periphery of the latently infected neuron in *in vivo* reactivation, but could also operate at some level in primary infection in the mouse eye. Here the reduced yield of virus would result in a lower inoculation of latent genomes into the neurons themselves.

The process of transcription and viral genome replication in both trigeminal neurons and in the cornea during the early stages of induced reactivation in rabbits is currently being carefully analyzed in our laboratories with the collaboration of J. M. Hill of LSU. Preliminary data rule out any readily measurable restriction of LAT($^-$) virus replication following reinfection of rabbits latently infected with *wt* virus, and similarly, no differences in LAT($^+$) and LAT($^-$) virus replication were detected *in vitro* in cultured corneal keritinocytes. These results suggest that any peripheral restriction in rabbits is subtle, at best. Hopefully, continued analysis will clarify the issues involved.

REFERENCES

1. Wagner, E. K., 1991, Herpesvirus transcription—General aspects, in: *Herpesvirus Transcription and its Regulation* (E. K. Wagner, ed.) CRC Press, Boca Raton, pp. 1–15.
2. Wagner, E. K., Guzowski, J. F., and Singh, J., 1995, Transcription of the herpes simplex virus genome during productive and latent infection, in: *Progress in Nucleic Acid Research and Molecular Biology* (W. E. Cohen and K. Moldave, eds.), Academic Press, San Diego, pp. 123–268.
3. Wagner, E. K., 1994, Herpes simplex virus—Molecular Biology, in: *Encyclopedia of Virology* (R. G. Webster and A. Granoff, eds.), Academic Press, London, pp. 593–603.
4. Margolis, T. P., Sedarati, F., Dobson, A. T., Feldman, L. T., and Stevens, J. G., 1992, Pathways of viral gene expression during acute neuronal infection with HSV-1, *Virology* **189:**150–160.
5. Speck, P. G., and Simmons, A. 1991, Divergent molecular pathways of productive and latent infection with a virulent strain of herpes simplex virus type 1, *J. Virol.* **65:**4001–4005.
6. Howard, M. K., Mailhos, C., Dent, C. L., and Latchman, D. S., 1993, Transactivation by the herpes simplex virus virion protein Vmw65 and viral permissivity in a neuronal cell line with reduced levels of the cellular transcription factor Oct-1, *Exp. Cell Res.* **207:**194–196.
7. Lillycrop, K. A., Dent, C. L., Wheatley, S., Beech, M. N., Ninkina, N. N., Wood J. N., and Latchman, D. S., 1991, The octamer-binding protein Oct-2 represses HSV immediate-early genes in cell lines derived from latently infectable sensory neuron, *Neuron* **7:**381–390.
8. Lillycrop, K. A., Howard, M. K., Estridge, J. K., and Latchman, D. S., 1994, Inhibition of herpes simplex virus infection by ectopic expression of neuronal splice variants of the Oct-2 transcription factor, *Nucleic Acids Res.* **22:**815–820.
9. Wheatley, S. C., Dent, C. L., Wood, J. N., and Latchman, D. S., 1992, Elevation of cyclic AMP levels in cell lines derived from latently infectable sensory neurons increases their permissivity for herpes virus infection by activating the viral immediate-early 1 gene promoter, *Mol. Brain Res.* **12:**149–154.
10. Valyi-Nagy, T., Deshmane, S. L., Spivack, J. G., Steiner, I., Ace, C. I., Preston, C. M. and

Fraser, N. W., 1991, Investigation of herpes simplex virus type 1 (HSV-1) gene expression and DNA synthesis during the establishment of latent infection by an HSV-1 mutant, *in* 1814, that does not replicate in mouse trigeminal ganglia, *J. Gen. Virol.* **72:**641–649.
11. Steiner, I., Spivack, J. G., Deshmane, S. L., Ace, C. I., Preston, C. M., and Fraser, N. W., 1990, A herpes simplex virus type 1 mutant containing a nontransinducing Vmw65 protein establishes latent infection *in vivo* in the absence of viral replication and reactivates efficiently from explanted trigeminal ganglia, *J. Virol,* **64:**1630–1638.
12. Sedarati, F., Margolis, T. P., and Stevens, J. G., 1993, Latent infection can be established with drastically restricted transcription and replication of the HSV-1 genome, *Virology* **192:**687–691.
13. Katz, J. P., Bodin, E. T., and Coen, D. M., 1990, Quantitative polymerase chain reaction analysis of herpes simplex virus DNA in ganglia of mice infected with replication-incompetent mutants, *J. Virol.* **64:**4288–4295.
14. Dobson, A. T., Margolis, T. P., Sedarati, F., Stevens, J. G., and Feldman, L. T., 1990, A latent, nonpathogenic HSV-1-derived vector stably expresses β-galactosidase in mouse neurons, *Neuron* **5:**353–360.
15. Harris, R. A., and Preston, C. M., 1991, Establishment of latency *in vitro* by the herpes simplex virus type 1 mutant *in* 1814, *J. Gen. Virol.* **72:**907–913.
16. Leib, D. A., Bogard, C. L., Kosz-Vnenchak, M., Hicks, K. A., Coen, D. M., Knipe, D. M., and Schaffer, P. A., 1989, A deletion mutant of the latency-associated transcript of herpes simplex virus type 1 reactivates from the latent state with reduced frequency, *J. Virol.* **63:**2893–2900.
17. Sedarati, F., Izumi, K. M., Wagner, E. K., and Stevens, J. G., 1989, Herpes simplex virus type 1 latency-associated transcription plays no role in establishment or maintenance of a latent infection in murine sensory neurons, *J. Virol.* **63:**4455–4458.
18. Steiner, I., Spivack, J. G., Lirette, R. P., Brown, S. J., MacLean, A. R., Subak-Sharpe, J. H., and Fraser, N. W., 1989, Herpes simplex virus type 1 latency-associated transcripts are evidently not essential for latent infection, *EMBO J.* **8:**505–511.
19. Hill, J. M., Sedarati, F., Javier, R. T., Wagner, E. K., and Stevens, J. G., 1990, Herpes simplex virus latent phase transcription facilitates *in vivo* reactivation, *Virology* **174:**117–125.
20. Krause, P. R., Stanberry, L. R., Boiurne, N., Connelly, B., Kurawadwala, J. F., Patel, A., and Straus, S. E., 1995, Expression of the herpes simplex virus type 2 latency-associated transcript enhances spontaneous reactivation of genital herpes in latently infected guinaea pigs, *J. Exp. Med.* **181:**297–306.
21. Javier, R. T., Stevens, J. G., Dissette, V. B., and Wagner, E. K., 1988, A herpes simplex virus transcript abundant in latently infected neurons is dispensable for establishment of the latent state, *Virology* **166:**254–257.
22. Devi-Rao, G. B., Bloom, D. C., Stevens, J. G., and Wagner, E. K., 1994, Herpes simplex virus type 1 DNA replication and gene expression during explant induced reactivation of latently infected murine sensory ganglia, *J. Virol.* **68:**1271–1282.
23. Bloom, D. C., Devi-Rao, G. B., Hill, J. M., Stevens, J. G., and Wagner, E. K., 1994, Molecular analysis of herpes simplex virus type 1 during epinephrine induced reactivation of latently infected rabbits *in vivo, J. Virol.* **68:**1283–1292.
24. Sawtell, N. M., and Thompson, R. L., 1992, Herpes simplex virus type 1 latency-associated transcription unit promotes anatomical site-dependent establishment and reactivation from latency, *J. Virol.* **66:**2157–2169.
25. Leib, D. A., Coen, D. M., Bogard, C. L., Hicks, K. A., Yager, D. R., Knipe, D. M., Tyler, K. L., and Schaffer, P. A., 1989, Immediate-early regulatory gene mutants define different stages in the establishment and reactivation of herpes simplex virus latency, *J. Virol.* **63:**759–768.
26. Izumi, K. M., McKelvey, A. M., Devi-Rao, G. B., Wagner, E. K., and Stevens, J. G., 1989, Molecular and biological characterization of a type 1 herpes simplex virus (HSV-1) specifically deleted for expression of the latency-associated transcript (LAT), *Microb. Pathog.* **7:**121–134.

27. Rock, D. L., and Fraser, N. W., 1983, Detection of the HSV-1 genome in the central nervous system of latently infected mice, *Nature* **302:**523–525.
28. Rock, D. L., and Fraser, N. W., 1985, Latent herpes simplex virus type 1 DNA contains two copies of the virion joint region, *J. Virol.* **62:**3820–3826.
29. Efstathiou, S., Minson, A. C., Field, H. J., Anderson, J. R., and Wildly, P., 1986, Detection of herpes simplex virus specific DNA sequences in latently infected mice and in humans, *J. Virol.* **57:**446–455.
30. Mellerick, D. M., and Fraser, N. W., 1987, Physical state of the latent simplex virus genome in a mouse model system: Evidence suggesting an episomal state, *Virology* **158:**254–275.
31. Cabrera, C. V., Wohlenberg, C., Openshaw, H., Rey-Mendez, M., Puga, A., and Notkins, A. L., 1980, Herpes simplex virus DNA sequences in the CNS of latently infected mice, *Nature* **288:**288–290.
32. Hill, J. M., Gebhardt, B. M., Wen, R. J., Bouterie, A. M., Thompson, H. W., O'Callaghan, R.J., Halford, W. P., and Kaufman, H. E., 1996, Quantitation of herpes simplex virus type 1 DNA and latency-associated transcripts in rabbit trigeminal ganglia demonstrates a stable reservoir of viral nucleic acids during latency, *J. Virol.* **70:**3137–3141.
33. Sequiera, L. W., Jennings, L. C., Carrasco, L. H., Lord, M. A., Curry, A., and Sutton, R. N., 1979, Detection of herpes-simplex viral genome in brain tissue, *Lancet* **2:**609–612.
34. Gressens, P., and Martin, J. R., 1994, *In situ* polymerase chain reaction: Localization of HSV-2 DNA sequences in infections of the nervous system, *J. Virol. Methods* **46:**61–83.
35. Nicoll, J. A. R., Love, S., and Kinrade, E., 1993, Distribution of herpes simplex virus DNA in the brains of human long-term survivors of encephalitis, *Neurosci. Lett.* **157:**215–218.
36. Mehta, A., Maggioncalda, J., Bagasra, O., Thikkavarapu, S., Saikumari, P., Valyi-Nagy, T., Fraser, N. W., and Block, T. M., 1995, *In situ* DNA PCR and RNA hybridization detection of herpes simplex virus sequences in trigeminal ganglia of latently infected mice, *Virology* **206:**633–640.
37. Miller, R. G., Fox, J. D., Thomas, P., Waite, J. C., Sharvell, Y., Gazzard, B. G., Harrison, M. J. G., and Brink, N. S., 1996, Acute lumbosacral polyradiculopathy due to cytomegalovirus in advanced HIV disease: CSF findings in 17 patients, *J. Neurol. Neurosug. Psychiatry* **61:**456–460.
38. Ecob-Prince, M. S., Hassan, K., Denheen, M. T., and Preston, C. M., 1995, Expression of β-galactosidase in neurons of dorsal root ganglia which are latently infected with herpes simplex virus type 1, *J. Gen. Virol.* **76:**1527–1532.
39. Ecob-Prince, M. S., Rixon, F. J., Preston, C. M., Hassan, K., and Kennedy, P. G. E., 1993, Reactivation *in vivo* and *in vitro* of herpes simplex virus from mouse dorsal root ganglia which contain different levels of latency-associated transcripts, *J. Gen. Virol.* **74:**995–1002.
40. Maggioncalda, J., Mehta, A., Su, Y. H., Fraser, N. W., and Block, T. M., 1996, Correlation between herpes simplex virus type 1 rate of reactivation from latent infection and the number of infected nerons in trigeminal ganglia, *Virology* **225:**72–81.
41. Ramakrishnan, R., Levine, M., and Fink, D. J., 1994, PCR-based analysis of herpes simplex virus type 1 latency in the rat trigeminal ganglion established with a ribonucleotide reductase-deficient mutant, *J. Virol.* **68:**7083–7091.
42. Ramakrishnan, R., Poliani, P. L., Levine, M., Glorioso, J. C., and Fink, D. J., 1996, Detection of herpes simplex virus type 1 latency-associated transcript expression in trigeminal ganglia by *in situ* reverse transcriptase PCR, *J. Virol.* **70:**6519–6523.
43. Kramer, M. F., and Coen, D. M., 1995, Quantification of transcripts from the ICP4 and thymidine kinase genes in mouse ganglia latently infected with herpes simplex virus, *J. Virol.* **69:**389–1399.
44. Wagner, E. K., Flanagan, W. M., Devi-Rao, G. B., Zhang, Y. F., Hill, J. M., Anderson, K. P., and Stevens, J. G., 1988, The herpes simplex virus latency-associated transcript is spliced during the latent phase of infection, *J. Virol.* **62:**4577–4585.

45. Zwaagstra, J. C., Ghiasi, H.,Slanina, S. M., Nesburn, A. B., Wheatley, S. C., Lillycrop, K., Woods, J., Latchman, D. S., Patel, K., and Wechsler, S. L., 1990, Activity of herpes simplex virus type 1 latency-associated transcript (LAT) promoter in neuron-derived cells: Evidence for neuron specificity and for a large LAT transcript, *J. Virol.* **64:**5019–5028.
46. Wechsler, S. L., Nesburn, A. B., Watson, R., Slanina, S., and Ghiasi, H., 1988, Fine mapping of the latency-related gene of herpes simplex virus type 1 in humans, *J. Gen. Virol.* **69:**3101–3106.
47. Wechsler, S. L., Nesburn, A. B., Watson, R., Slanina, S. M., and Ghiasi, H., 1988, Fine mapping of the latency-related gene of herpes simplex virus type 1: Alternative splicing produces distinct latency-related RNAs containing open reading frames, *J. Virol.* **62:**4051–4058.
48. Deatly, A. M., Spivack, J. G., Lavi, E., O'Boyle, D. R., and Fraser, N. W., 1988, Latent herpes simplex virus type 1 transcripts in peripheral and central nervous system tissues of mice map to similar regions of the viral genome, *J. Virol.* **62:**749–756.
49. Devi-Rao, G. B., Goodart, S. A., Hecht, L. B., Rochford, R., Rice, M. K., and Wagner, E. K., 1991, The relationship between polyadenylated and non-polyadenylated herpes simplex virus type 1 latency associated transcripts, *J. Virol.* **65:**2179–2190.
50. Dobson, A. T., Sedarati, F., Devi-Rao, B. G., Flanagan, W. M., Farrell, M. J., Stevens, J. G., Wagner,E. K., and Feldman, L. T., 1989, Identification of the latency-associated transcript promoter by expression of rabbit beta-globulin mRNA in mouse sensory nerve ganglia latently infected with a recombinant herpes simplex virus, *J. Virol.* **63:**3844–3851.
51. Mitchell, W. J., Deshmane, S. L., Dolan, A., McGeoch, D. J., and Fraser, N. W., 1990, Characterization of herpes simplex virus type 2 transcription during latent infection of mouse trigeminal ganglia, *J. Virol.* **64:**5342–5348.
52. Krause, P.R., Croen, K. D., Straus, S. E., and Ostrove, J. M., 1988, Detection and preliminary characterization of herpes simplex virus type 1 transcripts in latently infected human trigeminal ganglia, *J. Virol.* **62:**4819–4823.
53. Mitchell, W. J., Lirette, R. P., and Fraser, N. W., 1990, Mapping of low abundance latency-associated RNA in the trigeminal ganglia of mice latently infected with herpes simplex virus type 1, *J. Gen Virol.* **71:**125–132.
54. Spivack, J. G., and Fraser, N. W., 1988, Expression of herpes simplex virus type 1 (HSV-1) latency-associated transcripts and transcripts affected by the deletion in avirulent mutant HFEM: Evidence for a new class of HSV-1 genes, *J. Virol.* **62:**3281–3287.
55. Spivack, J. G., Woods, G. M., and Fraser, N. W., 1991, Identification of a novel latency-specific splice donor signal within the herpes simplex virus type 1 2.0-kilobase latency-associated transcript (LAT): Translation inhibition of LAT open reading frames by the intron within the 2.0-kilobase LAT, *J. Virol.* **65:**6800–6810.
56. Stevens, J. G., Wagner, E. K., Devi-Rao, G. B., Cook, M. L., and Feldman, L. T., 1987, RNA complementary to a herpesvirus alpha gene mRNA is prominent in latently infected neurons, *Science* **235:**1056–1059.
57. Suzuki, S., and Martin, J. R., 1989, Herpes simplex virus type 2 transcripts in trigeminal ganglia during acute and latent infection in mice, *J. Neurol. Sci.* **93:**239–251.
58. Wagner, E. K., Devi-Rao, G. B., Feldman, L. T., Dobson, A. T., Zhang, Y. F., Flanagan, W. M., and Stevens, J. G., 1988, Physical characterization of the herpes simplex virus latency-associated transcript in neurons, *J. Virol.* **62:**1194–1202.
59. Zwaagstra, J. C., Ghiasi, H., Nesburn, A. B., and Wechsler, S. L., 1991, Identification of a major regulatory sequence in the latency associated transcript (LAT) promoter of herpes simplex virus type 1 (HSV-1), *Virology* **182:**287–297.
60. Farrell, M. J., Dobson, A. T., and Feldman, L. T., 1991, Herpes simplex virus latency-associated transcript is a stable intron, *Proc. Natl. Acad. Sci. USA* **88:**790–794.
61. Wu, T. T., Su, Y. H., Block, T. M., and Taylor, J. M., 1996, Evidence that two latency-associated transcripts of herpes simplex virus type 1 are nonlinear, *J. Virol.* **70:**5962–5967.

62. Rodahl, E., and Haarr, L., 1997, Detection of a 2 kb LAT lariat in HSV-1 infected PC-12 cells, *J. Virol.*, in press.
63. Rice, M. K., Devi-Rao, G. B., and Wagner, E. K., 1993, Latent phase transcription by alpha herpesviruses, in: *Genome Research in Molecular Medicine and Virology* (K. W. Adolph, ed.), Academic Press, Orlando, pp. 305–324.
64. Doerig, C., Pizer, L. I., and Wilcox, C. L., 1991, An antigen encoded by the latency-associated transcript in neuronal cell cultures latently infected with herpes simplex virus type 1, *J. Virol.* **65:**2724–2729.
65. Nicosia, M., Zabolotny, J. M., Lirette, R. P., and Fraser, N. W., 1994, The HSV-1 2-kb latency-associaed transcript is found in the cytoplasm comigrating with ribosomal subunits during productive infection, *Virology* **204:**717–728.
66. Yeh, L., and Schaffer, P. A, 1993, A novel class of transcripts expressed with late kinetics in the absence of ICP4 spans the junction between the long and short segments of the herpes simplex virus type 1 genome, *J. Virol.* **67:**7373–7382.
67. Bohenzky, R. A., Lagunoff, M., Roizman, B., Wagner, E. K., and Silverstein, S., 1995, Two overlapping transcription units which extend across the L-S junction of herpes simplex virus type 1, *J. Virol.* **69:**2889–2899.
68. Snowden, B. W., Blair, E. D., and Wagner, E. K., 1988, Transcriptional activation with concurrent or nonconcurrent template replication has differential effects on transient expression from herpes simplex virus promoters, *Virus Genes* **2:**129–145.
69. Mitchell, W. J., Steiner, I., Brown, S. M., MacLean, A. R., Subak-Sharpe, J. H., and Fraser, N. W. 1990, A herpes simplex virus type 1 variant, deleted in the promoter region of the latency-associaed transcripts, does not produce any detectable minor RNA species during latency in the mouse trigeminal ganglion, *J. Gen. Virol.* **71:**953–957.
70. Krause, P. R., Ostrove, J. M., and Straus, S. E., 1991, The nucleotide sequence, 5′ end, promoter domain, and kinetics of expression of the gene encoding the herpes simplex virus type 2 latency-associated transcript, *J. Virol.* **65:**4519–5623.
71. Batchelor, A. H., and O'Hare, P. 1992, Localization of *cis*-acting sequence requirements in the promoter of the latency-associated transcript of herpes simplex virus type 1 required for cell-type-specific activity, *J. Virol.* **66:**3573–3582.
72. Batchelor, A. H., Wilcox, K. W., and O'Hare, P., 1994, Binding and repression of the latency-associated promoter of herpes simplex virus by the immeediate early 175K protein, *J. Gen. Virol.* **75:**753–767.
73. Kenny, J. J., Krebs, F. C., Hartle, H. T., Gartner, A. E., Chatton, B., Leiden, J. M., Hoeffler, J. P., Weber, P. C., and Wigdahl, B., 1994, Identification of a second ATF/CREB-like element in the herpes simplex virus type 1 (HSV-1) latency-associated transcript (LAT) promoter, *Virology* **200:**220–235.
74. Rivera-Gonzalez, R., Imbalzano, A. N., Gu, B., and DeLuca, N. A., 1994, The role of ICP4 repressor activity in temporal expression of the IE-3 and latency-associated transcript promoters during HSV-1 infection, *Virology* **202:**550–564.
75. Wechsler, S. L., Nesburn, A. B., Zwaagstra, J. C., and Chiasi, H., 1989, Sequence of the latency-related gene of herpes simplex virus type 1, *Virology* **168:**168–172.
76. Soares, K., Hwang, D. Y., Ramakrishnan, R., Schmidt, M. C., Fink, D. J., and Glorioso, J. C. 1996, *cis*-Acting elements involved in transcriptional regulation of the herpes simplex virus type 1 latency-associated promoter 1 (LAP1) *in vitro* and *in vivo*, *J. Virol.* **70:**5384–5394.
77. Leib, D. A., Nadeau, K. C., Rundle, S. A., and Schaffer, P. A., 1991, The promter of the latency-associated trnscripts of herpes simplex virus type 1 contains a functional cAMP response element: Role of the latency-associated transcripts and cAMP in reactivation of viral latency, *Proc. Natl. Acad. Sci. USA* **88:**48–52.
78. Frazier, D. P., Cox, D., Godshalk, E. M., and Schaffer, P. A., 1996, The herpes simplex virus

type 1 latency-associated transcript promoter is activated through Ras and Raf by nerve growth factor and sodium butyrate in PC12 cells, *J. Virol.* **70:**7424–7432.

79. Frazier, D. P., Cox, D., Godshalk,, E. M., and Schaffer, P. A., 1996, Identification of *cis*-acting sequences in the promoter of the herpes simplex virus type 1 latency-associated transcripts required for activation by nerve growth factor and sodium butyrate in PC12 cells, *J. Virol.* **70:**7433–7444.
80. Zwaagstra, J. C., Ghiasi, H., Nesburn, A. B., and Wechsler, S. L., 1989, In vitro promoter activity associated with the latency-associated transcript gene of herpes simplex virus type 1, *J. Gen. Virol.* **70:**2163–2169.
81. Batchelor, A. H., and O'Hare, P., 1990, Regulation and cell-type-specific activity of a promoter located upstream of the latency-associated transcript of herpes simplex virus type 1, *J. Virol.* **64:**3269–3279.
82. Mador, N., Panet, A., Latchman, D., and Steineer, I., 1995, Expression and splicing of the latency-associated transcripts of herpes simplex virus type 1 in neuronal and non-neuronal cell lines, *J. Biochem. (Tokyo)* **117:**1288–1297.
83. Margolic, T. P., Bloom, D. C., Dobson, A. T., Feldman, L. T., and Stevens, J. G., 1993, Decreased reporter gene expression during latent infection with HSV LAT promoter constructs, *Virology* **197:**585–592
84. Lokensgard, J. R., Bloom, D. C., Dobson, A. T., and Feldman, L. T., 1994, Long-term promoter activity during herpes simplex virus latency, *J. Virol.* **68:**7148–7158.
85. Rader, K. A., Ackland-Berglund, C. D., Miller, J. K., Pepose, J. S., and Leib, D. A., 1993, *In vivo* characterization of site-directed mutations in the promoter of the herpes simplex virus type 1 latency-associated transcrips, *J. Gen. Virol.* **74:**1859–1869.
86. Bloom, D. C., Stevens, J. G., Hill, J. M., and Tran, R. K., 1997, Mutagenesis of a cAMP response element within the latency-associated promoter of HSV-1 reduces adrenergic reactivation, *Virology* **236:**202–207.
87. Chen, X. W., Schmidt, M. C., Goins, W. F., and Glorioso, J. C., 1995, Two herpes simplex virus type 1 latency-active promoters differ in their contributions to latency-associated transcript expression during lytic and latent infections, *J. Virol.* **69:**7899–7908.
88. Deshmane, S. L., Nicosia, M., Valyi-Nagy, T., Feldman, L. T., Dillner, A., and Fraser, N. W., 1993, An HSV-1 mutant lacking the LAT TATA element reactivates normally in explant cocultivation, *Virology* **196:**868–872.
89. Ho, D. Y., and Mocarski, E. S., 1989, Herpes simplex virus latent RNA (LAT) is not required for latent infection in the mouse, *Proc. Natl. Acad. Sci. USA* **86:**7596–7600.
90. Goins, W. F., Sternberg, L. R., Croen, K. D., Krause, P. R., Hendricks, R. L., Fink D. J., Straus, S. E., Levine, M., and Glorioso, J. C., 1994, A novel latency-active promoter is contained within the herpes simplex virus type 1 U_L flanking repeats, *J. Virol.* **68:**2239–2252.
91. Fowler, S. L., Harrison, C. J., Meyers, M. G., and Stanberry, L. R., 1992, Outcome of herpes simplex virus type 2 infection in guinea pigs, *J. Med. Virol.* **36:**303–308.
92. Trousdale, M. D., Steiner, I., Spivack, J. G., Deshmane, S. L., Brown, S. M., MacLean, A. R., Subak-Sharpe, J. H., and Fraser, N. W., 1991, *In vivo* and *in vitro* reactivation impairment of a herpes simplex virus type 1 latency-associated transcript variant in a rabbit eye model, *J. Virol.* **65:**6989–6993.
93. Perng, G. C., Dunkel, E. C., Geary, P. A., Slanina, S. M., Ghiasi, H., Kaiwar, R., Nesburn, A. B., and Wechsler, S. L., 1994, The latency-associated transcript gene of herpes simplex virus type 1 (HSV-1) is required for efficient *in vivo* spontaneous reactivation of HSV-1 from latency, *J. Virol.* **68:**8045–8055.
94. Bloom, D. C., Hill, J. M., Devi-Rao, G., Wagner, E. K., Feldman, L. T., and Stevens, J. G., 1996, A 348-base-pair region in the latency-associated transcript facilitates herpes simplex virus type 1 reactivation, *J. Virol.* **70:**2449–2459.

95. Hill, J. M., Maggioncalda, J. B. Garza, H. H., Jr., Su, Y. H., Fraser, N. W., and Block, T. M., 1996, *In vivo* epinephrine reactivation of ocular herpes simplex virus type 1 in the rabbit is correlated to a 370-base-pair region located between the promoter and the 5′ end of the 2.0-kilobase latency-associated transcript, *J. Virol.* **70:**7270–7274.
96. Perng, G. C., Slanina, S. M., Ghiasi, H., Nesburn, A. B., and Wechsler, S. L., 1996, A 371-nucleotide region between the herpes simplex virus type 1 (HSV-1) LAT promoter and the 2-kilobase LAT is not essential for efficient spontaneous reactivation of latent HSV-1, *J. Virol.* **70:**2014–2018.
97. Bourne, N., Stanberry, L. R., Connelly, B. L., Kurawadwala, J., Straus, S. E., and Krause, P. R., 1994, Quantity of latency-associated transcript produced by herpes simplex virus is not predictive of the frequency of experimental recurrent genital herpes, *J. Infect. Dis.* **169:** 1084–1087.
98. Block, T. M., Spivack, J. G., Steiner, I., Deshmane, S., McIntosh, M. T., Lirette, R. P., and Fraser, N. W., 1990, A herpes simplex virus type 1 latency-associated transcript mutant reactivates with normal kinetics from latent infection, *J. Virol.* **64:**3417–3426.
99. Thompson, R. L., Cook, M. L., Devi-Rao, G. B., Wagner, E. K., and Stevens, J. G., 1986, Functional and molecular analysis of the avirulent wild-type herpes simplex virus type 1 strain KOS, *J. Virol.* **58:**203–211.
100. Sawtell, N. M., and Thompson, R. L., 1992, Rapid *in vivo* reactivation of herpes simplex virus in latently infected murine ganglionic neurons after transient hyperthermia, *J. Virol.* **66:**2150–2156.
101. Maggioncalda, J., Mehta, A., Fraser, N. W., and BLock, T. M., 1994, Analysis of a herpes simplex virus type 1 LAT mutant with a deletion between the putative promoter and the 5′ end of the 2.0—kilobase transcript, *J. Virol.* **68:**7816–7824.
102. Sandri-Goldin, R. M., Sekulovich, R. E., and Leary, K., 1990, The alpha protein ICP0 does not appear to play a major role in the regulation of herpes simplex virus gene expression during infection in tissue culture, *Nucleic Acids Res.* **15:**905–919.
103. Perng, G. C., Ghiasi, H., Slanina, S. M., Nesburn, A. B., and Weschsler, S. L., 1996, The spontaneous reactivation function of the herpes simplex virus type 1 LAT gene resides completely within the first 1.5 kilobases of the 8.3-kilobase primary transcript, *J. Virol.* **70:**976–984.
104. Farrell, M. J., Hill, J. M., Margolis, T. P., Stevens, J. G., Wagner, E. K., and Feldman, L. T., 1993, The herpes simplex virus type 1 reactivation function lies outside the latency-associated transcript open reading frame ORF-2, *J. Virol.* **67:**3653–3655.
105. Fareed, M. U., and Spivack, J. G., 1994, Two open reading frames (ORF1 and ORF2) within the 2.0-kilobase latency-associated transcript of herpes simplex virus type 1 are not essential for reactivation from latency, *J. Virol.* **68:**8071—8081.
106. Bolovan, C. A., Sawtell, N. M., and Thompson, R. L., 1994, ICP34.5 mutants of herpes simplex virus type 1 strain 17syn$^+$ are attenuated for neurovirulence in mice and for replication in confluent primary mouse embryo cell cultures, *J. Virol.* **68:**48—55.
107. Brown, S. M., Harland, J., MacLean, A. R. Podlech, J., and Clements, J. B., 1994, Cell type and cell state determine differential *in vitro* growth of non-neurovirulent ICP34.5-negative herpes simplex virus types 1 and 2, *J. Gen. Virol.* **75:**2367 –2377.
108. Chou, J., Kern, E. R., Whitley, R. J., and Roizman, B., 1990, Mapping of herpes simplex virus-1 neurovirulence to gamma$_1$34.5, a gene nonessential for growth in culture, *Science* **250:**1262–1266.
109. Perng, G. C., Chokephaibulkit, K., Thompson, R. L., Sawtell, N. M., Slanina, S. M., Ghiasi, H., Nesburn, A. B., and Wechsler, S. L., 1996, The region of the herpes simplex virus type 1 LAT gene that is colinear with the ICP34.5 gene is not involved in spontaneous reactivation, *J. Virol.* **70:**282–291.
110. Perng, G. C., Ghiasi, H., Slanina, S. M., Nesburn, A. B., and Wechsler, S. L., 1996, High-dose ocular infection with a herpes simplex virus type 1 ICP34.5 deletion mutant produces no corneal

disease or neurovirulence yet results in wild-type levels of spontaneous reactivation, *J. Virol.* **70:**2883–2893.

111. Valyi-Nagy, T., Fareed, M. U., O'Keefe, J. S., Gesser, R. M., MacLean, A. R., Brown, S. M., Spivack, J. G., and Fraser, N. W., 1994, The herpes simplex virus type 1 strain 17^+ gamma 34.5 deletion mutant 1716 is avirulent in SCID mice, *J. Gen. Virol.* **75:**2059–2063.
112. Lagunoff, M., Randall, G., and Roizman,. B., 1996, Phenotypic properties of herpes simplex virus 1 containing a derepressed open reading frame P gene. *J. Virol* **70:**1810–1817.
113. Natarajan, R., Deshmane, S., Valyi-Nagy, T., Everett, R. D., and Fraser, N.W ., 1991, A herpes simplex virus 1 mutant lacking the ICP0 introns reactivates with normal efficiency, *J. Virol.* **65:**5569–5573.
114. Clough, D. W., Kunkel, L. M., and Davidson, R. L., 1982, 5-Azacytidine-induced reactivation of a herpes simplex thymidine kinase gene, *Science* **216:**70–73.
115. Youssoufian, H., Hammer, S. M.,Hirsch, M. S., and Mulder, C., 1982, Methylation of the viral genome in an *in vitro* model of herpes simplex virus latency, *Proc. Natl. Acad. Sci. USA* **79:**2207–2210.
116. Dressler, G. R., Rock, D. L., and Fraser, N. W., 1987, Latent herpes simplex virus type 1 DNA is not extensively methylated *in vivo, J. Gen. Virol.* **68:**1761–1765.
117. Green, M. T., Courtney, R. J., and Dunkel, E. C., 1981, Detection of an immediate early herpes simplex virus type 1 polypeptide in trigeminalganglia from latently infected animals, *Infect. Immunol.* **34:**987–992.
118. Green, M. T., Dunkel, E.C.,,and Courtney, R. J., 1984, Detection of herpes simplex virus induced polypeptides in rabbit trigeminal ganglia, *Invest. Ophthalmol. Vis. Sci.* **25:**1436–1440.
119. Stevens, J. G., and Cook, M. L., 1974, Maintenance of latent herpetic infection: An apparent role for anti-viral IgG, *J. Immunol.* **113:**1685–1693.
120. Block, T. M., Deshmane, S., Masonis, J., Maggioncalda, J., Valyi-Nagi, T., and Fraser, N. W., 1993, An HSV LAT null mutant reactivates slowly from latent infection and makes small plaques on CV-1 monolayers, *Virology* **192:**618–630.
121. Rosen-Wolff, A., Scholz, J., and Darai, G., 1989, Organotropism of latent herpes simplex virus type 1 is correlated to the presence of a 1.5 kb RNA transcript mapped within the BamHI DNA fragment B (O.738 to 0.809 map units), *Virus Res.* **12:**43–51.

4

T Cell Activation and Lymphokine Induction in *Herpesvirus saimiri* Immortalized Cells

PETER GECK

1. INTRODUCTION

1.1. The Lymphotropic Gamma Herpesviruses

The dominant trend in the evolution of herpesviruses is organotropism, where a stage of the viral–host interaction becomes specific to a single permissive tissue. Organotropism and other biological features classify herpesviruses into three major subfamilies, alpha, beta, and gamma herpesviruses.[1] *Herpesvirus saimiri* belongs to the gamma herpesvirus subfamily that targets the lymphoid system. Only a subset of T lymphocytes can be transformed or immortalized by the virus.[2] Two factors are important for understanding the efficiency of lymphotropism by *Herpesvirus saimiri:* (1) coevolution of the virus with the immune system and (2) extensive molecular piracy by the virus.

PETER GECK • Department of Anatomy and Cellular Biology, Tufts University Health Science Schools, Boston, Massachusetts 02111.

Herpesviruses and Immunity, edited by Medveczky *et al.* Plenum Press, New York, 1998.

1.1.1. Coevolution of Gamma Herpesviruses with the Immune System

Ancient forms of herpesviruses evolved hundreds of millions of years ago in fish, reptiles, and birds.[3] Thus the ancestors of herpesviruses were already present when mammals appeared. During the explosive mammalian speciation about 60 million years ago, when the mammalian immune system achieved its present complexity, gamma herpesviruses became lymphotropic.[4] This subfamily was apparently very successful. It was shown that lymphocytes of several mammalian species carry gamma herpesviruses (see later). The coevolution process explains the highly specialized integration of the viral cycle into the immune environment.

1.1.2. Molecular Piracy by Gamma Herpesviruses

A set of unique molecular features of the virus poised *H. saimiri* for singularly effective molecular piracy.[5] (1) They are large virions with flexible spatial allowance for exoneous genetic material.[6] (2) The coding region of *H. saimiri* is flanked by long, repeated, noncoding sequences. Repetitive regions can serve as spacers and can easily buffer acquisitions of surplus DNA. Since every repetitive unit is a substrate for the packaging cleavage at the flanks, an increase in DNA size is automatically compensated for by losing the corresponding size in repeat units.[7] (3) The coding portion of the saimiri genome is uniquely high in A:T nucleotides.[5] This feature is apparently important for the virus because *H. saimiri* is the only virus that codes for the entire gene set of thymidine synthesis.[8] AT-rich sequences are known to be energetically favored for melting. Recombination events are initiated through partially denatured and cross-hybridized segments of DNA, e.g., crossing-over, Holliday junction.[9] Potentially, an AT-rich genome is a substrate for recombinations and gene acquisitions through low-energy melting. Another effect of high AT content is the low occurrence of CpG sequences that keeps inactivating methylation under control.[5]

1.2. The Biology of *H. saimiri*

Gamma herpesviruses (Epstein–Barr virus, *H. saimiri*, bovine herpesvirus 4, H. ateles, H. sylvilagus, alcelophine herpesvirus-1, and murine herpesvirus-68[5]) are distinguished by lymphotropism, capacity for immortalizing lymphocytes, and a common genomic structure.[6] *H. saimiri* is not pathogenic in its natural host, in squirrel monkeys. Infection of other New World monkeys results in progressive, fatal lymphoproliferative diseases.[10] Rabbits are also susceptible to *H. saimiri* infection and develop similar, rapidly progressing, fatal lymphomas.[11]

The virus infects several cell types such as fibroblasts and kidney cells, and produces progeny virions, but transformation takes place only in T lymphocytes.

In vitro immortalization is reported in T cells from marmosets,[12,13] from rabbits,[14] and from humans.[15,16] Immortalization was observed in various types and stages of T cell maturation, like TCR gamma/delta double negative ($CD4^-CD8^-$),[17] or immature double positive ($CD4^+CD8^+$) cells,[18] but the overwhelming majority of transformed cells are mature single positive $CD4^+$ or $CD8^+$ T lymphocytes (see later).

The immortalized cells produce virions in the initial phase, but productive infection is typically lost after prolonged culturing.[19] Analysis of the viral DNA in these nonproducer cell lines showed major deletions in the middle portion of the viral genome.[20] Notably, the left and right terminal 10–15 kb regions are always maintained, indicating the presence of potential oncogenes.[21]

The linear viral DNA genome is fully sequenced and contains a unique coding middle portion, flanked by repetitive sequences at both ends.[5] The unique middle region (L-DNA) has 113 kbp and is low in G+C nucleotides (34.5%). The repetitive sequence (H-DNA) has about 35 repeats of a noncoding tandem sequence of 1444 bp, and it is high in G+C nucleotides (70.8%). The L-DNA codes for 76 major open reading frames and 7 U-RNA genes. Arrangement of the genes is in overall colinearity with EBV, where conserved gene blocks are interspersed with unique sequences. Several genes exhibit great homology to human sequences, like thymidylate synthase,[22] dihydrofolate reductase,[8,23] complement control proteins,[24] CD59,[25] cyclin,[26] G protein-coupled receptors,[27] IL-17 homologue,[28] superantigenlike polypeptide,[29] etc.

The replicating viral DNA is circularized and follows the rolling circle model.[7] In productive infection, unit length linear genomes are cleaved and packaged by components of the capsid. In immortalized cells the viral DNA is maintained as an episomal circular element.

Analysis of naturally attenuated strains, which lost their ability to immortalize, demonstrated that the left terminal region is important in oncogenicity.[30] The region is nonhomologous between different strains, and three major hybridization subgroups were established, subgroups A, B, and C.[31] The subgroups also correlate with the *in vitro* transformation potential of strains. Group A and group C viruses transform common marmoset lymphocytes *in vitro*,[13] but only group C strains immortalize human cells.

Deletion mutations established that the leftmost open reading frame in subgroup A strains[32] or the leftmost two open reading frames in subgroup C virus strains[15] code for polypeptides involved in *H. saimiri* oncogenesis. The viral U-type small nuclear RNAs may also play a role in the phenotype of immortalized cells.[33]

Viral oncogenic factors have been extensively studied and are discussed later in detail. T cell activation and regulation are reviewed first to understand the role viral oncogenic factors play in T cell activation and cytokine induction.

2. THE MOLECULAR ENVIRONMENT OF *H. SAIMIRI* ONCOGENESIS: SIGNAL TRANSDUCTION IN T CELL ACTIVATION (OPERATIONAL PRINCIPLES, COMPETENCE PHASE, PROGRESSION PHASE)

2.1. Operational Principles of the Cellular Immune Response (Maintenance of Antigenic Integrity, T Cell Activation—the Target Pathway for *H. saimiri*)

2.1.1. Maintenance of the Antigenic Integrity of the Organism

Primitive immunoglobulins and lymphocytes evolved in early vertebrates. A complex immune system, however, in the form of a highly integrated network of anatomically disparate elements first developed in birds and mammals, relatively late in evolution.[34] This fact argues that the system came into existence to perform a broader function than simply to contain microbial pressure.

Along with the increasing complexity of an organism, dysfunctional molecules and cells that can escape proliferative control also emerge with increasing probability. Elimination of these biological entities is critical for the integrity of the individual. The immune system evolved to maintain the genetic and antigenic homeostasis of the organism through continuous survey and elimination of "non-self" structures.

The most salient feature of the immune response is the clear distinction between "self" and "non-self." The number of self antigens the organism entails is on the order of tens of thousands, whereas possible "non-self" antigenic structures that the organism may encounter are on the order of millions. Consequently, every organism must develop an entire repertoire of variable receptor sequences on T cells that recognizes and responds to all possible antigenic structures.[35] Maturation of the immune system eliminates clones that represent self-antigens by negative selection.[36] The rest of T cell clones are subject to positive selection and migrate to peripheral lymphoid compartments.[37] Proliferation of these cells is suspended, and only specific encounter with a cognate antigen elicits the response called T cell activation.

2.1.2. T Cell Activation, the Target of H. saimiri Oncogenesis

The collective repertoire of mature T lymphocytes constitutes the capacity to recognize all possible antigenic permutations. The number of individual cells, however, that can respond to a particular antigen is very low. As a result, these cells represent no defensive force for immediate deployment. Consequently, they need to increase rapidly in number to eliminate an emerging antigen. In fact, these cells are distinguished by an inherent capacity for rapid proliferative clone expansion by contact with specific antigens. The initial phase of this proliferative response is called T cell activation.[34] The mechanism is very efficient in raising a powerful defense on short notice, but in turn, it has to maintain a dangerous contingency of cells with explosive proliferative potential. Not surprisingly, these cells are targeted by a number of viruses (retroviruses, lentiviruses, adenoviruses, herpesviruses), including *H. saimiri.* To understand the pathways harnessed by *H. saimiri* to immortalize its host, T cell activation is reviewed in some detail in sections 2.2 and 2.3. The following paragraphs discuss a few general patterns in regulation.

2.1.3. *Uniform Patterns of Proliferative Control in T Cell Activation (Biphasic Induction, Multiple Signal Integration, Conserved Signal Pathway Hierarchy)*

2.1.3a. Biphasic Pattern in the Induction of Proliferation; Competence Phase/ Progression Phase. Mitotic induction from quiescence follows uniform strategies in most cells, with a few common basic patterns. T cells, like most eukaryotic cells, display a biphasic pattern in inducing proliferation.[38] Specific factors or stimuli prepare cells for mitosis. These factors, however, do not by themselves bring about proliferation. The activated cells are in a competence phase, and the factors are called competence factors.[34] The underlying molecular events are cell cycle related. Cells in quiescence are in the G0 phase, and competence factors induce changes in cell cycle regulation to bring cells into the G1 phase. During the G0 to G1 transition, the molecular basis of cell cycle regulation (cyclins, cyclin dependent kinases, and other regulators) are synthesized or activated. Cells in competence are in the G1 phase, where components of DNA replication are synthesized or activated.

A different set of factors or stimuli are needed to activate the regulation mechanism created in the G1 phase. These factors are called progression factors because they induce cells to progress through the G1–S checkpoint and enter the S phase.[34] In most cells a third set of conditions is also necessary to carry out a full cycle of mitosis. These factors regulate transit through the G2–M restriction point.

2.1.3b. Multiple Signal Pattern in Activation Phase. Activation to the competence phase cannot typically be achieved with a single signal in most cells. Fibroblasts, e.g., are activated by the combined effect of phorbol esters and ionophoric treatment, lymphocytes are activated with CD3 cross-linking and ionophoric treatment together, etc. Although T cells are activated by a single signal, i.e., the specific antigen, studies on antigenic presentation show that this process is also subject to the multiple-signal rule. The specific signal component is delivered by the T cell receptor (TCR), whereas the accessory signals are induced by coreceptors and accessory molecules in contact with their ligands on antigen-presenting cells (APC).[39]

Recent results on signal transduction have revealed the molecular basis of the phenomenon. The release of cells from proliferative arrest, even temporarily, represents some risk for the organism. Commitment for replication, therefore, requires reinforcements from several signal transduction pathways. Integration of multiple inputs is typically the function of multisubunit transcription transactivators. Transcription factors that render cells competent for replication are multisubunit ligands. The subunits are individually controlled by different signal transduction pathways, e.g., c-fos (lck–ras–MAPK pathway), c-jun (PKC–JNK1 pathway), and pNF-AT (PLC–Ca–calcineurin pathway) that together constitute AP1 and NF-AT, the major transcription regulators in IL-2 expression (see later).

2.1.3c. Conserved, Uniform Signal Pathway Hierarchy. Differences in biological response reflect a great variety of receptors and effectors, but intracellular signal pathways are remarkably similar. The apparently uniform hierarchy of signal transfer events evolved to amplify and distribute the signal from the receptor to the nucleus and to integrate highly complex signal patterns into cell proliferative control.

Signal transfer follows a conserved sequence of events[40] described later in detail. Briefly, the conformational change of the receptor induces dimerization, tyrosine phosphorylation, and activation of protein tyrosine kinases. The initial signal generates a wave of activation by phosphorylation that spreads two dimensionally along the cytoplasmic membrane surface through adapter and docking proteins. The signal is relayed to two parallel pathways of second-level signal processing: GTP-binding proteins and lipid second messengers, all of which eventually activate serine/threonine protein kinase cascades. The activated serine/threonine kinases translocate to the nucleus and translate the initial signal into modified gene-expression patterns.

2.2. Competence Phase in T Cell Activation (Antigen Recognition, T Cell Receptor Activation, Signal Transduction, Target Genes)

2.2.1. Antigen Recognition (Antigenic Presentation, T Cell Receptor Structure, Kinetic Proofreading in Antigen Binding)

2.2.1a. Antigenic Presentation. The molecular mechanisms of antigenic recognition are based on two major principles: (1) breakdown of large protein antigens into 8–10 residue peptides to limit antigenic sequence permutations and (2) peptide presentation on the surface of contact cells to confer further regulation through accessory proteins between two cells.

T lymphocytes that interact with antigen-presenting cells ($CD4^+$ and $CD8^+$ T cells) are also the overwhelming targets for *H. saimiri* immortalization and cytokine induction. To understand the changes that *H. saimiri* inflicts on these cells, antigenic presentation is discussed briefly.

There are two categories of antigens that the immune system encounters: (1) exogenous structures synthesized outside of the organism; and (2) viral or modified cellular (self) structures synthesized endogenously in the cells of the organism.[34] Cells that present endogenously generated heteroantigens are potentially dangerous because they carry the replicating genetic material (viral or altered self) and must be destroyed. On the other hand, cells presenting exogenous peptides are harmless. They merely display the antigen, so their destruction would be deleterious.

To deal with these widely different requirements, antigen-presenting cells (APC) can display the peptide on one of two different receptor structures (MHC class I or class II), which can assign a particular APC cell to a killer (MHC class I), or helper (MHC class II) T lymphocyte.[34] Helper T cells express the CD4 coreceptor that can engage only with the MHC class II structure on APC, which displays an exogenous peptide and signals no replicative genetic dangers within the cell. By activation, helper cells do not harm cells they contact, but recruit other cells for defense. On the other hand, endogenously synthesized peptides are displayed on MHC class I proteins, which are recognized by killer-specific CD8 receptors, and the ensuing contact with killer T cells destroys the antigen-presenting, virally infected, or altered tumor cells.

2.2.1b. The T Cell Receptor. H. saimiri interferes with T cell receptor activation pathways and interacts with several downstream components. The receptor on T lymphocytes that binds the MHC-bound peptide on antigen-presenting cells is a functional unit of the T cell receptor (TCR) and CD3.[40] It is a multisubunit complex consisting of eight membrane proteins. Except for the zeta chain, they belong to the immunoglobulin superfamily and display the characteristic variable and constant regions. Antigen recognition and binding is the function of the α and β chains (or in a subpopulation of lymphocytes, gamma and δ chains). They are in complex with elements of CD3: two heterodimers of gamma plus δ and gamma plus ϵ molecules and a dimer of the zeta chain. TCR α, β and CD3 gamma, δ and, ϵ chains are extracellular membrane-anchored proteins with short cytoplasmic domains. The CD3 zeta component is also a membrane protein, but it has no extracellular portion. Its active domain is entirely intracellular. CD3 gamma, δ, and ϵ cytoplasmic domains are targets for phosphorylation, interact with kinases, and are involved in signal generation. The major signal relay pathway, however, is the zeta chain.

2.2.1c. Specific Antigen Binding, the Kinetic Proofreading Model. The model of kinetic proofreading has been developed to explain a paradox in antigen recognition.[41] Antigen binding takes place with high sensitivity and specificity, and yet it is performed through surprisingly weak interactions between participating receptor–ligand systems. The dissociation constant of the TCR–peptide complex is on the order of 10^{-5}. By comparison, the cytokine–receptor Kd is about 10^{-11}.[42] The explanation is in the recruitment of multiple receptor–ligand contacts of individually weak forces that together establish a lasting complex and trigger the signal.

Lymphocytes are continuously in transient interactions with each other and other cells. These temporary contacts are realized through integrins like LFA1 and its ligand ICAM1, which are large molecules. The steric conformation is critical to the outcome of the interaction because as long as nonantigen-presenting cells are within range, the large integrin molecules rule out direct contact between the membrane-proximal MHC and TCR receptors.

When cells engage with potentially antigen-presenting cells that express CD58 (integrin LFA3), interactions progress to the next level. The receptor for CD58 is CD2 on lymphocytes, and the evolving molecular complex has the same dimensions (membrane-proximal) as the TCR–MHC complex (see Fig. 1). The significance of the similar steric conformation is that the CD2-CD58 ligation is ultimately the force that brings the TCR–MHC molecules in close contact. Once a T cell receptor binds a specific peptide on an opposing MHC molecule, it releases a cascade of events that quickly stabilizes and expands the complex. First the coreceptor CD4 or CD8 molecules bind the corresponding MHC class II or class I structures, respectively. This event stabilizes the complex and gives a chance for other molecules to interact. Eventually, several receptor–ligand systems join the core CD2–CD3–TCR–CD4–CD8 unit, involving CD5, CD26, CD28, CD29, CD40, CD44, and others. Stabilized by combined binding forces, the adhesion complex holds the two cells together long enough to activate the T cell receptor and trigger the antigen recognition signal.

2.2.2. T Cell Receptor Activation, the Target Pathway for H. saimiri Cytokine Induction and Immortalization

2.2.2a. Activation of the First Wave of Tyrosine Kinases, p56lck and Other src Family Kinases. During the gradual extracellular buildup of the adhesion complex, the intracellular cytoplasmic domains of its components are also engaged in the gradual formation of a signaling complex (see Fig. 1). The first participants to join the complex are src family kinases, most importantly p56lck, but p59fyn and p56lyn are also involved.[43] The p50csk kinase keeps lck inactive in quiescent cells by specifically phosphorylating a tyrosine at the regulatory C-terminus, a common domain in src family members.[40] A unique feature of src family kinases is the inability for autophosphorylation and autoactivation.[44] These kinases need to be positioned along another kinase to be activated. Because they associate with the cytoplasmic domains of TCR receptor elements, extracellular binding of a ligand by the TCR receptor contracts the intracellular domains (and the associated kinases) for mutual cross-reactivation. In quiescence, low amounts of lck activity are found in association with the cytoplasmic domains of CD4 and CD8,[40] and fyn activity is detected to be constitutively present on the zeta chain of CD3.[45]

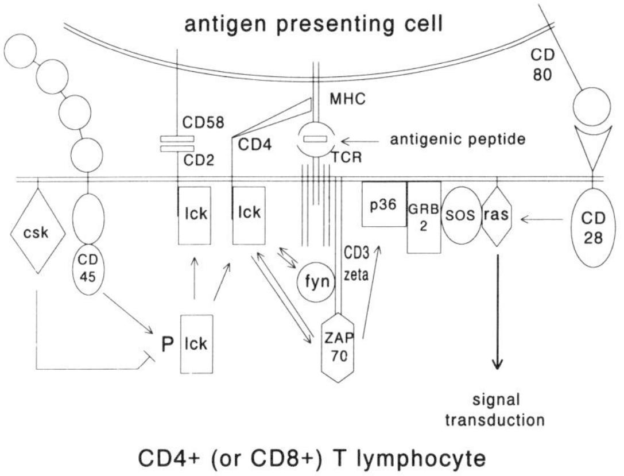

FIGURE 1. T cell receptor activation in native T lymphocytes. Components of the T cell receptor signal generation pathway are represented by symbols of various forms and sizes. The double line in the middle indicates the cell membrane. Intracellular components are below, and extracellular elements are above the lines. The curved double line at the top represents the cell membrane of an antigen-presenting cell. Of the large number of coreceptor and accessory molecular interactions, only those of MHC–TCR, CD58–CD2, and CD80–CD28 are depicted. The double lines in the TCR complex represent subunit molecules (not labeled). The rectangle between the TCR and MHC complexes is the antigenic peptide, as indicated. For details, see text.

The interaction is specific and realized through the membrane-proximal N-terminal domains of the kinases.[44] This activity is sufficient to increase tyrosine phosphorylation of CD2, CD3, CD4, and CD8 in the initial phase of TCR activation. Because lck and other src family kinases have SH2 domains that bind phosphotyrosines, increasing tyrosine phosphorylation recruits more lck to the complex, further increasing tyrosine phosphorylation.

CD2–CD58 binding is the turning point in the initial phase of activation. Before the sterically restrictive CD2–CD58 binding, the large CD45 tyrosine phosphatase is present in the complex[41] (see Fig. 1). Its function is critical. It can remove the inhibitory phosphate and activate lck and other kinases, but it can also deactivate other phosphorylated receptor elements. When sufficient lck activity is present and a large number of CD2 is phosphorylated, an increase in CD58 binding brings the membranes of the two cells in close proximity. This makes MHC–TCR contacts possible, and at the same time the large CD45 tyrosine phosphatase is excluded by steric incompatibility. This step is critical to stabilizing the lck-delivered tyrosine phosphates on CD2, CD3, CD4, or CD8, a prerequisite for subsequent steps in TCR activation. The multiple roles of CD2-CD58 engagement (phosphorylation and binding of lck, exclusion of CD45 phosphatase, and facilitation of MHC-TCR contacts) establish the initial complex. Beyond this point, rapid tyrosine phosphorylation of CD4, CD8, and elements of CD3 can take place and, through their SH2 domains, more and more lck and other src family kinases can join the complex. If TCR can recognize the specific MHC-presented peptide, the extracellular binding of the coreceptor CD4 or CD8 to MHC can bring an active lck in their intracellular domains to the zeta chain of CD3.[40] Tyrosine phosphorylation of the zeta chain is the first step in downstream transduction. The CD4/p56lck complex can also recruit the lipid kinase PI4K kinase that generates the substrate for the classical, lipid, second-messenger pathway.[44]

Although lck plays a decisive role in the process, tyrosine kinase activity of lck is dispensable for TCR activation.[46] The SH2 and SH3 domains, however, are important. SH3 domains associate with proteins through proline rich regions, so lck simply acts as an adaptor, holding separate protein complexes together in signal transduction.[47]

2.2.2b. Activation of the Second Wave of Tyrosine Kinases, the Syc/ZAP-70 Family The T cell-specific ZAP-70 is a cytoplasmic protein tyrosine kinase with double SH2 modules.[48] The binding substrate of ZAP-70 is the double phosphorylated ITAM sequence (immunoreceptor tyrosine based activation domain) with two tyrosines.[40] Activation requires double occupancy of the tyrosines. The CD3 zeta chain has three ITAM sequences, indicating that this domain is the nodal center for different signal pathways. After binding, the ZAP-70 molecules are activated by lck through tyrosine phosphorylation, as shown in Fig. 1, and can further phosphorylate their substrates in signal transduction.[48]

2.2.3. Signal Transduction Downstream of TCR Activation

2.2.3a. TCR Signal Pathway Through c-fos (RAS–MAPK–cfos). The substrate for ZAP-70 in the first channel of TCR signaling is a small adaptor molecule, the not well-characterized membrane-associated p36 protein[49] (see Fig. 1). Tyrosine phosphorylated p36 associates with the grb2 adaptor protein through its SH2 domain.[50] Grb2 is a multifunctional adaptor complexed with several other proteins through its two SH3 domains. The main signal pathway is through the SOS factor, but complexes with p115–120 kDa polypeptide (c-cbl proto-oncogen)[51] and the unknown p75 protein are also TCR activation specific.[38]

The membrane-associated SOS protein is a ras guanyl exchange factor.[52] Ras protein is a GTP-binding membrane-associated activator, inactive in GDP binding.[38] Ras guanyl exchange factors can facilitate GDP–GTP transition and can activate ras. The active conformation of ras plays a critical role in several transduction pathways. In TCR signal transduction it can activate raf-1, a serin/threonin kinase, at the level of mitogen-activated protein kinase kinase kinases (MAPKKK).[53] Through a middle level kinase (MAPKK), it can activate ERK1 and ERK2 serin/threonin kinases.[54] The phosphorylated ERK2 translocates to the nucleus and activates the Elk-1 transcription factor that induces c-fos expression.[55] Active c-fos protein is one component of the AP1 transcription factor complex that can induce IL-2 expression.

2.2.3b. TCR Signal Pathway Through c-jun (the p36–PLC–PKC–JNK1–cjun Pathway). The p36 membrane protein participates in an alternative pathway through phospholipase C that activates the classical, lipid second-messenger cascade (phosphorylation at the D4 position of inositol).[56] Phosphatidylinositol diphosphate is cleaved to generate inositol triphosphate (IP3) and diacylglycerol (DAG).[57] IP3-sensitive calcium channels regulate endoplasmic calcium pools, and IP3 induction results in calcium mobilization in the cytosol. Calcium and DAG can activate the conventional isoforms of protein kinase C (PKC), a cytosolic serin/threonin kinase.[58,59] PKC isoforms operate on a wide spectrum of substrates and can also activate a contingency of JNK1 serin/threonin kinases. Active JNK1 translocates to the nucleus and phosphorylates and activates c-jun.[38] c-jun is another component of AP1 on the promoter of the gene for IL-2.

Phosphorylation of c-jun is apparently important in TCR signalling because several parallel alternative pathways developed in evolution. Both the CD3 zeta chain and CD4 can complex and activate phosphatidylinositol 3 kinase (PI3K), probably through lck and ZAP70 binding.[47,60] This enzyme can generate another lipid messenger family, where phosphorylation takes place on the D3 position of the inositol ring, conferring resistance to phospholipase isoforms.[44,47] This novel class of second-messenger phospholipids can activate the nonconventional epsilon and zeta isoforms of PKC,[61] which participate in the c-jun activation pathway. Remarkably, TCR and CD4 recruitment of PI3K requires the adaptor function of lck but not the catalytic activity.[47] Furthermore, activated ras itself can also activate PI3K, so the ras pathway also contributes to c-jun activation.[62]

2.2.3c. TCR Signal Pathway Through NF-AT Activation. (p36–PLC–Calcium–Calcineurin–NF-AT Pathway). Calcium mobilization through the ZAP70-p36-PLC-lipid messenger pathway can activate a calcium/calmodulin binding serin/threonin phosphorylase, calcineurin.[63] The substrate for calcineurin is the cytoplasmic form of NF-AT, pNF-AT, a member of the c-rel family of transcription factors.[64] In its dephosphorylated form, pNF-AT can translocate to the nucleus, and in complex with AP1 (the cfos/cjun heterodimer), it participates in IL-2 expression.

2.2.3d. The Transient Nature of TCR-Activation, Inhibitory Pathways. The TCR-generated primary signal is a transient wave of activation along pathway elements and dissipates quickly in the activated cell. Various inhibitory mechanisms have been implicated in the rapid loss of signal: (1) ras is under twofold regulation. In addition to guanyl exchange factors like SOS that facilitate ras-GTP formation, other factors can induce GTPase activity (ras GTPase activating proteins, GAPs), like rasp120GAP protein that switches ras to an inactive conformation.[65] (2) TCR induction can activate a nuclear serine protein phosphatase, PAC1, that specifically dephosphorylates the nuclear form of ERK2, which, in turn, rapidly down-regulates c-fos expression.[38] (3) The mRNA for c-fos carries several repeats of the AUUUA destabilization signal on the 3′ untranslated region and is subject to rapid degradation.[66] (4) TCR activation upregulates p50csk kinase which can phosphorylate and inactivate lck, and thus a key element of the initial phase of TCR signal generation is rapidly withdrawn.[67]

2.2.4. *The Stabilizing Second Signal in TCR Activation; Permanent Activation through CD28*

It has recently been established that to maintain permanent activation of T cells, the second signal is the function of CD28 receptor[44] (see Fig. 1). CD28 is a homodimer of an 80-kDa transmembrane polypeptide with immunoglobulin superfamily features.[34] The ligand of CD28 is CD80 (B7/1) and CD86 (B7/2) on antigen-presenting cells, B lymphocytes, and macrophages.[47] Ligand binding facilitates tyrosine phosphorylation of the cytoplasmic domain of CD28 by p56lck and p59fyn.[47] The activated CD28 can recruit two downstream factors, phosphatidylinositide-3-kinase (PI3K) through the SH2 domain of its p85 subunit and the grb-2 adaptor protein; both of them target the ras factor. Later the IL-2 inducible T cell kinase (Itk) joins the complex.[68] CD28 pathways reach c-fos, c-jun, and pNF-AT, where costimulation by the second signal reinforces and stabilizes the effect of the primary signal. In addition, CD28 activates two more unique signal pathways: (1) the NF-kB signal pathway that specifically upregulates IL-2 receptor expression; and (2) Bcl-XL induction that directly regulates cell-death control and protects from apoptosis.

2.2.4a. The ras–c-fos Signal Stabilization Pathway. In this pathway PI3K also functions as an adaptor through its p85 subunit. The SH3 domain of PI3K interacts with the p62 factor (a CD28 specific protein), a ras GAP factor. The proline-rich domain of PI3K can bind the SH3 domain of grb-2.[47] These factors directly affect ras function, and through the raf–MAPK–ERK pathway the signal can stabilize c-fos activation.

2.2.4b. The JNK–c-jun Signal Stabilization Pathway. The PI3K pathway activates a ras-related small GTP-binding protein, Rac (CD47), probably through the ras–ralGTP pathway. This protein participates in upstream regulation of JNK kinase activation,[38] it is an effector of PI3K, and it contributes to c-jun stabilization.

2.2.4c. The PKCε–NF-AT Stabilization Pathway. The p110 catalytic subunit of activated PI3K generates D3 phosphorylated phosphatidylinositide derivatives, which can specifically activate the ε and zeta isoforms of PKC. The activated ε isoform is implicated in AP1 and NF-AT stabilization.[69]

2.2.4d. The PI3K–p70S6K–NF-kB Pathway. The activated PI3K also has protein kinase activity, and threonin phosphorylates the activator domain of p70S6K kinase directly.[70] The activated p70S6k is involved in down-regulation of the IkBα inhibitor and c-Rel translocation to the nucleus. These events are necessary for activating NF-kB. NF-kB contributes to IL-2 induction, and it is even more important in IL-2 receptor expression.

2.2.4e. The PI3K–BclXL–Antiapoptosis Pathway. It is reported that CD28 activation triggers T cell survival signals through PI3K involvement. The exact pathway is not known, but the catalytic activity of PI3K enhances the expression of BclXL, a Bcl-2 related membrane protein that protects T cells from apoptosis.[71]

2.2.5. Target Genes of TCR-Activated Pathways

Activated T cells are in the G1 phase of the cell cycle and assume competence for proliferation. By the end of the G1 phase, an entirely novel signaling circuitry is established that can trigger S phase entry and mitosis: the IL2, IL3, and IL4 signal pathways.

IL-2 expression is regulated in part at the transcriptional level by AP1 and NF-AT responsive elements on the IL-2 gene 5′ promoter region, as described above. Posttranscriptional regulation affects the stability of the transcript through 3′ untranslated AU-rich and AUUUA repeats.[66] T cell activation confers increased stability on the IL-2 mRNA and facilitates IL-2 expression within a 4–6 hour period following T cell induction.[34]

The regulation of IL-2 receptor expression is more complex. The receptor consists of three polypeptides, α (55k), β (70k), and gamma (64k) chains.[42] The trimolecular complex binds IL-2 with high affinity (10^{-11} M). Expression of the β and gamma chains is not TCR activation-dependent. The promoters contain IL-2 and IL-4 responsive elements.[72]

Expression of the α chain, however, is largely TCR activation-induced. Three relevant promoter regions regulate expression, NF-kB, Serum Response Factor, and IL2 response elements.[72] NF-kB regulation is the function of the CD28-PKCzeta-p70s6K signal pathway, as described above. SRF is regulated by the raf1–ERK2–Elk1/JAP1 pathway, similar to c-fos activation. The IL-2 (and IL-1) response elements can further enhance IL-2 receptor expression.[72]

IL-4 expression is not responsive to PMA (PKC) activation, but calcium mobilization can induce IL-4.[73] The pathway of signal transduction is not mediated by conventional isoforms of PKC. An alternative pathway functions through calcium/calmodulin-regulated calcineurin that dephosphorylates and activates pNF-AT.[73] The notion is confirmed by cyclosporin A treatment that interferes with calcineurin and inhibits IL-4 expression. The mRNA of IL-4 also contains AUUUA repeats in an AU-rich context on the 3′ untranslated region and is probably subject to regulation similar to that of IL-2.[66]

2.3. Progression Phase (Cytokine Systems, Cytokine Signal Transduction, Target Genes)

2.3.1. Cytokine Systems in H. saimiri Immortalization (Interleukin-2, Interleukin-2 Receptor, Other Growth-Related Cytokines)

2.3.1a. Interleukin-2. IL-2 is the principal inducer of cell-cycle progression through the G1–S checkpoint in T lymphocytes. It is a 14–17 kDa glycoprotein secreted largely by $CD4^+$ T cells, and to a lesser extent, by $CD8^+$ T cells.[34] It acts through paracrine and autocrine mechanisms to stimulate $CD4^+$, $CD8^+$, NK, B, and mononuclear cells.

2.3.1b. Interleukin-2 Receptor. The receptor of interleukin-2 is a complex of three membrane proteins, as mentioned, the α, β, and gamma subunits that together bind IL-2 with high affinity. The α subunit (p55) is specifically regulated by TCR activation. It has very low affinity to IL-2 (α+gamma complex, 10^{-8} M), it has no cytoplasmic domain, and does not generate signals by itself.[42] The β (p75) and gamma (p64) chains belong to the cytokine receptor family. The β+gamma complex binds IL-2 with intermediate affinity (10^{-9} M) and can trigger mitotic signals.[42] The cytoplasmic domain of the β chain has 286 residues and is composed of different subdomains that play different roles in signal transduction.[72] The membrane-proximal, serine-rich region is required for proliferative signals. The membrane-distal acidic region binds src family kinases and is involved in other effects of IL-2, e.g., differentiation.[74] The gamma chain has an 86-residue cytoplasmic domain, and participates in signal generation in cooperation with the β subunit.[72] The gamma chain is shared by several cytokine receptors. It is a component of IL-4R, IL-7R, IL-9R, and IL-13R.[42]

2.3.1c. Interleukin-4 and IL-4 Receptor. IL-4 is a 20-kDa glycoprotein, secreted overwhelmingly by a subset (Th2) of $CD4^+$ T lymphocytes. It stimulates B cells, mast cells, macrophages, and it exerts an autocrine effect with a set of $CD4^+$ cells.[34] The IL-4 receptor is a single chain of 145 kDa. It associates with the IL-2R gamma chain that increases binding affinity and constitutes a single class of high-affinity binding sites (10^{-10} M).[72] The cytoplasmic domain is composed of about 500 amino acids but does not contain receptor tyrosine kinase sequences. It is also part of the IL-13 receptor.

2.3.2. Signal Transduction through the IL-2 Receptor

2.3.2a. Lck–c-fos Channel. Similar to the TCR receptor, nonreceptor tyrosine kinases (JAK1 and JAK3) are involved in the first step of signal transduction. In the process of ligand (IL-2) acquisition, the β and gamma chains form a complex. Their associated intracellular domains create the binding site for JAK1 and JAK3, which only bind together and tyrosine phosphorylate the receptor.[42] Src family kinases are also involved in the initial phase of IL-2 receptor activation. Immunoprecipitation experiments showed that p56lck, p59fyn, and p56lyn are present in the activated β chain receptor complex.[72] The docking site of lck on the β chain is the membrane-distal acidic domain, but the serine-rich domain on the receptor is also required. Because the latter is the binding site for JAK1 and ZAP70 tyrosine kinases, they are probably involved in lck recruitment.[75] The binding site on lck is nonconventional and interferes with the catalytic activity of the kinase subdomain.[76] It is also observed that CD45 phosphatase is not involved in IL-2 induced proliferation,[72] and because it would activate lck, this further shows that the lck catalytic domain is not essential in this pathway. Apparently, lck is an adaptor through its SH2 and SH3 domains. The activated receptor complex also binds another adaptor, the shc factor,[38] through a tyrosine (Tyr338) of the receptor β chain. The complex further involves the SOS and p95vav factors[77] through grb-2 and functions as a ras GEF (activating) element. Interestingly, even as an adaptor, lck is not essential because p95vav activation is normal in lck-minus cells.[78] The pathway downstream of ras follows the raf–MAPK–ERK signal and up-regulates c-fos activity. This pathway is not essential because elimination of Tyr338 from the IL-2 receptor does not prevent IL-2 induced proliferation.[42]

2.3.2b. The G1/S Checkpoint Pathway (JAK/STAT–cdk2 Channel). As mentioned, recruitment of Janus family tyrosine kinases, JAK1 and JAK3, is the earliest event in IL-2 receptor activation. JAK1 binds the membrane-proximal, serine-rich domain of the β chain,[79] and JAK3 is complexed with the C-terminal portion of the gamma chain.[42] The kinases mutually phosphorylate each other and the receptor. The substrates for JAK kinases are STAT3 and STAT5 transcription factors. In their C-terminal portions STAT factors have SH2 and SH3 domains that recruit them to JAK kinases or to the proline-rich C-terminal domain of the receptor β chain.[72] Upon tyrosine and serine phosphorylation (by MAPKs), STAT factors dimerize, translocate to the nucleus, and through the N-terminal DNA-binding domains they can transactivate promoters of different genes.[42] The target genes are not known. An important pathway is down-regulation of p27kip1, an inhibitor of cdk2.[80] Activation of the cyclinE/cdk2 complex facilitates Rb protein phosphorylation, and the released E2F transactivator can induce the gene set involved in the G1/S transition.

2.3.2c. The G2/M Checkpoint Pathway (ZAP70/Syk–c-myc–cdc25). ZAP70 and Syk tyrosine kinases can bind the serine-rich domain of the activated IL-2R β chain. Activation of ZAP70 is probably the function of src kinases.[81] Downstream substrates in this pathway are not yet known, but the pathway targets c-myc expression and activation.[72] C-myc is complexed with c-max and acts as a transcriptional transactivator on the promoter of cdc25, a phosphotyrosine phosphatase.[82] Interestingly, c-myc and c-ras pathways converge on cdc25, because the c-myc induced cdc25 is active only after c-ras dependent raf serine/threonine phosphorylation.[83] Activated cdc25 dephosphorylates and activates cdc2, and the cdc2/cyclin B complex is involved in the G2/M transition in T cell proliferation.[72]

2.3.2d. The Cytoskeleton/Apoptosis Pathway (PI3K/PKC-zeta-bcl2). Unlike the TCR pathway, IL-2 receptor activation does not involve PLC-gamma.[72] Consequently, D4 phosphorylated lipid messengers (IP3 and DAG) are not generated, there is no calcium mobilization, and conventional PKCs (α, β, and gamma) are not activated. The nonclassical, D3 phosphorylated lipids, however, are essential elements in IL-2 signal transduction. PI3K, the enzyme that generates these lipids, is involved in IL-2 signal transduction. It is reported that three pathways recruit PI3K to the activated IL-2 receptor complex: (1) The p85 subunit SH2 domain of PI3K complexes with a phosphotyrosine on the β chain of the IL-2 receptor.[84] (2) It is activated through the ras pathway by direct interaction,[85] and (3) it is involved in the receptor complex through p56lck as a passive adaptor module.[42] PI3K activation in IL-2 stimulation occurs in two subsequent waves.[72] First, IL-2 receptor-associated tyrosine kinases phosphorylate the p85 subunit. The catalytic activity is transiently up-regulated and generates D3 phosphorylated lipid messengers that activate the zeta isoform of PKC. This isoform associates with the p110 subunit of PI3K, and serine phosphorylates the p85 subunit that permanently activates the complex. Activated PI3K regulates several signal transduction events. Through PKC zeta and ε isoforms and other pathways, it coordinates the reorganization of actin cytoskeleton, and through NF-kB, bcl-2, and bcl-XL, it supplies survival signals to the dividing cell.[74]

2.3.3. Unique Features of IL-4 Signal Transduction

The IL-4 receptor signal transduction pathways are not understood to the extent of the pathways of IL-2, but preliminary data indicate that several unique features are involved: (1) In the lck–c-fos pathway, lck and fyn coprecipitate with the activated IL-4 receptor.[86] Uniquely, a special adaptor protein, IRS-1, organizes IL-4 signal transduction.[72] This protein was originally found in the insulin receptor complex. It has about 20 phosphorylated tyrosines and can bind and activate a plethora of factors. Although grb-2 and SOS factors are involved, no ras activation was detected. (2) In the JAK/STAT pathway, the activated IL-4 receptor recruits a special transactivator, STAT6, and also associates with STAT3.[87] The IL-4 receptor also activates a STAT-related pathway, GAS signal transduction.[72] (3) In the PI3K/PKC pathway, no PLC gamma activation is detected, and no calcium mobilization occurs. It was also shown that even the PKC ε and zeta isoforms are not critical in IL-4 signal transduction.[72]

3. CHARACTERIZATION OF *H. SAIMIRI* TRANSFORMED LYMPHOCYTES

3.1. Surface Markers

The surface markers expressed on *H. saimiri* immortalized cells correspond to the pattern characteristic of mature, activated T cells. Components of the antigen-dependent signal generation are expressed: CD2, CD3, TCR, CD4, CD8, and CD45.[88] A key element in the stabilization of normal T cell activation, CD28, is not typically detected and is even down-regulated in $CD8^+$ cells, raising interesting questions about the stabilization of *H. saimiri* induced signals.[88,89] As mentioned earlier, $CD4^+$ and $CD8^+$ T cells are two entirely different classes of lymphocytes by lymphokine profile and by function. It is interesting to note, however, that *H. saimiri* does not display absolute coreceptor preference. In its original hosts, New World monkeys, $CD8^+$ cells are immortalized.[2] In humans, however, $CD4^+$ cells are more susceptible.[89,90] These data indicate that only mature, single-positive cells are immortalized, but double-positive $CD4^+CD8^+$ cell lines are also established in experiments with Old World monkeys.[18] Moreover, these coreceptors are not indispensable for immortalization because transformed cell lines with gamma/δ type TCR that express no CD4 or CD8 structures have also been immortalized.[17,91] CD25, the α chain of the IL-2 receptor is also expressed as a mediator for the progression phase of T cell activation.[88,88a,89] Accessory and costimulatory molecules are also present, and CD5,[89] CD26,[92] CD29,[18,89] and CD40[93] are expressed, together with the T cell activation marker CD30[16,89] and adhesion receptors, like CD11a (LFA-1), CD18 (ICAM-1), CD54, and CD58.[88] Some of the NK markers are detected, like CD16[15] and CD56,[16] although the latter was not present on $CD8^+$ cell lines.[88] Another NK marker, CD57, is not typically expressed.[16] B cell markers (CD19 and others) are never detected.[15]

3.2. Cytokine Profile of *H. saimiri* Transformed T Cells

3.2.1. Characterization of the Th1 and Th2 Subclasses of $CD4^+$ Cells

$CD4^+$ lymphocytes perform helper functions in two different directions. The Th1 subset can stimulate cellular elements of the immune system, and the TH2 subclass induces and regulates the humoral immune response. A third subset of $CD4^+$ T cells is not specialized, and this mixed phenotype is called Th0.[89] Because these diverse functions are mediated by humoral factors, cytokine expression profiles of the subsets are distinctly different.

Th1 lymphocytes secrete IL-2, IFN-gamma, and TNF-β (TNF-α is secreted in all three subclasses) upon antigen stimulation. This subset displays cytotoxicity for allogeneic cells, and in experimental conditions PHA (lectin, phytohemagglutinin) triggers cytotoxicity in 35–70% of TH1 cells. About 80% of TH1 cells also respond to antigen/APC treatment with proliferation.[89,94]

The Th2 subset of $CD4^+$ cells expresses only IL-4 and IL-5, in addition to the common TNF-α secretion, but neither of the TH1 cytokines. TH2 cells are not cytotoxic under normal conditions, and treatment with antigen/APC elicits proliferation at a rate similar to Th1.[89,94]

3.2.2. Cytokine Profile of H. saimiri Transformed TH1 Clones

A characteristic feature of *H. saimiri* induced changes in T cell activation is a nonspecific up-regulation of receptors and functions involved in T cell receptor function.[89] The unstimulated transformed Th1 cells spontaneously express high levels of IFN-gamma, TNF-α, and TNF-β and detectable amounts of an atypically induced cytokine, IL-3. The data on IL-2 expression are inconsistent and indicate heterogeneity of the various cell lines immortalized. Reports on IL-2 expression by unstimulated cells[95] are in contrast to observations that could not detect it, but even in this case low-level IL-2 mRNA expression was reported.[89] Because the IL-2 receptor α chain (CD25) is present at high levels in these cells, an autocrine mechanism was proposed where low amounts of IL-2 are immediately retained by the abundance of receptor on the secreting cell.[89] Neither IL-4 nor IL-4 receptor expression were detected in Th1 cells.

Expression patterns upon induction also show increased activation levels. SRBC (sheep red blood cell) treatment has no effect on normal T cells. In saimiri transformed Th1 cells, however, it increases the expression of the above cytokines and induces IL-2 and IL-3 expression.[89] PMA/anti-CD3 or antigen stimulation can further increase cytokine secretion. In summary, *H. saimiri* immortalization does not bring about major change in the cytokine profile of TH1 cells, except for the induction of IL-3.[89] Expression of the entire cytokine set is up-regulated, and IFN-gamma, IL-3, and membrane-TNF-α expression is constitutive.

3.2.3. Cytokines Expressed by the TH2 Subset of Transformed $CD4^+$ Cells

In contrast to Th1 cells, *H. saimiri* immortalization inflicts profound changes on the TH2 cytokine profile. Unstimulated, immortalized Th2 cells spontaneously secret TNF-α, but also other cytokines that are entirely uncharacteristic of this class: IFN-gamma and IL-3. IL-5 expression was not detected.[89] Reports on the expression of IL-4 in unstimulated cells are conflicting. Although spontaneous expression of IL-4 was never detected serologically, the biological test for IL-4 was positive in tissue culture supernatants.[95] Because IL-4 mRNA was detectable by RT-PCR,[95] low level IL-4 expression may take place without stimulation. IL-2 mRNA was also detected by RT-PCR in uninduced cells, but IL-2 was not present in secreted form.[89]

Stimulation experiments by SRBC also indicate that immortalization also increases nonspecific activation levels of Th2 cells; in addition, a shift in cytokine profile is observed. SRBC has no effect on the parent, untransformed Th2 cells, but in transformed cell lines it up-regulates TNF-α, IFN-gamma, and IL-3. Stimulation also induces detectable levels of IL-4 and IL-5 expression. Antigen/APC or PMA/antiCD3 treatment results in the same but also induces IL-2 secretion. Interestingly, the Th2 specific IL-4 and IL-5 induction in immortalized cells remains minimal, at a level much lower than what is observed in normal cells.[89] In summary, *H. saimiri* immortalization creates Th0 characteristics in parent Th2 cells. The induced expression of INF-gamma and IL-2 are clearly Th1 phenotypes, and IL-3 induction is unique to *H. saimiri* immortalization.[89] At the same time, induction of the Th2 specific IL-4 and IL-5 is marginal. As in Th1 cells, constitutive expression of IFN-gamma, IL-3, and TNFα is also observed with Th2 cells.

3.3. Functional Analysis of *H. saimiri* Transformed T Lymphocytes

3.3.1. Proliferative Response

H. saimiri transformed T cell lines proliferate spontaneously at a low rate, doubling in about 3–4 days.[90] IL-2 requirement is variable, with some cell lines being entirely IL-2 independent.[88] Typically, however, after an initial IL-2-free period, the cells require IL-2 for permanent growth. The proliferation rate is inducible with a very low threshold. In contrast to normal T cells that respond only to specific antigens on presenting cells, HVS-T cells are stimulated by SRBC or allogeneic cells without antigen.[89] Antigen recognition is functional, however, and elicits proliferative response.[96]

Receptor pathways that maintain permanent or induced proliferation have also been mapped. Monoclonal antibodies were used against an array of T cell surface components to establish their role in proliferation. The results indicated that the critical element in the spontaneous proliferation of HVS-T cells is autologous cell–cell contact mediated by the CD2–CD58 receptor–ligand system.[97] Notably, antibodies against CD2 domains that do not interact with CD58, are not effective. So, clearly, CD2–CD58 ligation is important.[97] Although this is not the predominant pathway in T cell proliferation, CD2 has the potential to trigger proliferative signals in normal T cells,[97] and is also involved in HTLV-1 transformed T cell signaling.[98] The degree of CD4 involvement is variable. mAB to CD4 inhibits proliferation with some cell lines but not with others.[88,99]

Antibodies (anti-TAC) against the α chain of IL-2 receptor (CD25) invariably inhibit proliferation, indicating the fundamental role of IL-2 in the autocrine growth of transformed T cells.[97] Other cytokines implicated in the proliferation of *H. saimiri* transformed T cells are IL-3 and IL-4.[89] Reports are incoherent on the effects of recombinant IL-3 or IL-4, or antibodies against these cytokines, and probably reflect differences between individual cell lines.[95]

3.3.2. *Cytotoxic Response*

Normally, the cytolytic, cytotoxic reaction against allogeneic target cells is the response of $CD8^+$ cells and the Th1 subset of $CD4^+$ cells.[34] In *H. saimiri* transformed cells, cytotoxicity, similarly to other tests, also displays a significant decrease in threshold. Nontransformed parental Th1 cells can be lectin-triggered for cytotoxicity in 35–70% of cells. *H. saimiri* transformed Th1 cells respond to PHA in 60–90%.[89] The difference is more manifest with Th2 lymphocytes. Cells of this subset are not cytotoxic normally, whereas in *H. saimiri* transformed Th2 cell lines lectin-induced cytotoxicity can be triggered in 40–77% of cells.[89]

Hyperactivity was also observed when individual surface receptors were investigated for cytotoxic signals in *H. saimiri* transformed cells. Interestingly, the signals were transduced by receptors not normally involved in the process. Results with monoclonal antibodies against surface receptors in transformed cells showed that cytotoxicity was inhibited by anti-CD2 and anti-CD58 antibodies.[92] Data on the inhibition of cytotoxicity by anti-CD11a indicate that LFA-1 integrin is also involved.[92] No inhibitory activity was observed with antibodies for CD4, CD8, CD28, B7, and HLA class I and class II receptors.[92]

To investigate the opposite direction, i.e., what receptors can actively trigger cytotoxic signal, mABs against individual receptors were cross-linked using secondary antibodies, Fc receptors on accessory cells, or Fc receptor coated plates. In normal, native T cells antibodies against CD3 and CD26 can trigger cytotoxicity by cross-linking.[92] In contrast, in *H. saimiri* transformed T cells, mABs against CD2, CD4, CD5, CD8, and CD26 were equally efficient to signal cytotoxicity. TCR–CD3 activity was also maintained, because both cross-linking anti-CD3 antibody and bacterial superantigen could elicit cytotoxicity. In addition, a number of reports indicate that the response for a specific antigen is also preserved and functional.[92]

3.3.3. *IL-2 Induction*

The majority of T lymphocytes is constitutively responsive to IL-2 because the β and gamma chains of the receptor are constitutively expressed and together form the intermediate affinity receptor for IL-2. In the initial phase of *H. saimiri* transformation, IL-2 induction is minimal. However, this low-level signaling is sufficient to induce expression of the α chain of the IL-2 receptor (CD25) and further induce IL-2 because both genes have IL-2 responsive elements in their promoters. Expression of CD25 and reconstitution of the high-affinity receptor further increases both IL-2 and IL-2 receptor expression. The phenomenon was clearly demonstrated on *H. saimiri* transformed $CD8^+$ T cell line. In the initial phase, at the 30th day postinfection, only 8% of the cells were CD25 positive. The number of positives increased gradually up to 50% at the 60th day and 90% at the 90th day.[100] Selective overgrowth of the $CD25^+$ population cannot be ruled out, however.

Induction of IL-2 is normally the function of the TCR–CD3 complex through pathways outlined above. In the case of *H. saimiri* transformed lymphocytes, cell–cell contact is the signal that maintains the activated state and IL-2 induction.[97] Transformed cells can be stimulated by contact with allogeneic cells, and if monoclonal antibodies are added to inhibit individual surface components, receptors that participate in signal generation can be identified. The results show that only anti-CD2 and anti-CD58 are inhibitory for IL-2 induction.[97] Notably, antibodies against components of regular antigenic signaling (TCR, CD3, MHC class I and class II) were ineffective, indicating that in the absence of specific antigen the system does not contribute to the signal. The TCR/CD3 system is operative, however, because specific antigenic response has been reported.[89,97,100] It is probably down-regulated, however, because crosslinking of anti-CD3 antibodies only weakly induces IL-2 production.[97]

3.4. The Immunological Profile of *H. saimiri* Transformed Cells and Implications in Signal Transduction

The data above clearly indicate that the key feature of *H. saimiri* immortalized cells is nonspecific, high-level activation. The activation reflects all of the characteristics of antigen-induced lymphocyte functions except for the antigen receptor TCR/CD3 system, which is not engaged and inactive. The latter feature and the polyclonal Vβ profile rule out a viral superantigen as an inducer. The source of the activation signal is the CD2–CD58 system. The CD2 receptor has long been recognized as a key factor in interaction with antigen-presenting cells, but in normal conditions it does not initiate signals. In addition, other accessory elements that are inactive in normal cells can also generate signals (CD4, CD4, CD8, CD26, etc.). These findings indicate an apparently indiscriminate, high-level activation of the intracytoplasmic domains of antigenic signal transduction elements.

The initial phase of T cell signaling is thoroughly studied and involves tyrosine phosphorylation by lck and other src family kinases. The receptors that are activated in *H. saimiri* immortalization are natural substrates for lck kinase activity. High-level, nonselective activation of these receptors clearly indicates lck

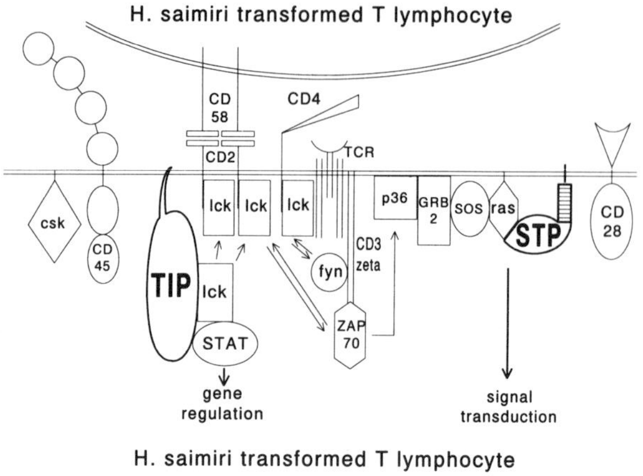

FIGURE 2. Activation of the T cell receptor signal transduction pathway by *H. saimiri* oncogenic proteins. The outline and symbols are the same as in Fig. 1. The TCR and CD4 complexes are unengaged. CD58–CD2 coupling between two *H. saimiri* transformed T lymphocytes is depicted. The oncogenic proteins of *H. saimiri* are in bold.

involvement in the process. Indeed, one of the viral proteins implicated in oncogenesis (tip) interacts with p56lck (see Fig. 2).

The TCR signal is a short-term, transient activation, and in normal T cells CD28 is involved to stabilize the signal. In *H. saimiri* transformed cells, however, CD28 is inactive, even down-regulated. Because the virally induced signal is apparently stable, an alternative stabilization pathway must be used by the virus. Another viral protein (stp-C488) forms a complex with c-ras, (Fig. 2), which is the central signal regulator in the c-fos/c-jun and possibly in the NF-AT pathway. These pathways converge on the IL-2 promoter and activate the IL-2, IL-2 receptor system. Other cytokine mechanisms (IL-3, IL-4) are also involved.

These data indicate that proliferation of *H. saimiri* transformed cells is based on autocrine mechanisms, but induction of the effector cytokines is secondary. The fundamental mechanisms that *H. saimiri* controls are part of the specific T cell activation circuitry. The means by which *H. saimiri* takes charge of T cell activation are discussed next.

4. *H. SAIMIRI* PATHWAYS FOR T CELL ACTIVATION AND CYTOKINE INDUCTION

4.1. *H. saimiri* Genes Involved in T Cell Activation

High homology with host sequences indicates that *H. saimiri* harbors a number of genes from its hosts. Some of these genes are related to cytokines, cytokine receptors, and other components of proliferative regulation, as outlined in the Introduction. Expression of these genes is documented in the replicative, lytic phase of viral infection, and some of them perform functions important in the viral cycle. In transformed, immortalized lymphocytes, however, these genes are silent.[101]

Two kinds of viral transcripts are detected in immortalized cells. The virus expresses one major mRNA[102] and a series of small transcripts.[103,104,105] A second, minor mRNA that codes for the dihydrofolate reductase open reading frame is also expressed in low abundance.[8] The only virally expressed polyA$^+$ mRNA that has been consistently demonstrated in immortalized cells represents the leftmost open reading frame(s) on the unique coding L-DNA portion of the virus.[101,102] The region corresponds to DNA sequences demonstrated by deletion analyses to be involved in oncogenesis.[32] In one subgroup of *H. saimiri* (subgroup A) this transcript codes for one open reading frame, called saimiri-transforming protein A (stpA).[106] In subgroup C the transcript is bicistronic.[102] The first open reading frame is similar to stpA, and it is called stpC.[106] The second open reading frame is dubbed tyrosine kinase interacting protein (tip)[101] and has no overall homology to other sequences. Both of these proteins are expressed in immortalized cells, and their involvement in immortalization is well documented (see later).

H. saimiri immortalization takes place exclusively in T lymphocytes. T cell specificity is controlled at two consecutive levels: regulation of the expression of the mRNA coding for stp and tip expression and by tip interaction that is specific to a tyrosine kinase expressed only in T cells, p56lck. The oncogenic potential of the other protein, stpC, is apparently not restricted to a particular cell type.

4.2. Regulation of the stp/tip mRNA Expression

Transcripts from the oncogenic region of *H. saimiri* show strain variations. A major 1.2-kb mRNA,[102] and a minor 5.3-kb transcript[8] were reported in strain 484-77 immortalized cells. In the related strain 488-77 the region is transcribed into several overlapping mRNAs. In lytic infection in owl monkey kidney cells, five transcripts were detected, 0.5, 1.7, 3.0, 6.8, and 7.6 kb in size.[107] In transformed simian cell lines, two transcripts were present with 1.7 and 3.0 kb. In immortalized human cells only the 1.7-kb mRNA was detected. Expression levels are low in human cell lines and vary from line to line.[107,108]

Baseline expression is inducible, however, and transcription is regulated.[107] Phorbol ester (TPA) at 1 ng/ml concentration increased mRNA expression after 25 min, and at 8 h after induction it reached 16-fold activation. Cycloheximide (10 ug/ml) also rapidly induced expression of the transcript, and at 32 h it reached 30-fold induction. The combined effect was additive, a 47-fold increase at 32 h. Detailed analysis of transcription start sites demonstrated that initiation for the baseline and induced transcripts are different. The baseline transcript starts with a CAAAAT site at −91 nucleotide upstream of the initiator codon, with no TATA boxlike sequence involved. The induced transcripts map to two different sites with traditional TATA and CAAAAT boxes at both, at −42 and −111 positions.[107]

Altogether, the expression pattern of the stp/tip transcript is similar to that of T cell activation genes. Strong inducibility by phorbol esters and mitogens and transcription enhancement in the absence of protein synthesis (cycloheximide) are typical features of T cell activation genes.[107] The resemblance is not complete, however. Cyclosporin A treatment down-regulated but did not abrogate expression of the stp/tip transcript, indicating that the NF-AT pathway is involved, but is not exclusive. The NF-kB pathway is probably not activated because response elements are not present in the promoter.[107]

4.3. The Function of the tip Protein

The first evidence that tip function is important in *H. saimiri* oncogenesis was generated by deleting the tip open reading frame in strain 484-77.[15] The knockout mutant had an intact stpC sequence, and the truncated mRNA was correctly expressed, but no IL-2 independent immortalization took place.

Analysis of the tip protein sequence indicated a C-terminal membrane-spanning domain.[109] Labeling, protease digestion, and immunofluorescence experiments demonstrated that the protein is localized in the inner cytoplasmic membrane.[110,111]

Features of the immortalized cells (T cells only, highly activated cytoplasmic domains of all TCR signal elements in membrane localization) focused the attention on a T cell-specific, membrane-localized, TCR signal associated tyrosine kinase, p56lck[101] (see Fig. 2). Immune precipitation with anti-lck antibody detected coprecipitation of a 40-kDa protein that is the *H. saimiri* tip protein.[101] Investigations with anti-tip antibody and GST-fusion product interactions also demonstrated complex formation between tip and p56lck.[111] Further analysis showed that tip is a substrate for p56lck kinase activity and is phosphorylated at a tyrosine residue.

Analysis of tip domains revealed two regions possibly involved in interactions.[101] At the 146–155 position a sequence was found with similarity to the negative regulatory C-terminal domain of src family kinases. At the 174–183 position a proline-rich sequence was located that fits the consensus SH3-binding motif. Deletion experiments with *in vitro* GST fusion product analysis and *in vivo* transfection assays indicated that both homologous regions are important in p56lck binding and tip phosphorylation.[111] In addition to lck, two more proteins (62 kDa and 110 kDa) were also detected in the immune complex with tip.[112,123,124] The data support the model that p56lck SH2 and SH3 domains are involved in a complex with a phosphotyrosine and the proline-rich region of the tip polypeptide, respectively.[111,112] In another report the SH3 domain by itself was sufficient for binding.[113]

There are three conflicting papers in the literature regarding the effect of the tip protein on p56lck. Jung *et al.* reported a down-regulation of tyrosine phosphorylation in cells expressing tip-488,[114] whereas Wiese *et al.* observed a marginal activation of p56lck by tip-488.[73] Data on tip-484[123] agree with the general conclusions of the work of Wiese *et al.*, who claimed an increase in p56lck activity in a cell-free system. Lund *et al.*[123] also presented evidence that this activation takes place in HVS transformed cells *in vivo*. Moreover, strain 484 tip is a much more potent activator of p56lck than strain-488 tip. Sequence comparison of tip-484 demonstrates a duplication of a C-terminal sequence in tip-488. This additional sequence predicts a larger protein, and this additional duplicated protein domain could interfere with the protein kinase activity of p56lck. Although Jung *et al.*[114] reported a decrease in tyrosine phosphorylation in cells expressing tip-488 and also a decrease in ZAP-70 phosphorylation after CD3 stimulation, this does not exclude the possibility that other proteins may interact with tip-488 and exert different effects on various signaling pathways.

Activation of p56lck by tip-484 correlates with the assembly and phosphorylation of a membrane-associated complex that includes tip, p56lck, and also two transcription factors, STAT-1 and STAT-3.[124] Signal transducers and activators of transcription (STATs) are transcription factors responsible for transducing signals from a variety of cytokine cell surface receptors, including IL-2, IL-10, and interferon-gamma (for a review, see [125]). STATs are inactive in the cytoplasm until a ligand-induced activation of cell surface receptors occurs. This leads to phosphorylation and dimerization of STATs on tyrosine residues. Phosphorylation of STATs can occur by either the receptor, or if the receptor lacks intrinsic kinase activity, by a receptor-associated member of the Janus kinase family (Jak-1,2,3 or Tyk2). Phosphorylated dimerized STATs then translocate to the nucleus where they directly activate transcription.[125]

Phosphorylation of STAT 1 and 3 proteins is also increased by the presence of tip and p56lck.[124] The unique interaction of tip and p56lck also leads to activation and constitutive up-regulation and translocation of STAT transcription factors to the nucleus, as detected by band-shift assays.[124] Further studies are required to understand which genes are activated by STATs in transformed T cells.

The data are clearly in striking contrast to observations in tip-transfected peripheral T lymphocytes. Tyrosine kinase inhibition is in line with an EBV transforming protein, LMP2A. This protein binds and inhibits B cell tyrosine kinases Lyn and Syk, it inhibits Ig receptor signaling, and contributes to the maintenance of latent infection.[114] In *H. saimiri* transformed T cells, however, receptor elements under lck control are highly activated, and this lends more credence to lck activation in a tip complex. Jurkat T cells are constitutively up-regulated and run at the highest activation level, as shown for several signal transduction pathways.[47] At this activation level, abundant expression of the transfected tip protein may only be inhibitory.

4.4. The Function of the stp Protein (stpA and stpC)

Early analysis of deletion mutants of H. simiri strain A identified a region on the viral genome required for immortalizing marmoset lymphocytes or for oncogenicity *in vivo*.[115,116] This region was not required for replication. Fine mapping of the region by restriction endonuclease-generated deletions identified a single open reading frame (later dubbed stpA) as the key component in immortalization.[32] Expression of the open reading frame, however, was never reported in subgroup A immortalized cells, probably because of a low level of transcription from the region. Transcription could be detected only in lytic infection by the virus in owl monkey kidney cells, where two overlapping polyA$^+$ mRNAs were shown, 2.3 and 4.9 kb in size.[117]

Subgroup A virus strains are weakly oncogenic, and they do not transform human lymphocytes, possibly because of the very low level of mRNA expression. Therefore, investigations turned to strains belonging to subgroup C, which are all strongly oncogenic, and high expression of oncogenic factors was expected. Indeed, the first transcript that expressed the stpC open reading frame was detected in a C strain, strain 484-77.[102] The transcript was a bicistronic mRNA and analysis of the first open reading frame identified a small protein with collagenlike sequences.

Retrovirus-mediated transfection of the isolated stpC open reading frame into rodent cells resulted in transformation.[109] The cells showed malignancy in terms of focus forming, growth at reduced serum concentrations, and invasive tumor growth in nude mice. The same experiment with stpA from a group A strain indicated much weaker oncogenicity. Transfection of the tip protein gene by itself did not result in a transformation phenotype.[109]

The stpC sequence was also expressed in transgenic mice.[118] Expression was regulated under various promoters, but tumor induction was observed only in epithelial tissues. Unfortunately, the promoters selected did not function in T cells, so no direct effects of stpC could be investigated on T cells. The results clearly show, however, that stpC is an efficient oncogene by itself and does not require the T cell background for activity.

In similar investigations, the stpA sequence was introduced into the murine genome to create transgenic mice and expressed from promoters induced in T lymphocytes.[119] T cell infiltrations and lymphomas were found in all animals, and the stpA protein was expressed in the proliferating cells that belonged to the CD4 subset. TCR-Vβ analysis showed polyclonality, so superantigenic effects were unlikely. The data indicate that stpA induces T cell proliferation by itself without the contribution of other genes of the viral genome.

The stpC protein was first detected and characterized using an antibody raised against an stpC-specific peptide antigen.[106] The protein was detected in immortalized primate lymphocytes, in transformed rodent cells, in transfected COS-1 cells, and from recombinant *E. coli* cells. In another report the protein was also shown in rabbit cell lines and in *in vitro* transformed human and monkey T cells.[108] Antibody titers indicated that the polypeptide was also expressed *in vivo* in tumor-bearing animals. Immunofluorescence tests and biochemical fractionation experiments indicated that stpC is membrane-associated in perinuclear localization. The protein was a substrate for collagenase, indicating that the predicted collagenlike domain was also expressed. Unlike cellular collagens, however, the viral stpC sequence showed no sign of glycosylation or cross-linking, and consensus motifs for glycosylation were also absent. Its perinuclear localization may indicate endoplasmic reticulum or Golgi association. The protein is phosphorylated at an N-terminal serine residue.[120] The phosphorylation site resembles a casein kinase II site. The serine is in the context of acidic residues. About 10% of the molecules are phosphorylated, but it is not essential for transformation. In rodent cells transformation was unaffected by non-phosphorylated mutants of the gene. The above features (nonessential, casein kinase II-like serine phosphorylation) were also found in another oncoprotein, in papillomavirus E7.[120]

Structural analysis of stpC established three major domains.[102,106] Domain I is the amino terminal sequence with an acidic net charge. Domain II has 18 complete collagenlike repeats, and Domain III is a hydrophobic transmembrane domain. Extensive mutational analyses were performed to identify what roles these domains play in oncogenesis.[121]

The acidic character of the N-terminal domain turned out to be important in oncogenesis.[121] Replacing glutamic acid residues with neutral or positively charged amino acids abrogated oncogenicity. Changing the net charge of this domain also affected subcellular localization. The normal diffuse expression was altered to punctuated staining.

Mutations in the middle collagenlike repeats also affected oncogenicity.[121] Insertion of three residues of noncollagen character between the 14th and 15th repeats eliminated oncogenicity but did not affect subcellular localization of the protein. Shortening the collagenlike domain by three repeats did not profoundly affect oncogenicity, but deletion of 15 repeats rendered the stpC protein nononcogenic.

Short deletions in the hydrophobic C-terminal domain drastically reduced the transforming capacity of the protein.[121] Deletion of a major part of the membrane-spanning domain abrogated oncogenicity, and the protein was diffusely cytoplasmic in localization. Point mutations did not function, but losing the C-terminal asparagine also reduced its transforming capacity.

The data indicate that the three domains are equally important in oncogenicity. The fact that the protein functions only in membrane-anchored localization points to a cellular target that is also in membrane localization. The role of the collagenlike domain is a hinge or spacer that brings the active domain into position for interaction with the target. The precise positioning may indicate a multifaceted target with multiple interactions where contact with a particular surface is effective only. The N-terminal acidic domain is the interface that delivers the oncogenic activation to its target.

In a search for the cellular target of the stpC protein, coprecipitation experiments were performed with antibodies against a number of proteins that could possibly match the characteristics outlined above.[122] Negative results were obtained in experiments with Shc, Nck, PI-3K, PLC gamma, EGF-R, PDGF-R, FGF-R, GAP factor, src, p62, and hsp70. Antibodies against p21 c-ras, however, successfully coprecipitated with stpC.[122] The ras-stpC interaction was also shown in transformed rodent cells, in stpC-transfected COS-1 cells, and in recombinant baculovirus infected insect cells. *In vitro* interaction with GST–stpC fusion protein was also demonstrated.

Experiments with mutant stpC-transformed cells indicated that only mutants which associate with ras retain transforming capacity.[122] StpC also competes for binding to ras with the raf-1 factor, an effector in the ras complex. Direct activation of the ras-GTP conformation was shown as a result of stpC binding, together with constitutively up-regulated MAP kinase activity in stpC-transformed rodent cells.[122] The biological effect of stpC-ras interaction was shown in stpC-expressing PC12 cells, where neurite outgrowth was induced, similarly to NGF induction that acts through the ras pathway.[122]

These data strongly indicate that stpC is complexed with p21 c-ras in transformed cells (see Fig. 2). The effect of stpC binding is similar to that of a GEF (GDP exchange factor), which promotes activation of ras through GTP binding. Alternatively, it can act as a GAP (GTPase activating protein) inhibitor, locking ras in the active GTP complex. Competition with raf-1 indicates, however, that the signal relay pathway is unique but eventually acts through the MAP kinase pathway.

Interestingly, similar assays with stpA have not been able to detect stpA-ras interaction.[122] Differences between stpA and stpC were outlined above. Apparently, stpA transforms T cells using pathways other than stpC.

5. CONCLUSIONS

Immortalization of T lymphocytes by *H. saimiri* is mediated by interleukins and interleukin receptors through autocrine mechanisms. Induction of soluble mediators, however, is secondary. The principal target for immortalization is the molecular mechanism of T cell activation. The virus eliminates (but not inactivates) the specific element of the machinery, the T cell receptor, and takes full control of regulation immediately downstream.

An entirely new phenomenon in the process is self-activation of T cells by direct contacts through the CD2 receptor–CD58 ligand. The CD2 receptor is a key player in signal generation, even in normal cells downstream of specific TCR/peptide/MHC complex formation between two cells. CD2 binding to CD58 generates conformational change in the CD2 cytoplasmic domain, it can associate with an lck kinase, if available, and contributes to the costimulatory signal.

The unique invention of *H. saimiri* is that it makes lck kinase available. One of the two transforming proteins of the virus, the tip polypeptide, is membrane-associated, and it can complex with lck. Because lck is the first-level tyrosine kinase that initiates the signal immediately downstream of TCR, supplying ample amounts of lck in the TCR context obviates the need for specific TCR involvement.

A special characteristic of src kinases, however, poses a problem for the virus. These kinases can not self-activate. They need to be in contact with another kinase to render them active, and this feature explains the need for CD2–CD58 contacts. Interaction of CD2 with CD58 of another cell performs the function of cross-ligation, pulls CD2 molecules together, and the cytoplasmic domains can cross-activate lck kinase molecules. Then local accumulation of lck activity can activate other players in TCR signal transduction and generate the described highly induced state of every element in this regulation.

The TCR-generated signal has a very short half-life in T cells, however, and cannot maintain permanent cell proliferation. Normally, the stabilizing second signal is the function of the CD28 receptor, but the virus does not directly interact with it, and activated lck does not affect its function. Instead, the other viral oncogene, the stpC protein, evolved to form a specific complex with the central signal processor immediately downstream of CD28, the ras protein. The way in which ras is activated by the viral stpC protein is not clear, but it can stabilize T cells in the activated state and maintain continuous proliferation.

The unique delicacy of the *H. saimiri* transformed cell lines is that the virus performs dramatic immortalization with minimal interference and leaves every function of the lymphocytes intact. From a practical point of view the importance of the system cannot be overemphasized. These cells offer the best, if not the only model, to study all of the important T cell functions and can be used as hosts to express lymphocyte-specific proteins. The system is already used as a host to maintain and study HIV infection. Herpesvirus saimiri based expression vectors are under development in several laboratories. The system is ideally suited for use in gene therapy because the patient's own lymphocytes can be efficiently transfected and transformed to express a therapeutic gene or, in the more distant future, to fight cancer.

REFERENCES

1. Roizman, B., Carmichael, L. E., Deinhardt, F., de-The, G., Nahmias, A. J., Plowright, W., Rapp, F., Sheldrick, P., Takahashi, M. and Wolf, K., 1981, *Herpesviridae, Intervirology* **16:**201–217.
2. Kiyotaki, M., Desrosiers, R. C., and Letvin, N. L., 1986, *Herpesvirus saimiri* strain 11 immortalizes a restricted marmoset T8 lymphocyte subpopulation *in vitro. J. Exp. Med.* **164:**926–931.
3. Davison, A. J., 1993, Origins of the herpesviruses, *Abstr.* Int. Herpesvirus Meet., Pittsburgh, PA, p. S-27.
4. Karlin, S., Mocarski, E. S. and Schachtel, G. A., 1994, Molecular evolution of Herpesviruses: genomic and protein sequence comparisons, *J. Virol.* **68:**1886–1902.
5. Albrecht, J.-C., Nicholas, J., Biller, D., Cameron, K. R., Biesinger, B., Newman, C., Wittmann, S., Craxton, M. A., Coleman, H., Fleckenstein, B., and Honess, R. W., 1992, Primary structure of the Herpesvirus saimiri genome, *J. Virol.* **66:**5047–5058.
6. Roizman, B., 1982, The family *Herpesviridae:* General description, taxonomy, and classification, in: *The Herpesviruses,* Volume 1, (B. Roizman, ed.), Plenum Press, New York, p. 1–23.
7. Medveczky, M. M., Szomolanyi, E., Hesselton, R., DeGrand, D., Geck, P., and Medveczky, P. G., 1989, Herpesvirus saimiri strains from three DNA subgroups have different oncogenic potentials in New Zealand white rabbits, *J. Virol.* **63:**3601–3611.
8. Whitaker, S., Geck, P., Medveczky, M. M., Cus, J., Kung, S.-H., Lund, T., Medveczky, P. G., 1995, A polycystronic transcript in transformed cells encodes the dihydrofolate reductase of Herpesvirus saimiri, *Virus Genes* **10:**163–172.
9. Shinagawa, H., and Iwasaki, H., 1996, Processing the Holiday junction in homologous recombination, *Trends Bio. Sci.* **21:**107–111.
10. Laufs, R. and Fleckenstein, B., 1973, Susceptibility to *herpesvirus saimiri* and antibody development in Old and New World monkeys. *Med. Microbiol. Immunol.* **158:**227–236.
11. Daniel, M. D., Mendelez, L. V., Hunt, R. D., King, N. W., Anver, M., Fraser, C. E. O., Barahona, H. H., and Bagss, R. B., 1974, Herpesvirus saimiri. VII. Induction of malignant lymphoma in New Zealand white rabbits, *J. Natl. Cancer Inst.* **53:**1803–1807.
12. Fleckenstein, B., and Desrosiers, R. C., 1982, *Herpesvirus saimiri* and *Herpesvirus ateles,* in *The Herpesviruses,* Volume 1, (B. Roizman, ed.), Plenum Publishing, New York, p. 253–332.
13. Szomolanyi, E., Medveczky, P., and Mulder, C., 1987, *In vitro* immortalization of marmoset cells with three subgroups of *Herpesvirus saimiri, J. Virol.* **61:**3485–3490.
14. Medveczky, P., Szomolanyi, E., Desrosiers, R. C., and Mulder, C., 1984, Classification of *Herpesvirus saimiri* into three groups based on extreme variation in a DNA region required for oncogenicity, *J. Virol.* **52:**938–944.
15. Medveczky, M. M., Geck, P., Sullivan, J. L., Serbousek, D., Djeu, J. Y., and Medveczky, P. G., 1993, IL-2 independent growth and cytotoxicity of Herpesvirus saimiri-infected human CD8 cells and involvement of two open reading frame sequences of the virus, *Virology* **196:**402–412.
16. Biesinger, B., Muller-Fleckenstein, I., Simmer, B., Lang, G., Wittmann, S., Platzer, E., Desrosiers, R. C., and Fleckenstein, B., 1992, Stable growth transformation of human T lymphocytes by *Herpesvirus saimiri. Proc. Natl. Acad. Sci. USA* **89:**3116–3119.
17. Yasukawa, M., Inoue, Y., Kimura, N., and Fujita, S., 1995, Immortalization of human T cells expressing T cell receptor gamma/delta by Herpesvirus saimiri, *J. Virol.* **69:**8114–8117.
18. Akari, H., Mori, K., Terao, K., Otani, I., Fukasawa, M., Mukai, R., and Yoshikawa, Y., 1996, *In vitro* immortalization of Old World monkey T lymphocytes with *Herpesvirus saimiri:* Its susceptibility to infection with simian immunodeficiency viruses, *Virology,* **218:**382–388.
19. Schirm, S., Muller, I., Desrosiers, R. C., and Fleckenstein, B., 1984, *Herpesvirus saimiri* DNA in a lymphoid cell line established by *in vitro* transformation, *J. Virol.* **49:**938–944.
20. Kaschka-Dierich, C., Werner, F. J., Bauer, I., and Fleckenstein, B., 1982, Structure of noninte-

grated circular *herpesvirus saimiri* and *herpesviruses ateles* genomes in tumor cell lines and *in vitro* transformed cells, *J. Virol.* **44:**295–310.

21. Desrosiers, R. C., 1981, Herpesvirus saimiri DNA in tumor cells—deleted sequences, and sequence rearrangements, *J. Virol.* **39:**497–509.
22. Honess, R. W., Bodamer, W., Cameron, K. R., Niller, H. H., Fleckenstein, B., and Randall, R. E., 1986, The A+T rich genome of *H. saimiri* contains a highly conserved gene for thymidilate synthase, *Proc. Natl. Acad. Sci. USA* **83:**3604–3608.
23. Trimble, J. J., Murthy, S. C. S., Bakker, A., Grassmann, R., and Desrosiers, R. C., 1988, A gene for dihydrofolate reductase in a herpesvirus, *Science,* **239:**1145–1147.
24. Albrecht, J.-C., and Fleckenstein, B., 1992, New member of the multigene family of complement control proteins in *herpesvirus saimiri, J. Virol.* **66:**3937–3940.
25. Albrecht, J.-C., Nicholas, J., Cameron, K. R., Newman, C., Fleckenstein, B., and Honess, R. W., 1992, Herpesvirus saimiri has a gene specifying a homologue of the cellular membrane glycoprotein CD59, *Virology* **190:**527–530.
26. Jung, J. U., Stager, M., and Desrosiers, R. C., 1994, Virus encoded cyclin, *Mol. Cell. Biol.* **14:**7235–7244.
27. Nicholas, J., Cameron, K. R., and Honess, R. W., 1992, Herpesvirus saimiri encodes homologs of G protein-coupled receptors and cyclins, *Nature (London)* **355:**362–365.
28. Yao, Z., Fanslow, W. C., Seldin, M. F., Rouseau, A. M., Painter, S. L., Comeau, M. R., Cohen, J. I., and Spriggs, M. K., 1995. Herpesvirus saimiri encodes a new cytokine, IL-17, which binds to a novel cytokine receptor, *Immunity* **3:**811–821.
29. Yao, Z., Maraskovsky, E., Spriggs, M. K., Cohen, J. I., Armitage, R. J., and Alderson, M. R., 1996, *Herpesvirus saimiri* open reading frame 14, a protein encoded by T lymphotropic herpesvirus, binds to MHC class II molecules and stimulates T cell proliferation, *J. Immunol.* **156:**3260–3266.
30. Koomey, J. M., Mulder, C., Burghoff, R. L., Fleckenstein, B., and Desrosiers, R. C., 1984, Deletion of DNA sequence in a nononcogenic variant of *Herpesvirus saimiri, J. Virol.* **50:**662–665.
31. Medveczky, P., Szomolanyi, E., Desrosiers, R. C., and Mulder, C., 1984, Classification of *herpesvirus saimiri* into three groups based on extreme variation in a DNA region required for oncogenicity, *J. Virol.* **52:**938–944.
32. Murthy, S. C. S., Trimble, J. J., and Desrosiers, R. C., 1989, Deletion mutants of *Herpesvirus saimiri* define an open reading frame necessary for transformation, *J. Virol.* **63:**3307–3314.
33. Geck, P., Medveczky, M. M., Chou, C.-S., Brown, A., Cus, J., and Medveczky, P. G., 1994, Herpesvirus saimiri small RNA and interleukin-4 mRNA AUUUA repeats compete for sequence-specific factors including a novel 70K protein, *J. Gen. Virol.* **75:**2293–2301.
34. Abbas, A. K., Lichtman, A. H., and Pober, J. S., 1991, *Cellular and molecular immunology.* W.B. Saunders, Philadelphia.
35. Davis, M. M., and Bjorkman, P. J., 1988, T cell antigen receptor genes and T cell recognition, *Nature* **334:**395–402.
36. Kappler, J. W., Roehm, N., and Marrack, P., 1987, T cell tolerance by clonal elimination in the thymus, *Cell* **49:**273–280.
37. von Boehmer, H., and Kisielow, P., 1990, Self-nonself discrimination by T cells, *Science* **248:**1369–1373.
38. Pastor, M. I., Reif, K., and Cantrell, D., 1995, The regulation and function of p21ras during T cell activation and growth, *Immunol. Today* **16:**159–164.
39. Crabtree, G. R., 1989, Contingent genetic regulatory events in T lymphocyte activation, *Science* **243:**355–361.
40. DeFranco, A. L., 1995, Transmembrane signaling by antigen receptors of B and T lymphocytes, *Curr. Opinion Cell Biol.* **7:**163–175.

41. Davis, S. J., and van der Merwe, P. A., 1996, The structure and ligand interactions of CD2: Implications for T cell functions, *Immunol. Today* **17:**177–187.
42. Theze, J., Alzari, P. M., and Bertoglio, J., 1996, Interleukin-2 and its receptors: Recent advances and new immunological functions, *Immunol. Today* **17:**481–486.
43. Burkhardt, A. L., Stealey, B., Rowley, R. B., Mahajan, S., Prendergast, M., Fargnoli, J., and Bolen, J. B., 1994, Temporal regulation of non-transmembrane protein tyrosine kinase enzyme activity following T cell antigen receptor engagement, *J. Biol. Chem.* **269:**23642–23647.
44. Rudd, C. E., Janssen, O., Cai, Y. C., da Silva, A. J., Raab, M., and Prasad, K. V. S., 1994, Two-step TCRzeta/CD3-CD4 and CD28 signaling in T cells: SH2/SH3 domains, protein-tyrosine and lipid kinases, *Immunol. Today* **15:**225–234.
45. Samelson, L., Phillips, A., Luong, E., and Klausner, R., 1990, Association of the fyn protein tyrosine kinase with the T cell antigen receptor, *Proc. Natl. Acad. Sci. USA* **87:**4358–4362.
46. Xu, H., and Littman, D. R., 1993, A kinase-independent function of lck in potentiating antigen-specific T cell activation, *Cell* **74:**633–644.
47. Ward, S. G., June, C. H., and Olive, D., 1996, PI 3-kinase: A pivotal pathway in T cell activation? *Immunol. Today* **17:**187–197.
48. Chan, A. C., Iwashima, M., Turck, C. W., and Weiss, A., 1992, ZAP-70, a 70 kd protein-tyrosine that associates with the TCR zeta chain, *Cell* **71:**649–662.
49. Buday, L., Egan, S. E., Viciana, P. R., Cantrell, D. A., and Downward, J., 1994, A complex of Grb2 adapter protein, Sos exchange factor, and a 36-kDa membrane-bound tyrosine phosphoprotein is implicated in ras activation in T cells, *J. Biol. Chem.* **269:**9019–9023.
50. McCormick, F., 1993, Signal transduction. How receptors turn ras on, *Nature* **363:**15–16.
51. Donovan, J. A., Wange, R. L., Langdon, W. Y., and Samelson, L. E., 1994, The protein product of the c-cbl proto-oncogene is the 120 kDa tyrosine phosphorylated protein in Jurkat cells activated via the T cell antigen receptor. *J. Biol. Chem.* **269:**22921–22924.
52. Boguski, M. S., and McCormick, F., 1993, Proteins regulating ras and its relatives. *Nature* **366:**643–654.
53. Avruch, J., Zhang, X.-F., and Kyriakis, J. M., 1994, Raf meets ras: Completing the framework of a signal transduction pathway, *Trends. Bio. Sci.* **19:**279–283.
54. Ahn, N. G., Seger, R., and Krebs, E. G., 1992, The mitogen activated protein kinase activation. *Curr. Opinion Cell. Biol.* **4:**992–999.
55. Marais, R., Wynne, J., and Treisman, R., 1993, The SRF accessory protein Elk-1, contains a growth factor regulated transcriptional activation domain, *Cell* **73:**381–393.
56. Sieh, M., Batzer, A., Schlessinger, J., and Weiss, A., 1994, GRB2 and phospholipase C-gamma1 associate with a 36- to 38-kilodalton phosphotyrosine protein after T cell receptor stimulation, *Mol. Cell. Biol.* **14:**4435–4442.
57. Lee, S. B., and Rhee, S. G., 1996, Significance of PIP2 hydrolysis and regulation of phospholipase C isozymes, *Curr. Opinion Cell. Biol.* **7:**183–189.
58. Weiss, A., and Imboden, J., 1987, Cell surface molecules and early events involved in human T lymphocyte activation, *Adv. Immunol.* **41:**1–38.
59. Berridge, M. J., 1993, Inositol triphosphate and calcium signaling, *Nature* **361:**315–325.
60. Pleiman, C. M., Hertz, W. M., and Cambier, J. C., 1994, Activation of phosphatidylinositol-3′ kinase by src family kinase SH3 binding to the p85 subunit, *Science* **263:**1609–1612.
61. Toker, A., Meyer, M., Reddy, K. K., Falck, J. R., Aneja, R., Aneja, S., Parra, A., Burns, D. J., Ballas, L. M., and Contley, L. C., 1994, Activation of protein kinase C family members by the novel polyphosphoinositides PtdIns-3,4,P2 and PtdIns-3,4,5-P3, *J. Biol. Chem.* **269:**32368–32377.
62. Feig, L. A., Urano, T., and Cantor, S., 1996, Evidence for a ras/ral signaling cascade, *Trends Biochem. Sci.* **21:**438–441.
63. Schreiber, S. L., and Crabtree, G. R., 1992, The mechanism of action of cyclosporin A and FK506, *Immunol. Today* **13,**136–142.

64. Rao, A., 1994, NF-ATp: A transcription factor required for the coordinate induction of several cytokines, *Immunol. Today* **15:**274–281.
65. Marshall, M. S., 1993, The effector interactions of p21ras, *Trends Biochem. Sci.* **18:**250–253.
66. Shaw, G., and Kamen, R., 1986, A conserved AU sequence from the 3′ untranslated region of GM-CSF mRNA mediates selective mRNA degradation, *Cell* **46:**659–667.
67. Oetken, C., Couture, C., Bergman, M., Bonnefoy-Berard, N., Williams, S., Alitalo, K., Burn, P., and Mustelin, T., 1994, TCR/CD3 triggering causes increased activity of the p50 csk tyrosine kinase and engagement of its SH2 domain. *Oncogene* **9:**1625–1631.
68. Raab, M., Cai, Y. C., Bunnell, S. C., Heyeck, S. D., Berg, L. J., and Rudd, C. E., 1995, p56lck and p59fyn regulate CD28 binding to phosphatidylinositol-3 kinase, growth factor receptor bound protein GRB-2, and T cell specific protein tyrosine kinase ITK: Implications for T cell costimulation. *Proc. Natl. Acad. Sci. USA* **92:**8891–8895.
69. Genot, E., Parker, P., and Cantrell, D. A., 1995, Analysis of the role of protein kinase C-alpha, -epsilon, and -zeta in T cell activation, *J. Biol. Chem.* **270:**9833–9839.
70. Weng, Q. P., Andrabi, K., Klippel, A., Kozlowski, M. T., Williams, L. T., and Avruch, J., 1995, Phosphatidylinositol-3 kinase signals activation of p70S6 kinase *in situ* through site specific p70 phosphorylation, *Proc. Natl. Acad. Sci. USA* **92:**5744–5748.
71. Boise, L. H., Minn, A. J., Noel, P. J., June, C. H., Accavitti, M. A., Lindsten, T., and Thompson, C. B., 1995, CD28 costimulation can promote T cell survival by enhancing the expression of Bcl-XL, *Immunity* **3:**87–98.
72. Rebollo, A., Gomez, J., and Martinez, A. C., 1996, Lessons from immunological, biochemical and molecular pathways of the activation mediated by IL-2 and IL-4. *Adv. Immunol.* **63:**127–196.
73. Paliogianni, F., Hama, N., Mavrothalassitis, G. J., Thyphronitis, G., and Boumpas, D. T., 1996, Signal requirements for interleukin-4 promoter activation in human T cells, *Cell. Immunol.* **168:**33–38.
74. Minami, Y., and Taniguchi, T., 1995, IL-2 signaling: Recruitment and activation of multiple protein tyrosine kinases by the components of the IL-2 receptor. *Curr. Opinion Cell. Biol.* **7:**156–162.
75. Minami, Y., Kono, T., Yamada, K., Kobayashi, N., Kawahara, A., Perlmutter, R. M., and Taniguchi, T., 1993, Association of p56lck with IL-2 receptor beta chain is critical for the IL-2 induced activation of p56lck, *EMBO J.* **12:**759–768.
76. Hatakeyama, M., Kono, T., Kobayashi, N., Kavahara, A., Lewin, S. D., Perlmutter, R. M., and Taniguchi, T., 1991, Interaction of the IL-2 receptor with the src-family kinase p56lck: Identification of novel intermolecular association, *Science* **252:**1523–1528.
77. Gulbins, E., Coggeshall, K. M., Gottfried, B., Katzav, S., Burn, P., and Altman, A., 1993, Tyrosine kinase stimulated guanine nucleotide exchange activity of Vav in T cell activation, *Science* **260:**822–825.
78. Evans, G. A., Howard, O. M., Erwin, R., and Farrar, W. L., 1993, Interleukin-2 induces tyrosine phosphorylation of the vav proto-oncogene product in human T cells: Lack of requirement for the tyrosine kinase lck. *Biochem. J.* **294:**339–342.
79. Hatakeyama, M., Mori, H., Doi, T., and Taniguchi, T., 1989, A restricted cytoplasmic region of IL-2 receptor beta chain is essential for growth signal transduction, but not for ligand binding and internalization. *Cell* **59:**837–845.
80. Firpo, E. J., Koff, A., Solomon, M. J., and Roberts, J. M., 1994, Inactivation of a cdk2 inhibitor during interleukin-2 induced proliferation of human T lymphocytes. *Mol. Cell. Biol.* **14:**4889–4901.
81. Minami, Y., Kono, T., Miyazaki, T., and Taniguchi, T., 1993, The IL-2 receptor complex: Its structure, function and target genes. *Annu. Rev. Immunol.* **11:**245–267.
82. Galaktionov, K., Chen, X.-C., and Beach, D., 1996, Cdc25 cell cycle phosphatase as a target of c-myc. *Nature* **382:**511–517.

83. Grana, X., and Reddy, E. P., 1995, Cell cycle control in mammalian cells: Role of cyclins, cyclin dependent kinases (CDKs), growth suppressor genes, and cycle dependent kinase inhibitors (CKIs), *Oncogene* **11:**211–224.
84. Truitt, K. E., Mills, G. B., Turck, C. W., and Imboden, J. B., 1994, SH2 dependent association of phosphatidylinositol-3′ kinase 85-kDa regulatory subunit with the interleukin-2 receptor beta chain, *J. Biol. Chem.* **269:**5937–5943.
85. Rodriguez-Viciana, P., Warne, P. H., Dhand, R., Vanhaesebroeck, B., Gout, I., Fry, M. J., Waterfield, M. D., and Downward, J., 1994, Phosphatidylinositol-3-oh kinase as a direct target of ras. *Nature* **370:**527–532.
86. Smerz-Bertling, C., and Duschl, A., 1995, Both interleukin-4 and interleukin-13 induce tyrosine phosphorylation of the 140 kDa subunit of the interleukin-4 receptor. *J. Biol. Chem.* **270:**966–970.
87. Callard, R. E., Matthews, D. J., and Hibbert, L., 1996, IL-4 and IL-13 receptors: Are they one and the same? *Immunol. Today* **17:**108–110.
88. Meinl, E., Hohlfeld, R., Wekerle, H., and Fleckenstein, B., 1995, Immortalization of human T cells by Herpesvirus saimiri, *Immunol. Today* **16:**55–58.
88a. Medveczky, P. G., and Medveczky, M. M., 1989, Expression of interleukin-2 receptors in T cells transformed by strains of Herpesvirus saimiri representing three DNA subgroups, *Intervirology* **30:**213–226.
89. De Carli, M., Berthold, S., Fickenscher, H., Muller-Fleckenstein, I., D'Elios, M. M., Gao, Q.-L., Biagiotti, R., Giudizi, M. G., Kalden, J. R., Fleckenstein, B., Romagnani, S., and Del Prete, G., 1993, Immortalization with *Herpesvirus saimiri* modulates the cytokine secretion profile of established Th1 and Th2 human T cell clones, *J. Immunol.* **151:**5022–5033.
90. Broker, B. M., Tsygankov, A. Y., Muller-Fleckenstein, I., Guse, A. H., Chitaev, N. A., Biesinger, B., Fleckenstein, B., and Emmrich, F., 1993, Immortalization of human T cell clones by Herpesvirus saimiri. Signal transduction analysis reveals functional CD3, CD4 and IL-2 receptors, *J. Immunol.* **151:**1184–1192.
91. Klein, J. L., Fickenscher, H., Holliday, J. E., Biesinger, B., and Fleckenstein, B., 1996, Herpesvirus saimiri immortalized gamma/delta T cell line activated by IL-12, *J. Immunol.* **156:**2754–2760.
92. Mittrucker, H.-W., Muller-Fleckenstein, I., Fleckenstein, B., and Fleischer, B., 1993, Herpesvirus saimiri transformed human T lymphocytes: Normal functional phenotype and preserved T cell receptor signalling, *Intern. Immunol.* **5:**985–990.
93. Hess, S., Kurrle, R., Lauffer, L., Reithmuller, G., and Engelmann, H., 1995, A cytotoxic CD40/p55 tumor necrosis factor receptor hybrid detects CD40 ligand on *Herpesvirus saimiri* transformed T cells, *Eur. J. Immunol.* **25:**80–86.
94. Maggi, E., Almerigogna, F., Del Prete, G., and Romagnani, S., 1993, Abnormal B cell helper activity by virus infected human CD4+ T cells, *Semin. Immunol.* **5:**449–455.
95. Chou, C.-S., Medveczky, M. M., Geck, P., Vercelli, D., and Medveczky, P. G. 1995, Expression of IL-2 and IL-4 in T lymphocytes transformed by Herpesvirus saimiri, *Virology* **208:**418–426.
96. Weber, F., Meinl, E., Drexler, K., Czlonkowska, A., Huber, S., Fickenscher, H., Muller-Fleckenstein, I., Fleckenstein, B., Wekerle, H., and Hohlfeld, R., 1993, Transformation of human T cell clones by Herpesvirus saimiri: Intact antigen recognition by autonomously growing myelin basic protein specific T cells, *Proc. Natl. Acad. Sci. USA* **90:**11049–11053.
97. Mittrucker, H.-W., Muller-Fleckenstein, I., Fleckenstein, B., and Fleischer, B., 1992, CD2-mediated autocrine growth of Herpesvirus saimiri-transformed human T lymphocytes, *J. Exp. Med.* **176:**909–913.
98. Del Prete, G., De Carli, M., D'Elios, M. M., Muller-Fleckenstein, I., Fickenscher, H., Fleckenstein, B., Almerigogna, F., and Romagnani, S., 1994, Polyclonal B cell activation induced by Herpesvirus saimiri transformed human CD4+ T cell clones, *J. Immunol.* **152:**4872–4879.

99. Broker, B. M., Tsygankov, A. Y., Fickenscher, H., Chitaev, N. A., Muller-Fleckenstein, I., Fleckenstein, B., Bolen, J. B., and Emmrich, F. 1994, Engagement of the CD4 receptor inhibits the interleukin-2 dependent proliferation of human T cells transformed by *Herpesvirus saimiri*, *Eur. J. Immunol.* **24:**843–850.
100. Berend, K. R., Jung, J. U., Boyle, T. J., DiMaio, J. M., Mungal, S. A., Desrosiers, R. C., and Lyerly, H. K., 1993, Phenotypic and functional consequences of *Herpesvirus saimiri* infection of human $CD8^+$ cytotoxic T lymphocytes, *J. Virol.* **67:**6317–6321.
101. Biesinger, B., Tsygankov, A. Y., Fickenscher, H., Emmrich, F., Fleckenstein, B., Bolen, J. B., and Broker, B. M., 1985, The product of the *Herpesvirus saimiri* open reading frame 1 (Tip) interacts with T cell-specific kinase p56lck in transformed cells, *J. Biol. Chem.* **270:**4729–4734.
102. Geck, P., Whitaker, S.C., Medveczky, M. M., and Medveczky, P. G., 1990, Expression of collagenlike sequences by a tumor virus, Herpes saimiri, *J. Virol.* **64:**3509–3515.
103. Murthy, S., Kamine, J., and Desrosiers, R. C., 1986, Viral encoded small RNAs in Herpesvirus saimiri induced tumors. *EMBO J.* **5:**1625–1632.
104. Lee, S. I., Murthy, S. C. S., Trimble, J., Desrosiers, R. C., and Steitz, J. A., 1988, Four novel U RNAs are encoded by a herpesvirus. *Cell* **54:**599–607.
105. Geck, P., Whitaker, S. A., Medveczky, M. M., Last, T. J., and Medveczky, P. G., 1994, Small RNA expression from the oncogenic region of a highly oncogenic strain of *Herpesvirus saimiri*, *Virus Genes* **8:**25–34.
106. Jung, J. U., and Desrosiers, R. C., 1991, Identification and characterization of the *Herpesvirus saimiri* oncoprotein STP-C488, *J. Virol.* **65:**6953–6960.
107. Fickenscher, H., Biesinger, B., Knappe, A., Wittman, S., and Fleckenstein, B., 1996, Regulation of the *Herpesvirus saimiri* oncogene stpC, similar to that of T cell activation genes, in growth-transformed human T lymphocytes, *J. Virol.* **70:**6012–6019.
108. Medveczky, M. M., Geck, P., Vassallo, R., Medveczky, P. G., 1993, Expression of the collagen-like putative oncoprotein of Herpesvirus saimiri in transformed T cells, *Virus Genes* **7:**349–365.
109. Jung, J. U., Trimble, J. J., King, N. W., Biesinger, B., Fleckenstein, B. W., and Desrosiers, R. C., 1991, Identification of transforming genes of subgroup A and C strains of *Herpesvirus saimiri*, *Proc. Natl. Acad. Sci. USA* **88:**7051–7055.
110. Lund, T., Medveczky, M. M., Geck, P., and Medveczky, P. G., 1995, A *Herpesvirus saimiri* protein required for interleukin-2 independence is associated with membranes of transformed T cells, *J. Virol.* **69:**4495–4499.
111. Lund, T., Medveczky, M. M., Neame, P. J., and Medveczky, P. G., 1996, A *Herpesvirus saimiri* membrane protein required for interleukin-2 independence forms a stable complex with p56lck, *J. Virol.* **70:**600–606.
112. Jung, J. U., Lang, S. M., Friedrich, U., Jun, T., Roberts, T. M., Desrosiers, R. C., and Biesinger, B., 1995, Identification of lck-binding elements in tip of *Herpesvirus saimiri*, *J. Biol. Chem.* **270:**20660–20667.
113. Wiese, N., Tsygankov, A. Y., Klauenberg, U., Bolens, J. B., Fleischer, B., and Broker, B. M., 1996, Selective activation of T cell kinase p56lck by *Herpesvirus saimiri* protein tip, *J. Biol. Chem.* **271:**847–852.
114. Jung, J. U., Lang, S. M., Jun, T., Roberts, T. M., Veillette, A., and Desrosiers, R. C., 1995, Down-regulation of Lck-mediated signal transduction by tip of *Herpesvirus saimiri*, *J. Virol.* **69:**7814–7822.
115. Desrosiers, R. C., Bakker, A., Kamine, J., Falk, L. A., Hunt, R. D., and King, N. W., 1985, A region of the Herpesvirus saimiri genome required for oncogenicity, *Science* **228:**184–187.
116. Desrosiers, R. C., Silva, D. P., Waldron, L. M., and Letvin, N. L., 1986, Nononcogenic deletion mutants of *Herpesvirus saimiri* are defective for *in vitro* immortalization, *J. Virol.* **57:**701–705.
117. Kamine, J., Bakker, A., and Desrosiers, R. C., 1984, Mapping of RNA transcribed from a region of the Herpesvirus saimiri genome required for oncogenicity, *J. Virol.* **52:**532–540.

118. Murphy, C., Kretschmer, C., Biesinger, B., Beckers, J., Jung, J., Desrosiers, R. C., Muller-Hermelink, H. K., Fleckenstein, B. W., and Ruther, U., 1994, Epithelial tumours induced by a Herpesvirus oncogene in transgenic mice, *Oncogene* **9:**221–226.
119. Kretschmer, C., Murphy, C., Biesinger, B., Beckers, J., Fickenscher, H., Kirchner, T., Fleckenstein, B., and Ruther, U., 1996, A Herpes saimiri oncogene causing peripheral T cell lymphoma in transgenic mice, *Oncogene* **12:**1609–1616.
120. Jung, J. U., and Desrosiers, R. C., 1992, Herpesvirus saimiri oncogene STP-C488 encodes a phosphoprotein, *J. Virol.* **66:**1777–1780.
121. Jung, J. U., and Desrosiers, R. C., 1994, Distinct functional domains of STP-C488 of Herpesvirus saimiri, *Virology* **204:**751–758.
122. Jung, J. U., and Desrosiers, R. C., 1995, Association of the viral oncoprotein STP-C488 with cellular ras, *Mol. Cell. Biol.* **15:**6506–6512.
123. Lund, T., Medveczky, M. M., and Medveczky, P. G., 1997, A Herpesvirus saimiri Tip-484 membrane protein markedly increases $p56^{lck}$ activity in T cells, *J. Virol.* **71:**378–382.
124. Lund, T., Garcia, R., Medveczky, M. M., Jove, R., and Medveczky, P. G., 1997, Activation of STAT transcription factors by a viral protein requires p56lck, *J. Virol.* **71** in press.
125. Ihle, J. N., 1996, STATs: Signal transducers and activators of transcription. *Cell* **84:**331–334.

5

Kaposi's Sarcoma-Associated Herpesvirus (KSHV/HHV8) and the Etiology of KS

SONJA J. OLSEN and PATRICK S. MOORE

1. INTRODUCTION

Kaposi's sarcoma-associated herpesvirus (KSHV) is a newly discovered herpesvirus initially identified in Kaposi's sarcoma (KS) lesions from an AIDS patient.[1] Using representational difference analysis (RDA)[2] to search diseased tissue for unique DNA fragments, Chang and colleagues isolated two sequences, KS330Bam and KS631Bam. Initial analysis of these fragments identified high homology to two lymphotrophic gamma herpesviruses, Herpesvirus saimiri (HVS) and Epstein–Barr virus (EBV).[1] Further phylogenetic analysis established KSHV as a novel gamma herpesvirus belonging to the genus *Rhadinovirus* with HVS as its closest relative.[3]

KSHV is the eighth herpesvirus to be isolated in humans, and unless it has a primarily nonhuman host, it will be classified as HHV8. Early speculation about the epidemiology of KSHV was based on our knowledge of other human herpesviruses. Many herpesviruses, such as Epstein–Barr virus (EBV), spread widely through populations, cause asymptomatic or mild infection early in life, and

SONJA J. OLSEN and PATRICK S. MOORE • Division of Epidemiology, Columbia University School of Public Health, New York, New York 10032.

Herpesviruses and Immunity, edited by Medveczky *et al.* Plenum Press, New York, 1998.

achieve seroprevalences greater than 90% by adulthood.[4] However, there are exceptions. Herpes simplex virus 2 (HSV2) which is spread through venereal contact usually does not affect children and reaches high prevalences only in select risk groups.[4] Those scientists who are convinced that all human herpesviruses are ubiquitous are likely to conclude that the recognition of a newly identified herpesvirus in KS lesions is coincidental and not causal.[5,6] Others who demonstrate the absence of KSHV sequences in most control tissues suggest that perhaps there is a causal link. A careful examination of the literature should help to clarify the role, if any, of KSHV in the pathogenesis of KS.

Irrespective of this debate, the discovery of KSHV should also be viewed in a broader public health perspective. The recent increased attention to the threat of emerging diseases[7] has highlighted the need for molecular approaches that rapidly identify microbes. Although RDA detected KSHV nucleic acids in KS lesions in 1994, its use since this has been limited to the new identification of hepatitis G virus and neurotropic strains of HHV6 in multiple sclerosis patient brain tissue. Nonetheless, RDA and related techniques hold great promise in the future study of new and emerging pathogens.[8] Furthermore, the association between KSHV and KS provides additional evidence of the increased role of infectious agents in chronic diseases. As traditional disease paradigms begin to shift, so too must our approach to understanding and preventing them.

2. EPIDEMIOLOGY

2.1. Kaposi's Sarcoma

Kaposi's sarcoma (KS) was first described by Moritz Kaposi in 1872 and based on epidemiological and clinical differences, it has since been classified into four subtypes: classical (sporadic) KS, African (endemic) KS, posttransplant or immunosuppression-related KS, and AIDS (epidemic) KS (for review, see [9]). Despite differences in clinical and epidemiological features, all subtypes of KS are indistinguishable histologically, suggesting a common cause. KS typically appears as cutaneous lesions although mucosal and lymph node involvement occur. The classical form of the disease typically occurs in elderly men of Mediterranean descent and generally follows an indolent course. In Eastern and Central Africa, where it is endemic, KS is similar in adults but in children the lymphadenopathic form predominates. KS also occurs after immunosuppressive therapy (e.g., after organ transplantation) but is unique in that complete regression is often achieved following cessation of immunosuppression. The most recently described subtype, AIDS-KS, is very aggressive and often involves oral, genital, and intestinal mucosa and lymph nodes and skin.

High incidence rates and clustering of KS in Africa led to speculation in the

1960s that KS is caused by an infectious agent,[10] and in the 1970s cytomegalovirus (CMV) was suspected as a prime candidate.[11] However it was not until the AIDS epidemic, when KS incidence increased dramatically in America and Europe, that the infectious etiological hypothesis received serious attention.[12–15] Identification of behavioral and geographic risk factors pointed toward a sexually transmitted agent.[12–19] Gay and bisexual men with AIDS are approximately 20 times more likely to develop KS than hemophiliac AIDS patients, despite comparable levels of immunosuppression.[15,16] Furthermore, HIV seropositive gay and bisexual men living in AIDS epicenters[13,20,21] and their sex partners[13,20] are at a higher risk for KS than those living elsewhere, and the HIV-positive female sex partners of bisexual men are more likely to develop KS than other women with AIDS.[12] The recent discovery of KSHV in a KS lesion has led to much speculation about its role in the disease. However, establishing causality between an exposure and a disease requires a robust analysis of the existing data.

2.2. Evidence for Causality

The first criteria for causal inference in infectious disease were the result of the seminal work of Jakob Henle and Robert Koch in the 1880s. Koch's postulates, as they are more commonly known, consist of the following three criteria: (1) the parasite occurs in every case of the disease in question and under circumstances which can account for the pathological changes and clinical course of the disease; (2) the parasite occurs in no other disease as a fortuitous and nonpathogenic parasite; and (3) after being fully isolated from the body and repeatedly grown in pure culture, the parasite can induce the disease anew. These criteria have many obvious shortcomings because they do not account for pathogen carriage, multiple outcomes, or inability to cultivate the agent *in vitro*. Similar arguments have been used to suggest that HIV does not cause AIDS,[22,23] which modern epidemiologic theories find inappropriate. Koch and others rapidly realized the problems with "Koch's postulates"[24,25] and more appropriate standards have since been developed.[24–28] The criteria now widely used to distinguish a causal from a noncausal association include (1) strength of association, (2) specificity, (3) temporality, (4) consistency, (5) biological gradient, (6) biological plausibility, and (7) experimental evidence. These criteria are used to weigh the current evidence linking KSHV and KS.

2.2.1. *Strength of Association*

The best indicator of the strength of an association is given by the rate ratio (RR), a comparison of the incidence of disease in the exposed to the unexposed. The larger the magnitude of the association over unity, the more likely the association is real and not caused by confounding or bias. Rate ratios less than one indicate a protective or inverse association between the agent and disease.[27] Calculating a RR requires following a cohort of KSHV-positive and KSHV-negative individuals over time to see who develops KS. Although methodologically superior, cohort studies are expensive and time-consuming. Most of the studies on KSHV to date have been case-control or nested case-control (i.e., within a cohort) which select subjects on the basis of disease and then assesses exposure to KSHV. (The establishment of numerous cohorts in the 1980s to study HIV/AIDS has allowed more robust retrospective analyses of KSHV and KS.) Case-control studies measure the strength of an association by calculating the odds ratios (OR), an estimate of the RR.

We have examined major studies published to date (December 1994–January 1997) to survey results of the association between KSHV and KS. Tables 1–3 are a compilation of the major studies which examine the association between KSHV and KS: studies using PCR to detect KSHV in KS lesions (Table 1), PBMCs (Table 2), and serologic assays to detect antibodies to latent and lytic phase KSHV antigens (Table 3). In evaluating these data, one should be aware that technical differences in detecting KSHV infection are clearly a major barrier to epidemiological studies. Most DNA-based studies use PCR which is highly susceptible to contamination, producing false-positive results. This problem is nearly intractable for nested PCR despite extreme precautions[29] and is highly likely in studies which report aberrant or irreproducible results. Southern hybridization is less sensitive, but provides adequate detection rates (~70%) to confirm the presence of the virus in truly infected tissues. Similarly, unnested PCR has low sensitivity for detecting viral DNA in fixed tissues. Finally, serological studies are currently contentious on the issue of sensitivity for antibody detection (among latent antigen tests) and cross-reactivity to other herpesviruses (for lytic antigen tests).

Twenty-one studies have examined a total of 540 KS tissues from all forms of KS. KSHV DNA was found in 95% of these KS samples (Table 1). However, these results are meaningful only when compared to the percentage of KSHV-positive tissues found among control specimens. Because cases and controls theoretically arise from the same joint-source population,[30] the prevalence of KSHV in the controls should represent a background level of infection. In practice, however, selection of controls for these studies was highly variable and in no case was a truly population-based comparison group employed. Although this quantitatively affects the presumed risk of KSHV in developing KS in a general population, the magnitude and consistency of the results from various studies is striking. To give the reader a rough idea of the strength of the association, we computed ORs from the available data (Table 1). The reader should be aware that most of these studies have small sample sizes and controls were selected with unknown stringency criteria. Therefore, these calculations should be reviewed cautiously. Four studies did not use a control group required to calculate an OR,[31–34] making it difficult to draw conclusions about the strength of association from these studies. However, because these studies also tested normal skin from the KS patients, they can be used to examine the biological gradient between KSHV and KS (see below). For the 17 studies in which an odds ratio can be calculated, 12 have OR greater than 100 and the rest lie between 37 and 69. This is an extraordinarily high rate of association compared with most epidemiological exposures likely to have a causal association. For comparison, ORs for cigarette smoking and lung cancer generally range between 2 and 10 for most case-control studies.[27]

Most of the differences observed among the studies can be explained by the selection criteria used for KS and control tissues. Studies that extracted DNA from paraffin-embedded tissue have lower PCR detection rates. In addition, HIV serostatus among control patients appears to be important. KSHV is detected at a higher rate in tissues from HIV-infected than HIV-uninfected persons without KS. Because it has been hypothesized that HIV plays a major role in KS,[14,35] to examine the role of KSHV in the development of KS independently of HIV, one must control for HIV status either by restricting the selection criteria or by controlling for HIV in the analysis. Only a few studies have attempted to do this.[1,36–38] DNA-based detection studies should also be examined in the light of the fact that major sites of latency are largely unknown. One could hypothesize, for example, that KS cases and controls are infected at equal rates and that the virus present in a tissue type is not examined in these studies. Examination of KSHV in peripheral blood and serological studies, however, addresses this question to a limited extent. Nevertheless, the overwhelming conclusion from these analyses is that KSHV is likely to be present in almost all samples from the various forms of KS.

2.2.1a. Potential Sites of Viral Latency. A defining characteristic of herpesviruses is their ability to remain latent in their host. However, the site of viral latency varies with the specific herpesvirus. It has been hypothesized that several sites harbor latent KSHV. In a study investigating the distribution of KSHV DNA in tissue from AIDS-KS, Corbellino and colleagues found the virus in all lymphoid organs and all prostate glands examined by both PCR and nested PCR.[38] The virus was also found to a lesser degree in bone marrow and in normal skin of these same AIDS-KS patients whereas all control tissue samples from AIDS cases with KS were negative.[38] The same group has also suggested that perhaps like some alpha herpesviruses, KSHV could persist in the peripheral nervous system in viral latency. They found KSHV DNA in the sensory ganglia of seven patients with AIDS-associated KS by both nested and unnested PCR whereas the sensory ganglia from four AIDS and three HIV-negative patients without KS were negative.[39]

TABLE I
KSHV DNA Detection Rates by PCR Amplification of KS Lesions

Reference	KS type	No.+	Total	(%+)	Control tissues	No.+	Total	(%+)	Odds ratio (95% CI)[a]
Chang et al. (1994)[1]	AIDS-KS	25	27	(93%)	AIDS				68.8 (11.2–670)
•Fresh frozen					Lymphoma	3	27	(11%)	
•Direct Southern blot					Lymph node	3	12	(25%)	
•Sequencing done					Total	6	39	(15%)	
					Non-AIDS				901 (80.4–22026)
					Lymphoma	0	29	(0%)	
					Lymph node	0	7	(0%)	
					Vascular tumor	0	5	(0%)	
					Opportunistic inf.	0	13	(0%)	
					Surgical biopsy	0	49	(0%)	
					Total	0	103	(0%)	
Su et al. (1995)[121]	AIDS-KS	4	4	(100%)	AIDS lymph node	0	5	(0%)	133 (8.2–6128)
•KS lesions + lymph nodes	Non-AIDS KS	2	3	(67%)	Benign hyperplasia	0	10	(0%)	
•PCR Southern blot	Total	6	7	(86%)	B-cell lymphoma	0	12	(0%)	
					T-cell lymphoma	0	10	(0%)	
					Total	0	37	(0%)	
Dupin et al. (1995)[31]	Classical KS	5	5	(100%)	None				NA
•Snap-frozen	AIDS-KS	4	4	(100%)					
•83% of adjacent skin was PCR$^+$	Total	9	9	(100%)					
Boshoff et al. (1995)[66]	Classical KS	16	17	(94%)	Angioma/angiosarcoma	0	4	(0%)	240 (16.3–10473)
•Fresh-frozen	Transplant KS	8	8	(100%)	Skin nevi	0	3	(0%)	
•Paraffin-embedded (nested PCR)	AIDS-KS	14	14	(100%)	Granulation tissue	0	4	(0%)	
	HIV$^-$ gay men	1	1	(100%)	Total	0	11	(0%)	
•Sequencing done	Total	39	40	(98%)					
Ambroziak et al. (1995)[32]	AIDS-KS	12	12	(100%)	None				NA
•Sequencing done	HIV$^-$ gay men	1	1	(100%)					
•33% of nearby skin was PCR$^+$	Total	13	13	(100%)					

•Fresh-frozen	Classical KS	6	6	(100%)	PBMCs from healthy subjects	0	10	(0%)	
•PCR Southern blot	HIV⁻ gay men	4	4	(100%)	Total	1	21	(5%)	
•21% KS normal skin was PCR⁺	Total	20	21	(95%)					
Lebbé et al. (1995)[33]	Immuno-								
•33% KS normal skin was PCR⁺	suppressed KS	1	1	(100%)	None				NA
	Classical KS	10	10	(100%)					
	African KS	3	3	(100%)					
	AIDS-KS	2	2	(100%)					
	Total	16	16	(100%)					
Schalling et al. (1995)[36]	AIDS-KS	25	25	(100%)	HIV^+				650 (31.3–22026)
•PCR Southern blot	African KS	18	18	(100%)	PBL (PBMNC)	0	13	(0%)	
	Classical KS	3	3	(100%)	PBL-KS	0	3	(0%)	
	Total	46	46	(100%)	Lymph node	0	8	(0%)	
					Total	0	24	(0%)	
					HIV^-				396 (18.5–17142)
					PBL (PBMNC)	0	12	(0%)	
					Skin, non-KS patient	0	1	(0%)	
					Skin, KS patient	0	2	(0%)	
					Hemangioma	0	1	(0%)	
					Pyogenic granuloma	0	1	(0%)	OR (MH)
					Total	0	17	(0%)	508 (50.8–3831)
Chang et al. (1996)[37]	AIDS-KS	22	24	(92%)	HIV^+	1	7	(14%)	66.0 (3.8–3161)
•Paraffin-embedded	African KS	17	20	(85%)	HIV^-	2	15	(13%)	36.8 (4.3–428)
•Negatives retested by Nested PCR	Total	39	44	(89%)	Total	3	22	(14%)	OR MH 44.8 (7.9–259)
Chuck et al. (1996)[34]	African KS	4	4	(100%)	None				NA
•Fresh-frozen (Endemic)	HIV⁻ gay men	1	2	(50%)					
•Paraffin-embedded (HIV⁻ gay men)	Total	5	6	(83%)					
O'Neill et al. (1996)[123]	AIDS KS	7	7	(100%)	HIV^-	0	1	(0%)	NA
•Nested PCR									
•Sequencing done									
•0% KS normal skin was PCR⁺									

(*continued*)

TABLE I (*Continued*)

Reference	KS type	No.+	Total	(%+)	Control tissues	No.+	Total	(%+)	Odds ratio (95% CI)[a]
Buonaguro et al. (1996)[124]	African KS	12	12	(100%)	Reduction mammoplasties	0	3	(0%)	2046 (103–22026)
•Snap-frozen	Classical KS	28	28	(100%)	Penile carcinoma	0	4	(0%)	
•PCR Southern blot	Immuno-				XP pigmentosum skin cancer	0	5	(0%)	
•Sequencing done	suppressed KS	2	2	(100%)	XP autologous normal skin	0	5	(0%)	
•67% KS normal skin was	AIDS-KS	19	19	(100%)	PBMC of HIV$^+$	0	15	(0%)	
PCR$^+$	Total	61	61	(100%)	Total	0	32	(0%)	
Cathomas et al. (1996)[125]	AIDS-KS	9	9	(100%)	HIV$^+$ other skin lesions	0	4	(0%)	345 (15.9–15062)
•Paraffin-embedded	Classical KS	12	12	(100%)	HIV$^-$ other skin lesions	0	10	(0%)	
•Nested PCR	Transplant KS	1	1	(100%)	Total	0	14	(0%)	
	Total	22	22	(100%)					
Gaidano et al. (1996)[84]	AIDS-KS	35	35	(100%)	NHL (3 BCBLs)[b]	3	31	(9%)	549 (55.0–22026)
•Fresh-frozen (a few paraffin-					HD[b]	0	3	(0%)	
embedded)					PGL[b]	0	15	(0%)	
•Direct Southern blot					Anogenital neoplasia	0	14	(0%)	
•All 3 BCBLs were PCR$^+$					Total	3	63	(5%)	
•100% KS normal skin was PCR$^+$									
•50% KS normal skin was Southern blot +									
Jin et al. (1996)[68]	AIDS-KS	5	5	(100%)	Hemangiosarcoma	0	15	(0%)	1566 (76.4–22026)
•Paraffin-embedded	Non-AIDS-KS	12	12	(100%)	Hemangioma	0	75	(0%)	
	Total	17	17	(100%)	Lymphangioma	0	15	(0%)	
					Lymphangiomatosis	0	2	(0%)	
					Pyogenic granuloma	0	25	(0%)	
					Hemangiopericytoma	0	3	(0%)	
					Kimura's disease	0	2	(0%)	
					Lymphangiomyomatosis	0	1	(0%)	
					Total	0	138	(0%)	

Dictor et al. (1996)[62]	Classical KS	35	40	(88%)	Endothelial lesions	0	86	(0%)	725 (84.3–22026)
•Paraffin-embedded	AIDS-KS	14	14	(100%)					
	Total	49	54	(91%)					
Marchiolo et al. (1996)[126]	AIDS-KS	28	28	(100%)	Normal skin	0	10	(0%)	612 (66.9–22026)
•Fresh-frozen and paraffin-embedded	Classical KS	7	8	(88%)	Pediatric lymphomas	0	8	(0%)	
	African KS	7	10	(70%)	Adult lymphomas	0	37	(0%)	
•23% KS normal skin was PCR+	HIV^- gay men	2	2	(100%)	Carcinomas	0	12	(0%)	
	Total	44	48	(92%)	Total	0	67	(0%)	
Luppi et al. (1996)[127]	Classical KS	15	22	(68%)	Normal PBMCs	0	13	(0%)	41.6 (7.1–404)
•Paraffin-embedded	AIDS-KS	3	4	(75%)	Normal salivary glands	0	9	(0%)	
	Total	18	26	(69%)	Normal saliva samples	0	6	(0%)	
					Hyperplastic tonsils	2	11	(18%)	
					Total	2	39	(5%)	
McDonagh et al. (1996)[65]	KS	9	9	(100%)	Angiosarcoma	7	24	(29%)	47.8 (5.2–2119)
•Fresh-frozen and paraffin-embedded					Hemangioma	1	20	(5%)	
					Hemangiopericytoma	0	6	(0%)	
					Total	8	50	(16%)	
Corbellino et al. (1996)[38]	AIDS-KS	7	7	(100%)	AIDS, no KS	0	6	(0%)	56.0 (2.1–2822)
•Snap-frozen									
•PCR Southern blot									
•60% KS normal skin was PCR^+									
Lebbé et al. (1997)[128]	Classic	16	16	(100%)	HIV^-				416 (19.4–18005)
•Fresh-frozen	African	3	3	(100%)	Dermatology biopsies	0	10	(0%)	
•Nested PCR	CD[b]	1	1	(100%)	Reduction mammaplasties	0	5	(0%)	
•24% KS normal skin was PCR^+ (50–100ng)	HIV^- gay men	3	3	(100%)	Total	0	15	(0%)	
•70% KS normal skin was PCR^+ (250–500ng)	Immuno-suppressed/Transplant KS	2	2	(100%)					
	Total	25	25	(100%)					

[a]A value of one was added to each cell when zero cells were encountered; exact 95% confidence intervals were computed.
[b]NHL = Non Hodgkin's lymphoma; BCBL = body cavity-based lymphoma; HD = Hodgkin's disease; PGL = persistent generalized lymphadenopathy; CD = Castleman's disease.

TABLE II

KSHV DNA Detection Rates by PCR Amplification of Peripheral Blood Mononuclear Cells (PBMCs)

Reference	KS type	No.+	Total	(%+)	Control tissues	No.+	Total	(%+)	Odds ratio (95% CI)[a]
Collandre et al. (1995)[128]	AIDS-KS	2	10	(20%)	HIV^+, no KS	0	9	(0%)	3.3 (0.2–192)
•PCR Southern blot									
Ambroziak et al. (1995)[32]	AIDS-KS	7	7	(100%)	HIV^+, no KS	0	6	(0%)	231 (10.3–10354)
	HIV^-, KS	3	3	(100%)	HIV^-, no KS	0	14	(0%)	
	Total	10	10	(100%)	Total	0	20	(0%)	
Whitby et al. (1996)[82]	AIDS-KS	24	46	(52%)	HIV^+	11	143	(8%)	13.1 (5.2–33.5)
•Nested PCR					Oncology patients	0	26	(0%)	29.4 (4.0–1248)
					Blood donors	0	134	(0%)	147 (21.1–6066)
					Total	11	303	(4%)	
Moore et al. (1996)[81]	AIDS-KS	11	21	(52%)	Gay/bisexual AIDS,	3	23	(13%)	7.3 (1.4–47.9)
•Nested PCR					no KS	0	19	(0%)	21.8 (2.4–978)
					Hemophiliac AIDS,	3	42	(7%)	
					no KS				
					Total				

Marchiolo et al. (1996)[126]	AIDS-KS	46	99	(46%)	HIV$^+$	0	64	(0%)	191 (30.8–7749)
•PCR Southern blot	HIV$^-$ gay men	0	2	(0%)	HIV$^-$	0	163	(0%)	
	Total	46	101	(46%)	Total	0	227	(0%)	
Humphrey et al. (1996)[130]	AIDS-KS	34	98	(35%)	HIV$^+$, no KS	12	64	(19%)	2.3 (1.0–5.4)
•PCR Southern blot					HIV$^-$, no KS	0	11	(0%)	6.5 (0.9–284)
					Total	12	75	(16%)	
Decker et al. (1996)[83]	KS	8	9	(89%)	Allograft patients	4	5	(80%)	3.4 (0.2–202)
•Multiple samples tested to obtain positives in controls					Healthy donors	3	5	(60%)	
					Total	7	10	(70%)	
Lebbé et al. (1997)[128]	Classical KS	9	18	(50%)	Blood donors	0	20	(0%)	16.1 (1.9–714)
•PCR Southern blot	African KS	2	3	(67%)					
•Nested PCR	CD[b]	0	1	(0%)					
•Unnested 32% of KS subjects PCR$^+$	HIV$^-$ gay men	1	4	(25%)					
	Immuno-suppressed/ Transplant KS	0	2	(0%)					
	Total	12	28	(43%)					

[a]A value of one was added to each cell when zero cells were encountered; exact 95% confidence intervals were computed.
[b]CD = Castleman's disease.

TABLE III
KSHV Antibody Detection Rates

					Various populations							
Reference	KS Type	No.$^+$	Total	(%+)	HIV$^+$ Gay/Bi	HIV$^+$ IDU	HIV$^+$ Hemo	Women	HIV$^+$	HIV$^-$	Blood donors	EBV$^+$ pts.
Miller et al. (1996)[76]												
BC-1 Western blot •Butyrate induced •p40	AIDS-KS	32	48	(67%)	7/54 (13%)[a]	–	–	–	–	–	–	–
BC-1 IFA •Butyrate induced •1:10 dilution	AIDS-KS	31	48	(65%)	7/54 (13%)[a]	–	–	–	–	–	–	–
Gao et al. (1996),[53] (1996)[77]												
BC-1 Western blot •Uninduced •p234 & p226 (LNA)	US AIDS-KS	32	40	(80%)	7/40 (18%)	–	0/20	–	–	–	0/122	0/69
•Controls include HIV$^+$ and HIV$^-$ cancer patients from Uganda	Italy AIDS-KS	11	14	(79%)	–	–	–	–	–	–	4/107 (4%)	–
	Classical KS	11	11	(100%)	–	–	–	–	25/35 (71%)	29/47 (62%)	–	–
	Uganda AIDS-KS	16	18	(89%)								
	African KS	1	1	(100%)								
BCP-1 IFA •Uninduced •1:160 dilution	US AIDS-KS	35	40	(88%)	12/40 (30%)	–	0/20	–	–	–	0/122	0/69

•Controls include HIV$^+$ and HIV$^-$ cancer patients from Uganda	Italy AIDS-KS	10	14	(71%)	–	–	–	–	–	–	4/107 (4%)	–
	Classical KS	11	11	(100%)	–	–	–	–	18/35 (51%)	24/47 (51%)	–	–
	Uganda AIDS-KS	14	18	(78%)								
	African KS	1	1	(100%)								
Kedes et al. (1996)[52]												
BCBL-1 (nuclei) IFA •Uninduced	AIDS-KS	37	45	(82%)	13/37 (35%)	–	9/300 (3%)	5/59 (8%)[b]	–	7/107 (7%)[b]	2/141 (1%)[c]	–
•1:40 dilution	HIV-KS	1	1	(100%)								
Lennette et al. (1996)[54]												
BCBL-1 IFA •Uninduced	AIDS-KS	47	91	(52%)	19/94 (20%)	0/13	–	0/87[d]	–	–	0/44	0/40
•1:10 dilution	African KS	28	28	(100%)								
BCBL-1 IFA •TPA induced •1:10 dilution	AIDS-KS	87	91	(96%)	87/94 (93%)	3/13 (23%)	–	22/87 (25)	–	–	9/44 (20%)	8/40 (20%)
	African KS	28	28	(100%)								
Simpson et al. (1996)[79]												
BCP-1 IFA •Uninduced •1:150 dilution	AIDS-KS, U.S.	84	103	(82%)	10/33 (30%)	0/38	0/26	3/15 (20%)[b]	–	14/166 (8%)[b]	4/257 (1%)[e]	–
	Classical KS	17	18	(94%)								
ORF65 ELISA •Recombinant	AIDS-KS, U.S.	46	57	(81%)	5/16 (31%)	2/38 (1%)	0/28	–	–	–	9/291 (3%)[e]	–
	AIDS-KS, Uganda	14	17	(82%)								
	Classical KS	17	18	(94%)								

[a] 89% were homo/bisexual HIV$^+$ men.
[b] STD clinic attenders.
[c] HIV$^-$; 30% HIV$^+$ blood donors were seropositive.
[d] IFA uninduced: 0/33 HIV$^+$ and 0/54 HIV$^-$; IFA induced: 7/33 HIV$^+$ and 1/54 HIV$^-$.
[e] IFA: 3% U.K. blood donors, 0% U.S. blood donors; ELISA: 2% U.K. blood donors, 5% U.S. blood donors.

2.2.2. *Specificity*

Specificity of cause and effect implies that a cause leads to a single effect. It is well established, however, that nearly all pathogens cause a spectrum of clinical disorders. This is most clearly seen in AIDS where the variety of clinical outcomes from AIDS depend on HIV-induced immunosuppression. Therefore, the possibility that KSHV is associated with a variety of disorders does not in itself eliminate a potential causal association with KS. Lymphomas occur more frequently as second cancers in patients with KS than other primary malignancies.[40,41] Further, several nonmalignant lymphoproliferative diseases, such as giant lymph node hyperplasia (Castleman's disease) and angioimmunoblastic lymphadenopathy with dysproteinemia (AILD), occur at high rates in KS patients.[42,43] Since the initial discovery of KSHV in KS lesions, the virus has now been consistently linked to two other diseases, body-cavity-based lymphomas (BCBLs) and Multicentric Castleman's disease (MCD). There is compelling evidence to suggest that the virus may cause or contribute to the pathogenesis of these diseases.

KSHV is present in high copy number in some non-Hodgkin's lymphomas (NHL) localized to body cavities.[1,44,45] Body-cavity-based lymphomas (BCBLs) or primary effusion lymphomas (PEL) differ from most NHLs in that they grow predominantly in pleural, pericardial, and abdominal cavities as effusions and are usually not associated with identifiable tumor mass. Unlike many B cell NHLs occurring in AIDS patients, it is now recognized that those harboring KSHV represent a subset of BCBLs characterized by distinct morphological, immunophenotypical, and molecular features.[46,47] A phenotypically similar lymphoma, pyothorax-associated lymphoma (PAL), is not infected with KSHV but generally is infected with EBV and has *c-myc* rearrangements.[48] BCBL cells are highly pleomorphic, have prominent, irregular convoluted nuclei, and display a high mitotic rate. They usually lack expression of any lineage-associated antigens but demonstrate light and heavy chain immunoglobulin gene rearrangements by molecular analysis, indicating their B cell origin. BCBL cells express epithelial membrane antigen (EMA) and a variety of activation markers including CD30, CD38, HLA-DR, and CD71. Unlike EBV-associated Burkitt's lymphoma and PAL, no *c-myc* rearrangements are present. BCBLs, however, are frequently dually infected with KSHV and EBV. Cases infected by KSHV alone also exist,[49–51] and cell lines generated from these EBV-tumors are used in several current serological assays.[52–54] The clinical course of BCBL in the setting of AIDS follows an aggressive course, whereas HIV seronegative, non-immunosuppressed patients may have a more indolent course, homologous to KS in adult patients with and without AIDS.

Multicentric Castleman's disease (MCD) is an atypical lymphoproliferative disorder epidemiologically associated with KS and lymphoma. Two subtypes are differentiated on the basis of clinical and pathological features: hyaline-vascular variant and plasma-cell variant. The hyaline-vascular variant is usually seen as a solitary lesion typically presenting in the mediastinum of young adults. Generally no systemic symptoms complicate the clinical course, and surgical excision is curative. In contrast, the plasma-cell variant of Castleman's disease usually presents as generalized (multicentric) lymphadenopathy in an older population, is accompanied by systemic symptoms of fever and dysproteinemia, and does not respond well to surgical or medical intervention. Corbellino and colleagues have identified a strong correlation between the plasma-cell variant and KSHV.[55] Patients with KSHV-associated Castleman's disease are at high risk for KS and frequently experience an aggressive and rapidly fatal clinical course with a high incidence of autoimmune hemolytic anemia.

In a study by Soulier and colleagues, KSHV DNA from tumor tissue samples was detected by PCR and Southern hybridization in all patients with HIV-associated MCD and less frequently (41% by PCR and 13% by Southern blotting) in HIV-negative patients with MCD.[56] Among the patients without detectable KS, 100% of HIV-positive and 38% of HIV-negative patients were PCR positive, whereas only one of 34 (3%) control lymph nodes from non-KS, non-MCD, HIV-negative patients was positive. PBMCs were examined in one case and KSHV was detected only by PCR. Dupin and colleagues also detected KSHV DNA by PCR in PBMCs of two AIDS patients with MCD, one with KS and one without.[57] A subsequent study found 75% of HIV-infected persons positive for KSHV DNA by PCR of lymph node biopsies. However, all three positives developed KS.[58] In this study 17% percent of HIV-negative, KS-negative patients with MCD were positive for KSHV by PCR. However, the sample size was very small. Additional studies with larger samples are needed to further understand the role of KSHV in MCD.

KSHV has also been inconsistently linked to a few others diseases. Rady and colleagues examined various non-KS skin lesions from four HIV-negative transplant patients and found KSHV DNA by PCR in a wide variety of non-KS skin tumors.[59] Several groups using larger sample sizes have not been able to replicate these findings.[60–62] Another group detected KSHV DNA by PCR in an HIV-negative patient with angiosarcoma of the face[63] and in four HIV-negative patients with angiolymphoid hyperplasia and eosinophilia.[64] A larger study found seven of 24 (29%) angiosarcomas positive for KSHV DNA by PCR, and one was positive by direct Southern hybridization.[65] However, numerous other studies have failed to demonstrate KSHV DNA in angiosarcomas.[1,62,66–68]

A few studies have detected KSHV in healthy populations. Lin and colleagues were unable to isolate KSHV DNA by unnested PCR from the semen of healthy blood donors. However, 23% became positive when the analysis was repeated with nested PCR.[69] Because nested PCR is an extremely sensitive technique prone to contamination, this inconsistency should be viewed with caution. A similar study found an even higher prevalence of KSHV by nested PCR in the semen and prostatic tissue from HIV-negative Italian patients undergoing surgery.[70] These findings have not been confirmed by others.[71–74] Interestingly, the rates of detection were significantly lower when tested in a blinded fashion rather than an unblinded examination.[75] It is possible that these contradictory results are due to regional variation, but the possibility of PCR contamination cannot be excluded, especially when using nested PCR. Contamination can occur even in the most stringent and meticulous laboratories. The bulk of studies now suggest that KSHV is generally not found in semen or prostatic tissues from healthy adult men.

2.2.2a. Serological Assays. DNA-based studies provide evidence that KSHV is present in KS lesions. However, this is only a good marker for present infection. It is possible that the virus exists in persons without KS or is present in a site of latency not generally sampled by these tissue-based studies. Serological studies allow a better estimate of the true prevalence of KSHV in different populations because they detect antibodies to past and present infection and are independent of the tissue site examined.

The first serological assay to detect antibodies to KSHV in serum used BC-1, a B cell line derived from a BCBL that was dually infected with KSHV and EBV, requiring preadsorption of cross-reacting antibodies.[3] Using immunofluorescence and immunoblot assays, Miller and colleagues tested sera from AIDS-KS patients for antibodies against a lytic phase polypeptide, p40, in sodium butyrate induced BC-1 cells.[76] Among HIV-positive patients, 65–67% with KS were KSHV-seropositive compared to 13% without KS (Table 3). Gao and colleagues identified two KSHV-related latent nuclear antigens (LANA) of high molecular weight (p226 and p234) by immunoblot using uninduced BC-1 cell lysates.[77] Similar results were obtained: 80% of U.S. AIDS-associated KS patients were seropositive compared to 18% of homosexual control patients with AIDS (Table 3). None of 20 hemophiliacs with AIDS, 122 HIV-negative blood donors, or 69 HIV-negative patients with high EBV titers were positive. Further, these authors demonstrated evidence of seroconversion to KSHV positivity in a group of HIV-positive gay men who were followed until developing KS.[77]

The subsequent development of two KSHV-positive but EBV-negative cell lines, BCP-1 and BCBL-1, helped to strengthen the seroepidemiological findings. Two groups independently developed an immunofluorescence assay (whole-cell[53] and nuclei[52]) to detect antibodies to latent nuclear antigens. The results are highly concordant, showing that the seroprevalence of KSHV antibodies to latent antigens is high in U.S. KS patients (83–88%), moderate in HIV-infected homo/bisexual men (30–35%), and very low in U.S. blood donors (0–1%) (Table 3).[52,53] The IFA antigen detected by these assays has recently been found to be identical to the immunoblot LANA antigen and is encoded by ORF73.[78]

Recently, several groups developed serological assays to detect antibodies to lytic-phase antigens. Lennette and colleagues, using TPA-induced BCBL-1 cell preparations, found higher rates of seropositivity in low risk populations even though their whole cell, uninduced IFA produced results similar to previously published studies.[54] However, because of the use of low serum dilutions (1:10), antibody cross-reactivity to other herpesviruses, such as EBV or CMV, cannot be excluded. Gao and colleagues demonstrated that at serum dilutions less than 1:160 both AIDS-KS and non-KS patients had diffuse cross-reactive cytoplasmic staining of BCP-1 cells that could be eliminated by preadsorption with EBV-positive cells.[53] Kedes and colleagues found similar nonspecific staining with the BCBL-1 cell line and used isolated nuclei to avoid cross-reactivity.[52] In contrast to lytic IFA, a recombinant lytic antigen generated from a truncated KSHV open reading frame (ORF) 65.2 capsid protein and tested by enzyme-linked immunosorbent assay (ELISA) found rates of infection comparable to the latent, whole-cell immunofluorescence antigen assay.[79] Combined use of the ELISA and the IFA increased sensitivity to 94%.

These assays have also detected a fair amount of geographic variation in seroprevalence. This is not surprising because the incidence and prevalence of KS varies widely.[14] In Uganda, the seroprevalence of KSHV among 82 HIV-seropositive and seronegative subjects with and without cancer and all without KS was 66% by immunoblot LANA and 51% by IFA.[53] Furthermore, 4% of northern Italian blood donors were seropositive by both assays, a significantly higher prevalence than found among North American blood donors.[53] Lennette and colleagues found that the seroprevalence obtained from their lytic-phase IFA tracked loosely with KS prevalence in several African countries.[54] All were in the range of 32–100% seropositivity. In contrast, the seroprevalence obtained in these countries from their uninduced IFA was much lower (a range of 6–43%). Countries with a low prevalence of KS, including Haiti, Dominican Republic, and Guatemala, demonstrated a low to moderate prevalence with the induced IFA (a range of 10–29%) and complete absence of the virus with the uninduced IFA.[54] The recombinant lytic-phase assay developed by Simpson and colleagues found 43% seroprevalence among the same Ugandan subjects surveyed by Gao and colleagues.[79] Despite the different seroprevalences obtained from these assays, certain geographic regions such as Central Africa and perhaps Italy have a higher seroprevalence than North America.

Serological studies are now beginning to examine risk factors for KSHV infection. Kedes and colleagues looked at broad risk groups and found that compared to HIV-negative blood donors (1%), KSHV seropositivity was significantly greater in HIV-positive gay and bisexual men (35%) and in HIV-negative STD attendees (8%) and no different in recipients of HIV-positive blood (5%) or in hemophiliacs (3%).[52] The similarity between these seroprevalences and the prevalence of KS in comparable risk groups[12,14] is quite striking.

A controversy remains over the true prevalence of this virus in the human population. One argument is that, compared with other herpesviruses, KSHV is ubiquitous, and the reason that serological assays are not detecting higher rates of infection in low risk populations, i.e., greater than 75%, is that current assays can detect antibodies generated only during viral reactivation and not primary infection. Reactivation could manifest as direct viral replication or, as with EBV,[80] through expression of cell-transforming antigens which otherwise would be detected by immune surveillance in immunocompetant individuals. However, both DNA and serological evidence suggest otherwise. In a study comparing AIDS patients before and after development of KS with gay and bisexual AIDS patients with comparable $CD4^+$ levels, Moore and colleagues found that AIDS-KS patients were significantly more likely to be KSHV-positive than controls by PCR at both times.[81] Although it has been suggested that viral reactivation may occur as a result of immunosuppression (e.g., posttransplant KS), these results suggests that immunosuppression alone does not increase detection rates. Additional evidence comes from antibody kinetic studies. If current serological assays can detect antibodies only during viral reactivation, then it follows that antibody titers should vary with antigen load. Gao and colleagues demonstrated that in six AIDS-KS patients there was an initial rise in immunoglobulin-g (IgG) titers after seroconversion followed by a prolonged, elevated antibody response over 36–93 months until KS onset.[53]

Second-generation serological assays and additional large-scale seroprevalence studies are needed to help clarify this issue. As in the case of HIV, a combination of assays is most likely necessary to obtain optimal sensitivity and specificity.

2.2.3. *Temporality*

In a causal association it is necessary for the cause to precede the effect (i.e., infection with KSHV precedes onset of KS). Although this criterion may appear intuitive, it is not easy to determine by a case-control study which generally cannot distinguish the temporal sequence of occurrence. The first group to circumvent this limitation was Whitby and colleagues,[82] who prospectively examined serial blood samples by nested PCR from subjects enrolled in an ongoing cohort study. One hundred forty-three HIV-positive subjects without KS were followed for a median of 30 months. At the initial time, the PBMCs were KSHV DNA-positive in 11 subjects, 55% of whom went on to develop KS, whereas only 9% of KSHV-negative subjects subsequently developed KS (RR = 6, 95% C.I. = 2.8–12.9). A second study, using a nested case-control design, has confirmed these findings.[81] Using nested PCR on paired PBMCs taken before and after KS development, Moore and colleagues showed that after controlling for HIV status, subjects with AIDS-KS were significantly more likely to have KSHV before KS onset than either a high or low risk AIDS control group (Table 2). KS patients were also more frequently positive near the time of KS onset in serial samples, consistent with either denovo infection or selective reactivation.

The majority of the studies suggest that PCR is only 20–50% sensitive in detecting KSHV infection of PBMCs from KS patients (Table 2). This is not unexpected because this technique measures the current presence of viral DNA and less than 1 in 10^6 PBMCs are likely to have detectable virus during active infection.[83] One intriguing study suggests that virus infection may be present in PBMCs in most healthy adults at extremely low levels and that the virus is in a linear, actively replicating form.[83] However, repetition with adequate PCR safeguards (e.g., nonoverlapping primers) is needed to verify this finding.

Serological data lend additional evidence that infection with KSHV precedes KS onset.[53,77] The results of two assays which detect antibodies to KSHV-related latent antigens demonstrate KSHV infection before KS onset in most patients; 50% of the patients seroconvert between 33 and 46 months before KS diagnosis.[53,77] Furthermore, these data show that the rate of infection is constant, suggesting that the risk of KS may be independent of the duration of infection and is instead under strict immunologic control.[53,77] Taken together, these DNA and serological data provide evidence that KSHV infection precedes KS onset, a necessary requirement for causality.

2.2.4. *Consistency*

Consistency of the findings is the ability of different groups to achieve similar findings through alternate study designs and populations. KSHV DNA was originally identified in KS lesions in persons with AIDS,[1] but, as Table 1 indicates, over 95% of KS lesions from all forms of KS including AIDS-KS, classical KS, African KS, and posttransplantation or immunosuppression-associated KS have detectable KSHV DNA. In addition, studies have now been performed in North America, Europe, Asia, and Africa. Although the majority of the early studies were of case-control design, the development of serological assays permits larger population-based studies to detect KSHV-specific antibodies in different populations.

Most detection methods show that KSHV is generally absent or rare from healthy control populations without KS (Tables 1–3). Some diseases, such as skin cancer and angiosarcoma, generally lack KSHV DNA, as assessed by PCR in most studies,[61,62,66] although contradictory studies exist.[59,63–65] A few studies have found high prevalence of KSHV in healthy populations. As previously mentioned, KSHV DNA was detected at a high rate in the semen of healthy men.[69,70] However, others have not been able to confirm these findings.[71–74].

2.2.5. *Biological Gradient*

A dose response between the amount of exposure to a risk factor and the risk of subsequently developing the disease is generally consistent with a causal relationship. This has been looked at in the association between KSHV and KS by examining KS lesions, adjacent tissue, and normal distant skin all from subjects with KS. If a gradient exists, one might expect to see a decease in viral copy number per cell, as ascertained by semiquantitative PCR and/or Southern blot analysis, as the distance from the KS lesion increases. This was first examined by Chang and colleagues in which KS lesions from four AIDS-KS patients were all positive by PCR and direct Southern hybridization, but unaffected tissue from these same patients was largely negative.[1] Dupin and colleagues were also able to show by semiquantitative PCR that the copy number of KSHV DNA was higher in KS lesions than in normal tissue from the same patients, whereas there was no gradient for detecting HIV proviral DNA,[31] and Gaidano and colleagues estimated that the amount of KSHV was 10–30 times greater in KS lesions than in adjacent uninvolved tissues.[84]

Studies have also used *in situ* hybridization (ISH) to locate specific KSHV nucleic acid sequences within cells. Using PCR-ISH, a sensitive technique but difficult to reproduce, Boshoff and colleagues showed that the majority of nuclei from spindle cells within KS lesions were KSHV-positive, whereas the surrounding normal skin was negative.[85] Positive cells were generally also positive for CD34 by immunohistochemistry which is a marker for KS spindle cells. Li and colleagues found similar results. KSHV DNA localized mostly to nuclei of endothelial cells and to some spindle cells.[86] Most recently, Staskus and colleagues used two viral transcripts, T0.7 and T1.1, to probe for KSHV latent and lytic gene expression, respectively, in KS tumor cells.[87] T0.7, the latent phase probe, was expressed in both the nucleus and the cytoplasm of the majority of spindle cells in KS lesions, and T1.1 was expressed in the nucleus of approximately 10% of these. The identification of active viral replication in some spindle cells within KS tumors suggests that this tumor may be receptive to antiviral therapy. Interestingly, they also examined prostatic tissue from 16 HIV-positive and HIV-negative patients without KS and found T0.7 expression in 75%. Recently, PCR studies have failed to detect KSHV DNA in prostatic tissues from large numbers of healthy adult men.[88,89]

2.2.6. *Biological Plausibility*

Phylogenetic analyses link KSHV to two gamma herpesviruses with oncogenic potential, EBV and HVS.[1,3] EBV is a human herpesvirus that transforms B lymphocytes and is associated with Burkitt's lymphoma and other neoplastic disorders in immunocompetent and immunocompromised patients (for review, see [80]). HVS transforms human T cells and causes lymphomas in some species of New World monkeys.[90] Although this phylogenetic association does not in itself demonstrate that KSHV has cell-transforming capacity, sequencing and functional studies demonstrate that KSHV encodes a number of potential oncogenes.[91] These homologues to known human oncogenes[91] include cyclin D, *bcl-2,* interleukin 6 (vIL-6), interferon regulatory factors (vIRF), and an Il-8-like receptor (ORF 74). The KSHV D-type cyclin is expressed in KS lesions and has conserved sequence homology in the functional "cyclin box" region.[92] The human cyclin D proteins initiate cell progression through the G1 checkpoint by inhibiting the tumor suppressor protein pRb in combination with cellular cyclin-dependent kinases (for a review [93]). Overexpression of D-type cyclins have been associated with mantle cell lymphomas and parathyroid tumors.[94] The KSHV cyclin encoded by ORF72[91,92] initiates cell cycling in SAOS-2 cells and phosphorylates pRb at authentic sites.[95] This protein may have less substrate specifically than cellular D-type cyclins because it also phosphorylates histone H1 in association with cdk6.[96,97]

Unopposed inhibition of pRb can lead to cell apoptosis through both p53-dependent and p53-independent pathways.[98] KSHV also encodes several genes which can potentially inhibit apoptosis and are potential oncogenes. Like EBV, KSHV possesses a functional *bcl-2* gene[99,100] that inhibits apoptosis induced by the related members of the bax proapoptotic protein family. These proteins are activated through homodimerization to induce apoptosis by p53 or through other apoptotic pathways.[101] A KSHV-encoded cytokine, vIL-6 (ORFK2[102]), is functionally active in preventing B cell apoptosis[102,103] and like human IL-6, activates the Jak-STAT pathway.[104] However, vIL-6 is probably not expressed in KS lesions and is unlikely to contribute to KS pathogenesis.[102]

The KSHV vIRF (ORFK9[102]) also has the unique potential to induce tumorigenic cell transformation. This protein has a low but significant degree of sequence similarity to interferon regulatory factors (IRF) involved in interferon signal transduction and control. Injection of vIFR-transformed NIH3T3 cells into nude mice rapidly forms tumors.[104a] This gene, which is unique to KSHV, also inhibits β-interferon signal transduction which may allow the virus to escape from interferon-mediated host defenses. Other potential oncogenes include genes which are expressed during virus latency, such as ORF 73[78] and the Il-8-like receptor, which is constitutively activated and induces cellular proliferation.[105]

The molecular piracy of cellular regulatory and signaling genes by KSHV[102] may reflect a requirement of the virus to overcome cellular antiviral defenses. Many of these pathways (e.g., pRb and p53 pathways) are involved both in preventing tumor cell replication and replicating infecting viruses. Like other human tumor viruses, such as papillomaviruses, KSHV may possess specific genes active in inhibiting cellular pathways responsible for eliminating virally infected cells and controlling cell replication. Despite the functional similarity of putative oncogenes encoded by KSHV to those of other human DNA tumor viruses, these viruses are distantly related to each other and achieve their effects through fundamentally different mechanisms.

The identification of specific genes capable of inducing tumorigenesis in

KSHV provides biologically plausible support for the concept that KSHV is causal in the genesis of KS and related neoplasias.

2.2.7. *Experimental Evidence*

Animal and human studies perhaps provide the most concrete proof of a causal association, but human experiments are obviously unethical. However, it is possible to obtain indirect evidence from drug trials. If KSHV is causally associated with KS, it has been hypothesized that antiherpetic drugs should have an effect on KS development. However, current antiherpetic drugs affect only the lytic phase of viral replication by inhibiting the viral DNA polymerase. In a large retrospective study using data from the Adult/Adolescent Spectrum of Disease cohort, Jones and colleagues found that only foscarnet had a significant protective effect (OR = 0.3, 95% C.I. = 0.1–0.6) on the risk of KS development whereas no protective effect was seen for acyclovir and ganciclovir.[106] Additional studies have found similar results. Ganciclovir and foscarnet had an insignificant but protective effect on the risk of KS among men with cytomegalovirus disease,[107] and foscarnet significantly decreased the KS risk among HIV-positive patients.[108] Anecdotal studies suggest a possible effect of foscarnet on established AIDS-KS disease.[109] Recently, Kedes and colleagues examined the ability of antiviral drugs to block the release of KSHV DNA in TPA-induced BCBL-1 cells *in vitro*.[110] Foscarnet and cidofovir, but not acyclovir or ganciclovir, were found to be effective.

Other experimental evidence comes from preliminary studies on mice. Nude mice injected with vIFR-transformed NIH3T3 cells rapidly form tumors,[104a] and injections of BCP-1 and HBL-6 cells produces ascites in Nod/SCID mice.[111] These studies suggest that KSHV or specific oncogenes may mimic KSHV-induced neoplasia in animals.

2.2.8. *Current Evidence for Causality*

As seen in Tables 1–3, there is overwhelming and consistent evidence that KSHV is present in nearly all KS tumors examined. Within the limits of current detection methods, KS patients become infected with KSHV months to years before KS development. These findings strongly suggest a causal role for KSHV in KS development. If KSHV were a "passenger" virus (for which no other virus has a clear model), one would expect that a substantial fraction of KS lesions would be negative for the virus.

The major controversy regarding KSHV and KS involves the specificity of the association. There is increasing evidence that KSHV is not a ubiquitous infection and instead, like HSV2, is a sexually or parasexually transmitted infection. Although the prevalence in North American populations is likely to be low, even studies with high serological sensitivity[54] indicate that only a minority of U.S. adults are infected with KSHV. Detection of KSHV DNA by PCR in the semen of subjects without KS[69,70] has sparked further discussion on the prevalence of KSHV.[71,72,74]

KS is not the only disease associated with KSHV. Not surprisingly it is related to two other lymphoproliferative disorders. However, KSHV DNA has also been found to a varying degree in non-KS skin tumors and angiosarcomas. Although KSHV is consistently absent in control tissue from non-KS subjects both with and without HIV infection (including angiosarcomas), it may be that KSHV is present in a subset of high risk individuals with angiosarcoma. Additional studies are needed to further understand the association between KSHV and angiosarcomas.

Clearly, other noninfectious cofactors, e.g., immunosuppression, play a role in the pathogenesis of KS. However, the bulk of the studies indicate that KSHV is indeed the causative infectious agent for KS and related neoplasms.

3. TRANSMISSION

Based primarily on epidemiological evidence gathered on KS, several modes of transmission for KSHV have been hypothesized. These include sexual transmission, either through contact with infected semen or feces, oral transmission through contact with infected saliva, vertical transmission occurring perinatally or through breast feeding, and transmission through organ transplantation. The evidence for each is discussed in turn.

3.1. Sexual

In an early examination of semen among men dually infected with KS and HIV, KSHV DNA was not detectable by PCR.[32] However, the sample size was small, and the use of unnested PCR probably yielded low sensitivity. As mentioned above, using blinded, unnested PCR, Lin and colleagues[69] detected KSHV DNA in the semen of 64% of HIV-seropositive, gay men and 0% of healthy blood donors. However, examination by nested PCR was more sensitive, with 91% and 23% positive, respectively. Similar high rates of nested PCR detection in semen were found by Monini and colleagues,[70] but the much higher prevalence obtained from nested PCR remains controversial.

Attempts to replicate these findings have failed,[71,72,74] and it has since been reasoned that these discrepancies may result from geographic differences in KSHV prevalence or may be artifactual as a result of nested PCR contamination.[71,72,75]

Additional studies have detected KSHV DNA in prostatic tissue[38] and the semen[73] of patients with AIDS-associated KS but not in HIV-infected patients without KS. Furthermore, an examination of serial semen samples suggests that viral shedding may be intermittent and may play some role in varying transmission rates.[73] A more recent study demonstrated the presence of KSHV in the semen of AIDS-KS subjects by *in situ* hybridization.[112] Thus, it appears that transmission of KSHV through infected semen is possible, although the risk is probably restricted to high-risk populations.

Gay and bisexual men exposed to fecal matter during sexual intercourse are at a greater risk for KS compared to those not exposed.[15] Only one study attempted but was unable to detect the virus in feces.[82] KSHV DNA was isolated by PCR from 13 of 24 (54%) duodenal and 11 of 24 (46%) rectal biopsies, and from 7 of 7 duodenal aspirates of HIV-seropositive patients without KS, suggesting that perhaps viral shedding into the bowel facilitates sexual transmission of KSHV through fecal contact.[113] One case report suggests heterosexual transmission of KSHV from husband to wife. However, viral reactivation cannot be ruled out.[114]

3.2. Nonsexual

Preliminary data suggest that there may be nonsexual modes of transmission in Africa. We recently conducted a KSHV seroprevalence study[114a] in Lusaka, Zambia, using sera drawn in 1985 for an HIV seroprevalence study.[115] Our findings demonstrate that the epidemiology of HIV and KSHV was markedly different in 1985. HIV seroprevalence peaked in persons aged 20–29 and was entirely absent in persons over 50 years of age, consistent with a cohort effect due to the recent introduction of HIV into the sexually active population (Fig. 1).[115] In contrast, KSHV infection increased linearly with age among the study subjects (Fig. 1). KSHV seroprevalence was high among persons reaching sexual debut (45% in 14–19 year olds) consistent with virus transmission occurring through nonsexual modes. The linear increase with age implies a constant rate of, or steady-state, transmission during adulthood, possibly through sexual transmission. Antibodies to KSHV are long-lived,[53] and thus Figure 1 probably reflects the cumulative prevalence of KSHV infection in this population.

3.3. Saliva

Several groups have examined the potential for transmitting KSHV through saliva because this route of transmission is common in other herpesviruses, such as EBV and CMV. Two early studies detected an absence[32] or rare[82] presence of KSHV DNA in the saliva of KS patients by PCR, while two others found a moderate (33%) prevalence among HIV-positive persons[116] and high (74%)[117]

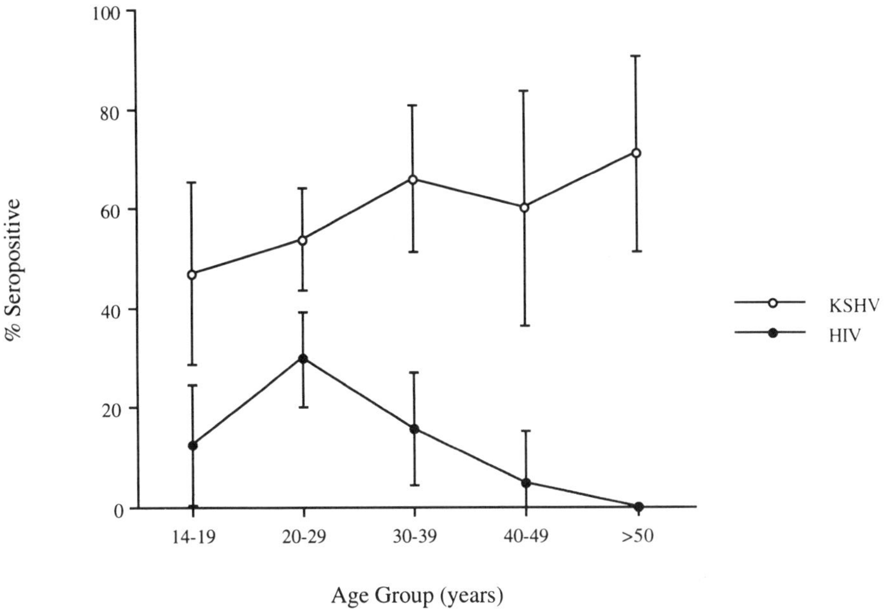

FIGURE 1. KSHV and HIV seroprevalence and 95% confidence intervals by age group in Lusaka, Zambia, 1985 (ref. 114a).

prevalence among HIV-positive KS patients. The latter study is distinguished by the use of direct culture to detect the virus.[118]

3.4. Vertical

A retrospective analysis of 100 childhood cases of KS in Uganda led Ziegler and colleagues to hypothesize that the potential exists for perinatal transmission of KSHV or transmission through breast feeding.[119] An estimation of age of KS onset emphasized birth or early infancy and presumably if KS were due to KSHV, infection would occur prior to clinical signs and symptoms. A high concordance between mother and child HIV seropositivity could also imply simultaneous transmission of HIV and KSHV. Neither child nor maternal KSHV serostatus was determined in this study, and only eight KS tissues from children were tested and found to be KSHV DNA positive by PCR.[1] It is of some interest that none of the mothers in this study had KS. However, because sexual transmission in this age group is unlikely, vertical transmission remains a potential cause of KSHV transmission in children.

3.5. Organ Transplantation

Using a case-control study design of Italian organ allograft recipients matched to their donors, Parravicini and colleagues found that 91% of KS cases compared to 12% of controls were KSHV-seropositive by two or more assays (latent and lytic) before transplantation (OR = 74, 95% C.I.. = 4.7–3457).[120] This suggests that the risk of KSHV transmission through organ transplantation is rare and that KS onset in these cases may be due to virus reactivation. However, one case independent of the case-control study was documented where transmission from a living, related donor to a recipient was likely to have occurred. A 24-year-old female kidney transplant patient was seronegative two months before and at the time of transplantation but developed Castleman's disease 13 months and KS 27 months posttransplantation. A blood sample taken at 13 months was seropositive for KSHV antibodies as was a sample from the living, related kidney donor. These data suggest that allograft transmission is uncommon (even in a setting of relatively high prevalence) but can occur.

4. CONCLUSION

New molecular techniques now play a tremendous role in the field of emerging infectious disease, as evidenced by the discovery of KSHV in 1994. Within only two years the entire genome of this new human herpesvirus was sequenced. Although initial studies were inconclusive about the role of KSHV in KS, the accumulation of more methodologically sound studies provides strong evidence of the causal association. Other noninfectious cofactors are certain to contribute to KS pathogenesis, however, as the current epidemiological evidence demonstrates KSHV is almost certainly the causative infectious agent for KS.

REFERENCES

1. Chang, Y., Cesarman, E., Pessin, M. S. *et al.*, 1994, Identification of herpesvirus-like DNA sequences in AIDS-associated Kaposi's sarcoma, *Science* **265:**1865–1869.
2. Lisitsyn, N. A., Lisitsina, N. M., Dalbagni, G. *et al.*, 1995 Comparative genomic analysis of tumors: Detection of DNA losses and amplification, *Proc. Natl. Acad. Sci. USA* **92**(1)**:**151–155.
3. Moore, P. S., Gao, S.-J., Dominguez, G. *et al.*, 1996, Primary characterization of a herpesvirus-like agent associated with Kaposi's sarcoma, *J. Virol.* **70:**549–558.
4. Roizman, B., 1993, The family Herpesviridae, in: *The Family Herpesviridae,* (B. Roizman, R. J. Whitley, and C. Lopez, eds.), Raven Press, New York, pp. 1–9.
5. Cohen, J., 1994, Controversy: Is KS really caused by new herpesvirus?, *Science* **268:**1847–1848.
6. Levy, J. A., 1995, A new human herpesvirus: KSHV or HHV8?, *Lancet* **346:**786.
7. Anonymous, 1994, Addressing emerging infectious disease threats: A prevention stategy for the United States. Executive Summary, *MMWR* **43**(RR-5)**:**1–18.

8. Gao, S.-J., Moore, P. S., 1996, Molecular identification of unculturable infectious agents, *Emerging Infect. Dis.* **2:**159–167.
9. Tappero, J. W., Conant, M. A., Wolfe, S. F., and Berger, T. G., 1993, Kaposi's sarcoma: Epidemiology, pathogenesis, histology, clinical spectrum, staging criteria and therapy, *J. Am. Acad. Dermatol.* **28:**371–395.
10. Oettle, A. G. (ed.), 1962, *Geographic and Racial Differences in the Frequency of Kaposi's Sarcoma as Evidence of Environmental or Genetic Causes*, Karger, Basel, p. 18.
11. Giraldo, G., Beth, E., and Haguenau, F., 1972, Herpes-type particles in tissue culture of Kaposi's sarcoma from different geographic regions. *J. Natl. Cancer Inst.* **49:**1509–1526.
12. Beral, V., Peterman, T. A., Berkelman, R. L., and Jaffe, H. W., 1990, Kaposi's sarcoma among persons with AIDS: A sexually transmitted infection?, *Lancet* **335:**123–128.
13. Beral, V., Bull, D., Jaffe, H. *et al.*, 1991, Is risk of Kaposi's sarcoma in AIDS patients in Britain increased if sexual partners came from United States or Africa?, *BMJ* **302:**624–625.
14. Beral, V., 1991, Epidemiology of Kaposi's sarcoma, in: *Epidemiology of Kaposi's Sarcoma, Vol. 10*, Imperial Cancer Research Fund, London, pp. 5–22.
15. Beral, V., Bull, D., Darby, S. *et al.*, 1992, Risk of Kaposi's sarcoma and sexual practices associated with faecal contact in homosexual or bisexual men with AIDS, *Lancet* **339:**632–636.
16. Archibald, C. P., Schechter, M. T., Le, T. N., Craib, K. J. P., Montaner, J. S. G., and O'Shaughnessy, M. V., 1992, Evidence for a sexually transmitted cofactor for AIDS-related Kaposi's sarcoma in a cohort of homosexual men, *Epidemiology* **3:**203–209.
17. Hoover, D. R., Black, C., Jacobson, L. P. *et al.*, 1993, Epidemiologic analysis of Kaposi's sarcoma as an early and later AIDS outcome in homosexual men. *Am. J. Epidemiol.* **138:**266–278.
18. Albrecht, H., Helm, E. B., Plettenberg, A. *et al.*, 1994, Kaposi's sarcoma in HIV infected women in Germany: More evidence for sexual transmission. A report of 10 cases and review of the literature [Review], *Genitourinary Med.* **70**(6)**:**394–398.
19. Armenian, H. K., Hoover, D. R., Rubb, S. *et al.*, 1993, Composite risk score for Kaposi's sarcoma based on a case-control and longitudinal study in the Multicenter AIDS Cohort Study (MACS) population, *Am. J. Epidemiol.* **138:**256–265.
20. Archibald, C. P., Schechter, M. T., Craib, K. J. *et al.*, 1990, Risk factors for Kaposi's sarcoma in the Vancouver Lymphadenopathy-AIDS Study, *J. AIDS* **3(suppl):**S18–23.
21. Schechter, M. T., Marion, S. A., Elmslie, K. D., Ricketts, M. N., Nault, P., and Archibald, C. P., 1991, Geographic and birth cohort associations of Kaposi's sarcoma among homosexual men in Canada, *Am. J. Epidemiol.* **134:**485–488.
22. Duesberg, P., 1988, HIV is not the cause of AIDS, *Science* **241:**514–517.
23. Cohen, J., 1994, Could drugs, rather than a virus, be the cause of AIDS?, *Science* **266:**1648–1649.
24. Evans, A. S., 1978, Causation and disease: A chronological journey, *Am. J. Epidemiol.* **108:**249–258.
25. Fredericks, D. N., and Relman, D. A., 1996, Sequence-based identification of microbial pathogens: A reconsideration of Koch's postulates, *Clin. Microbiol. Rev.* **9:**18–33.
26. Hill, A. B., 1965, Environment and disease: Association or causation?, *Proc. R. Soc. Med.* **58:**295–300.
27. Schlesselman, J. J., 1982, *Case Control Studies: Design, Conduct Analysis*, Oxford University Press, New York.
28. Susser, M., 1977, Judgment and causal inference: Criteria in epidemiologic studies, *Am. J. Epidemiol.* **105:**1–15.
29. Parry, J. P., and Moore, P. S., 1997, Corrected prevalence of Kaposi's sarcoma (KS)-associated herpesvirus infection prior to onset of KS, *AIDS* **11:**127–128.
30. Rothman, K. J., 1986, *Modern Epidemiology*, Little, Brown, Boston.

31. Dupin, N., Grandadam, M., Calvez, V. *et al.*, 1995, Herpesvirus-like DNA in patients with Mediterranean Kaposi's sarcoma, *Lancet* **345:**761–762.
32. Ambroziak, J. A., Blackbourn, D. J., Herndier, B. G. *et al.*, 1995, Herpes-like sequences in HIV-infected and uninfected Kaposi's sarcoma patients. *Science* **268:**582–583.
33. Lebbé, C., de Crémoux, P., Rybojad, M., Costa da Cunha, C., Morel, P., and Calvo, F., 1995, Kaposi's sarcoma and new herpesvirus, *Lancet* **345:**1180.
34. Chuck, S., Grant, R. M., Katongole-Mbidde, E., Conant, M., and Ganem, D., 1996, Frequent presence of a novel herpesvirus genome in lesions of human immunodeficiency virus-negative Kaposi's sarcoma, *J. Infect. Dis.* **173:**248–251.
35. Peterman, T. A., Jaffe, H. W., Friedman-Kien, A. E., and Weiss, R. A., 1991, The aetiology of Kaposi's sarcoma, in: *The Aetiology of Kaposi's Sarcoma, Vol. 10,* Imperial Cancer Research Fund, London pp. 23–37.
36. Schalling, M., Ekman, M., Kaaya, E. E., Linde, A., and Biberfeld, P., 1995, A role for a new herpesvirus (KSHV) in different forms of Kaposi's sarcoma, *Nature Med.* **1:**707–708.
37. Chang, Y., Ziegler, J. L., Wabinga, H. *et al.*, 1996, Kaposi's sarcoma-associated herpesvirus and Kaposi's sarcoma in Africa, *Arch. Int. Med.* **156:**202–204.
38. Corbellino, M., Poirel, L., Bestetti, G. *et al.*, 1996, Restricted tissue distribution of extralesional Kaposi's sarcoma-associated herpesvirus-like DNA sequences in AIDS patients with Kaposi's sarcoma, *AIDS Res. Hum. Retrovirus* **12:**651–657.
39. Corbellino, M., Parravincini, C., Aubin, J. T., and Berti, E., 1996, Kaposi's sarcoma and herpesvirus-like DNA sequences in sensory ganglia, *N. Engl. J. Med.* **334:**1341–1342.
40. Safai, B., Miké, V., Giraldo, G., Beth, E., and Good, R. A., 1980, Association of Kaposi's sarcoma with second primary malignancies, *Cancer* **45:**1472–1479.
41. Biggar, R. J., Curtis, R. E., Cote, T. R., Rabkin, C. S., and Melbye, M., 1994, Risk of other cancers following Kaposi's sarcoma: Relation to acquired immunodeficiency syndrome. *Am. J. Epidemiol.* **139:**362–368.
42. Frizzera, G., Banks, P. M., Massarelli, G., and Rosai, J., 1983, A systemic lymphoproliferative disorder with morphologic features of Castleman's disease. Pathological findings in 15 patients. *Am. J. Surg. Pathol.* **7:**211–231.
43. Varsano, S., Manor, Y., Steiner, Z., Griffel, B., and Klajman, A., 1984, Kaposi's sarcoma and angioimmunoblastic lymphadenopathy, *Cancer* **54**(8)**:**1582–1585.
44. Karcher, D. S., and Alkan, S., 1995, Herpes-like DNA sequences, AIDS-related tumors, and Castleman's disease, *New Eng. J. Med.* **333**(12)**:**797–798.
45. Cesarman, E., Chang, Y., Moore, P. S., Said, J. W., and Knowles, D. M., 1995, Kaposi's sarcoma-associated herpesvirus-like DNA sequences are present in AIDS-related body cavity based lymphomas, *N. Engl. J. Med.* **332:**1186–1191.
46. Nador, R. G., Cesarman, E., Chadburn, A. *et al.*, 1996, Primary effusion lymphoma: A distinct clinicopathologic entity associated with the Kaposi's sarcoma-associated herpes virus, *Blood* **88**(2)**:**645–656.
47. Knowles, D. M., Inghirami, G., Ubriaco, A., and Dalla-Favera, R., 1989, Molecular genetic analysis of three AIDS-associated neoplasms of uncertain lineage demonstrates their B-cell derivation and the possible pathogenetic role of Esptein–Barr virus, *Blood* **73:**792–799.
48. Cesarman, E., Nador, R. G., Aozasa, K., Delsol, G., Said, J. W., and Knowles, D. M., 1996, Kaposi's sarcoma-associated herpesvirus in non-AIDS-related lymphomas occurring in body cavities, *Am. J. Pathol.* **149**(1)**:**53–57.
49. Strauchen, J. A., Hauser, A. D., Burstein, D. A., Jiminez, R., Moore, P. S., and Chang, Y., 1996, Body cavity-based malignant lymphoma containing Kaposi sarcoma-associated herpesvirus in an HIV-negative man with previous Kaposi sarcoma, *Ann. Intern. Med.* **125:** 822–825.
50. Nador, R. G., Cesarman, E., Knowles, D. M., and Said, J. W., 1995, Herpes-like DNA

sequences in a body cavity-based lymphoma in an HIV-negative patient, *N. Engl. J. Med.* **333:**943.

51. Renne, R., Zhong, W., Herndier, B., McGrath, M., Abbey, N., and Ganem, D., 1996, Lytic growth of Kaposi's sarcoma-associated herpesvirus (human herpesvirus 8) in culture, *Nature Med.* **2:**342–346.
52. Kedes, D. H., Operskalski, E., Busche, M., Flood, J., Kohn, R., and Ganem, D., 1996, The seroepidemiology of human herpesvirus 8 (Kaposi's sarcoma-associated herpesvirus): Distribution of infection in KS risk groups for sexual transmission, *Nature Med.* **2:**918–924.
53. Gao, S.-J., Kingsley, L., Li, M. *et al.*, 1996, KSHV antibodies among Americans, Italians and Ugandans with and without Kaposi's sarcoma, *Nature Med.* **2:**925–928.
54. Lennette, E. T., Blackbourne, D. J., and Levy, J. A., 1996, Antibodies to human herpesvirus type 8 in the general population and in Kaposi's sarcoma patients, *Lancet* **348:**858–861.
55. Corbellino, M., Poirel, L., Aubin, J. T. *et al.*, 1996, The role of human herpesvirus 8 and Epstein–Barr virus in the pathogenesis of giant lymph node hyperplasia (Castleman's disease), *Clin. Infect. Dis.* **22:**1120–1121.
56. Soulier, J., Grollet, L., Oskenhendler, E. *et al.*, 1995, Kaposi's sarcoma-associated herpesvirus-like DNA sequences in multicentric Castleman's disease, *Blood* **86:**1276–1280.
57. Dupin, N., Gorin, I., Deleuze, J., Agut, H., Huraux, J.-M., and Escande, J.-P., 1995, Herpes-like DNA sequences, AIDS-related tumors, and Castleman's disease, *N. Engl. J. Med.* **333:**797–798.
58. Gessain, A., Sudaka, A., Briére, J. *et al.*, 1996, Kaposi's sarcoma-associated herpes-like virus (human herpesvirus type 8) DNA sequences in multicentric Castleman's disease: Is there any relevant association in non-human immunodeficiency virus-infected patients?, *Blood* **87**(1)**:**414–416.
59. Rady, P. L., Yen, A., Rollefson, J. L. *et al.*, 1995, Herpesvirus-like DNA sequences in non-Kaposi's sarcoma skin lesions of transplant patients, *Lancet* **345:**1339–1340.
60. Boshoff, C., Talbot, S., Kennedy, M. *et al.*, 1996, New herpesvirus and immunosuppressed skin cancers, *Lancet* **347:**338–339.
61. Adams, V., Kempf, W., Schmid, M., Müller, B., Briner, J., and Burg, G., 1995. Absence of herpesvirus-like DNA sequences in skin cancers of non-immunosuppressed patients, *Lancet* **346:**1715–1716.
62. Dictor, M., Rambech, E., Way, D., Witte, M., and Bendsöe, N., 1996, Human herpesvirus 8 (Kaposi's sarcoma-associated herpesvirus) DNA in Kaposi's sarcoma lesions, AIDS Kaposi's sarcoma cell lines, endothelial Kaposi's sarcoma stimulators, and the skin of immunosuppressed patients, *Am. J. Pathol.* **148:**2009–2016.
63. Gyulai, R., Kemeny, L., Kiss, M., Adam, E., Nagy, F., and Dobozy, A., 1996, Herpesvirus-like DNA sequence in angiosarcoma in a patient without HIV infection. *N. Engl. J. Med.* **334:**540–541.
64. Gyulai, R., Kemeny, L., Adam, E., Nagy, F. and Dobozy, A., 1996, HHV8 DNA in angiolymphoid hyperplasia of the skin, *Lancet* **347:**1837.
65. McDonagh, D. P., Liu, J., Gaffey, M. J., Layfield, L. J., Azumi, N., and Traweek, S. T., 1996, Detection of Kaposi's sarcoma-associated herpesvirus-like DNA sequences in angiosarcoma, *Am. J. Pathol.* **149:**1363–1368.
66. Boshoff, C., Whitby, D., Hatziionnou, T. *et al.*, 1995, Kaposi's sarcoma-associated herpesvirus in HIV-negative Kaposi's sarcoma, *Lancet* **345:**1043–1044.
67. Tomita, Y., Naka, N., Aozasa, K., Cesarman, E., and Knowles, D. M., 1996, Absence of Kaposi's sarcoma-associated herpesvirus-like DNA sequences (KSHV) in angiosarcomas developing in body-cavity and other sites, *Int. J. Cancer* **66:**141–142.
68. Jin, Y.-T., Tsai, S.-T., Yan, J.-J., Hsiao, J.-H., Lee, Y.-Y., and Su, I.-J., 1996, Detection of Kaposi's sarcoma-associated herpesvirus-like DNA sequence in vascular lesions: A reliable diagnostic marker for Kaposi's sarcoma, *Am. J. Clin. Pathol.* **105:**360–363.

69. Lin, J.-C., Lin, S.-C., Mar, E.-C. *et al.*, 1995, Is Kaposi's sarcoma-associated herpesvirus detectable in semen of HIV-infected homosexual men?, *Lancet* **346:**1601–1602.
70. Monini, P., de Lellis, L., Fabris, M., Rigolin, F., and Cassai, E., 1996, Kaposi's sarcoma-associated herpesvirus DNA sequences in prostate tissue and human semen, *N. Engl. J. Med.* **334:**1168–1172.
71. Corbellino, M., Bestetti, G., Galli, M., and Parravincini, C., 1996, Absence of HHV-8 in prostate and semen, *N. Engl. J. Med.* **335:**1237.
72. Tasaka, T., Said, J. W., Koeffler, H. P., 1996, Absence of HHV-8 in prostate and semen, *N. Engl. J. Med.* **335:**1237–1238.
73. Gupta, P., Mandaleshwar, D. S., Rinaldo, C. *et al.*, 1996, Detection of Kaposi's sarcoma herpesvirus DNA in semen of homosexual men with Kaposi's sarcoma, *AIDS* **10:**1596–1598.
74. Howard, M. R., Whitby, D., Bahadur, G. *et al.*, 1997, Detection of human herpesvirus 8 DNA in semen from HIV-infected individuals but not healthy semen donors, *AIDS* **11:**F15–F19.
75. Monini, P., de Lellis, L., and Cassai, E., 1996, Absence of HHV-8 in prostate and semen, *N. Engl. J. Med.* **335:**1238–1239.
76. Miller, G., Rigsby, M., Heston, L. *et al.*, 1996, Antibodies to butyrate inducible antigens of Kaposi's sarcoma-associated herpesvirus in HIV-1 infected patients., *N. Engl. J. Med.* **334:**1292–1297.
77. Gao, S.-J., Kingsley, L., Hoover, D. R. *et al.*, 1996, Seroconversion of antibodies to Kaposi's sarcoma-associated herpesvirus-related latent nuclear antigens prior to onset of Kaposi's sarcoma, *N. Engl. J. Med.* **335:**233–241.
78. Rainbow, L., Platt, G. M., Simpson, G. R. *et al.*, 1997, The 226- to 234-kilodalton latent nuclear protein (LNA) of Kaposi's sarcoma-associated herpesvirus (KSHV/HHV8) is encoded by ORF73 and is a component of the latency-associated nuclear antigen, *J. Virol.* **71:** 5919–5921.
79. Simpson, G. R., Schulz, T. F., Whitby, D. *et al.*, 1996, Prevalence of Kaposi's sarcoma-associated herpesvirus infection measured by antibodies to recombinant capsid protein and latent immunofluorescence antigen, *Lancet* **348:**1133–1138.
80. Klein, G., 1994, Epstein–Barr Virus strategy in normal and neoplastic B cells, *Cell* **77:**791–793.
81. Moore, P. S., Kingsley, L., Holmberg, S. D. *et al.*, 1996, Kaposi's sarcoma-associated herpesvirus infection prior to onset of Kaposi's sarcoma, *AIDS* **10:**175–180.
82. Whitby, D., Howard, M.R., Tenant-Flowers, M. *et al.*, 1995, Detection of Kaposi's sarcoma-associated herpesvirus (KSHV) in peripheral blood of HIV-infected individuals predicts progression to Kaposi's sarcoma, *Lancet* **364:**799–802.
83. Decker, L. L., Shankar, P., Khan, G. *et al.*, 1996, The Kaposi's sarcoma-associated herpesvirus (KSHV) is present as an intact latent genome in KS tissue but replicates in the peripheral blood mononuclear cells of KS patients, *J. Exp. Med.* **184:**283–288.
84. Gaidano, G., Pastore, C., Gloghini, A. *et al.*, 1996, Distribution of human herpesvirus-8 sequences throughout the spectrum of AIDS-related neoplasia, *AIDS* **10:**941–949.
85. Boshoff, C., Schulz, T. F., Kennedy, M. M. *et al.*, 1995, Kaposi's sarcoma-associated herpes virus (KSHV) infects endothelial and spindle cells, *Nature Med.* **1:**1274–1278.
86. Li, J. J., Huang, Y. Q., Cockerell, C. J., and Fliedman-Kien, A. E., 1996, Localization of human herpes-like virus type 8 in vascular endothelial cells and perivascular spindle-shaped cells of Kaposi's sarcoma lesions by *in situ* hybridization, *Am. J. Pathol.* **148:**1741–1748.
87. Staskus, K. A., Zhong, W., Gebhard, K. *et al.*, 1997, Kaposi's sarcoma-associated herpesvirus gene expression in endothelial (spindle) tumor cells, *J. Virol* **71:**715–719.
88. Lebbé, C., Pellet, C., Tatoud, R. *et al.*, 1997, Absence of human herpesvirus 8 sequences in prostate specimens, *AIDS* **11:**270.
89. Rubin, M. A., Parry, J. P., and Singh, B., 1997, Kaposi's sarcoma-associated herpesvirus deoxyribonucleic acid sequences: Lack of expression in prostatic tissue of human immunodeficiency virus-negative immunocompetent adults, *J. Urol.* **159:**146–148.

90. Fleckenstein, B., and Desrosiers, R. C., 1982, Herpesvirus saimiri and herpesvirus ateles, in: *Herpesvirus saimiri and Herpesvirus ateles*, (B. Roizman, ed.), Raven Press, New York.
91. Russo, J. J., Bohenzky, R. A., Chien, M. *et al.*, 1996, Nucleotide sequence of Kaposi's sarcoma-associated herpesvirus (HHV8), *Proc. Natl. Acad. Sci. USA* **93:**14862–14867.
92. Cesarman, E., Nador, R. G., Bai, F. *et al.*, 1996, Kaposi's sarcoma-associated herpesvirus contains G protein-coupled receptor and cyclin D homologs which are expressed in Kaposi's sarcoma and malignant lymphoma, *J. Virol.* **70:**8218–8223.
93. Peters, G., 1994, The D-type cyclins and their role in tumorigenesis, *J. Cell Sci.* **Suppl. 18:** 89–96.
94. Yatabe, Y., Nakamura, S., Seto, M. *et al.*, 1996, Clinicopathologic study of PRAD1/cyclin D1 overexpressing lymphoma with special reference to mantle cell lymphoma. A distinct pathologic entity, *Am. J. Surg. Pathol.* **20:**1110–1122.
95. Chang, Y., Moore, P. S., Talbot, S. J. *et al.*, 1996, Cyclin encoded by KS herpesvirus, *Nature* **382:**410.
96. Li, M., Lee, H., Yoon, D.-W., *et al.*, 1997, Kaposi's sarcoma-associated herpesvirus encodes a functional cyclin, *J. Virol.* **71:**1984–1991.
97. Godden-Kent, D., Talbot, S. J., Boshoff, C. *et al.*, 1997, The cyclin encoded by Kaposi's sarcoma associated herpesvirus (KSHV) stimulates cdk6 to phosphorylate the retinoblastoma protein and Histone H1, *J. Virol.* **71:**4193–4198.
98. White, E., 1994, Tumour biology. p53, guardian of Rb, *Nature* **371**(6492)**:**21–22.
99. Sarid, R., Sato, T., Bohenzky, R. A., Russo, J. J., and Chang, Y., 1997, Kaposi's sarcoma-associated herpesvirus encodes a functional Bcl-2 homolog, *Nature Med.* **3:**1–6.
100. Cheng, E.H.-Y., Nicholas, J., Bellows, D. S. *et al.*, 1997, A Bcl-2 homolog encoded by Kaposi sarcoma-associated virus, human herpesvirus 8, inhibits apoptosis but does not heterodimerize with Bax or Bak, *Proc. Natl. Acad. Sci. USA* **94:**690–694.
101. Reed, J. C., Miyashita, T., Krajewski, S. *et al.*, 1996, Bcl-2 family proteins and the regulation of programmed cell death in leukemia and lymphoma, *Cancer Treatment Res.* **84:**31–72.
102. Moore, P. S., Boshoff, C., Weiss, R. A., and Chang, Y., 1996, Molecular mimicry of human cytokine and cytokine-response pathway genes by KSHV, *Science* **274:**1739–1744.
103. Nicholas, J., Ruvolo, V. R., Burns, W. H. *et al.*, 1997, Kaposi's sarcoma-associated human herpesvirus-8 encodes homologues of macrophage inflammatory protein-1 and interleukin-6, *Nature Med.* **3:** 287–292.
104. Molden, J., Chang, Y., Yun, Y., Moore, P. S., and Goldsmith, M. A., 1997, A KSHV-encoded cytokine homologue (vIl-6) activates signaling through the shared gp130 receptor subunit, *J. Biol. Chem.* **272:**19625–19631.
104a. Gao, S.-J., Boshoff, C., Jayachandra, S., Weiss, R. A., Chang, Y., and Moore, P. S., 1997, KSHV ORF K9 (vIRF) is an oncogene which inhibits the interferon signaling pathway, *Oncogene* **15:** 1979–1985.
105. Arvanitakis, L., Geras-Raaka, E., Varma, A., Gershengorn, M. C., and Cesarman, E., 1997, Human herpesvirus KSHV encodes a constitutively active G-protein-coupled receptor linked to cell proliferation, *Nature* **385:**347–349.
106. Jones, J., Peterman, T., Chu, S., and Jaffe, H., 1995, AIDS-associated Kaposi's sarcoma, *Science* **267:**1078–1079.
107. Glesby, M. J., Hoover, D. R., Weng, S. *et al.*, 1996, Use of antiherpes drugs and the risk of Kaposi's sarcoma: Data from the Multicenter AIDS Cohort Study, *JID* **173:**1477–1480.
108. Mocroft, A., Youle, M., Gazzard, B., Morcinek, J., Halai, R., and Phillips, A. N., 1996, Anti herpesvirus treatment and risk of Kaposi's sarcoma in HIV infection, *AIDS* **10:**1101–1105.
109. Morfeldt, L., and Torsander, J., 1994, Long-term remission of Kaposi's sarcoma following foscarnet treatment in HIV-infected patients, *Scand. J. Infect. Dis.* **26:**749.
110. Kedes, D. H., and Ganem, D., 1997, Susceptibility of KSHV (HHV8) to antiviral drugs in culture. *J. Clin. Invest.* **99:**2082–2086.

111. Boshoff, C., Gao, S.-J., Healy, L. *et al.*, 1997, *In vivo* characterization of KSHV positive primary effusion lymphoma (PEL) cells, *Blood,* in press.
112. Huang, Y.-Q., Li, J. J., Poiesz, B. J., Kaplan, M. H., and Friedman-Kien, A. E., 1997, Detection of the herpesvirus-like DNA sequences in matched specimens of semen and blood from patients with AIDS-related Kaposi's sarcoma by polymerase chain reaction *in situ* hybridization, *Amer. J. Pathol.* **150:** 147–153.
113. Thomas, J. A., Brookes, L. A., McGowan, I., Weller, I., and Crawford, D. H., 1996, HHV8 DNA in normal gastrointestinal mucosa from HIV seropositive people, *Lancet* **347:**1337–1338.
114. Tirelli, U., Gaidano, G., Errante, D., and Carbone, A., 1996, Potential heterosexual Kaposi's sarcoma-associated herpesvirus transmission in a couple with HIV-induced immunodepression and with Kaposi's sarcoma and multicentric Castleman's disease, *AIDS* **10**(11)**:**1291–1292.
114a. Olsen, S. J., Chang, Y., Moore, P. S., Biggar, R. J., and Melbye, M., 1998, Increasing Kaposi's sarcoma-associated herpesvirus seroprevalence with age in a highly Kaposi's sarcoma endemic region, Zambia in 1985, *AIDS* **12** (in press).
115. Melbye, M., Bayley, A., Manuwele, J., *et al.*, 1986, Evidence for heterosexual transmission and clinical manifestations of human immunodeficiency virus infection and related conditions in Lusaka, Zambia, *Lancet* **15 Nov.:** 109–111.
116. Boldough, I., Szaniszlo, P., Bresnahan, W. A., Flaitz, C. M., Nichols M. C., and Albrecht, T., 1996, Kaposi's sarcoma herpesvirus-like DNA sequences in the saliva of individuals infected with human immunodeficiency virus, *Clin. Infect. Dis.* **23:**406–407.
117. Koelle, D. M., Huang, M.-L., Vieira, J., Berger, D., Piepkorn, M., and Corey L. Detection of Kaposi's sarcoma-associated herpesvirus in saliva of human immunodeficiency virus infected individuals with and without Kaposi's sarcoma. *ICAAC,* New Orleans, ASM, 1996.
118. Vieira, J., Koelle, D., Huang, M.-L., and Corey, L., Transmissible Kaposi's associated herpesvirus (KSHV) in saliva, *Fourth Conference on Retroviruses and Opportunistic Infections,* Washington, D. C., 1997.
119. Ziegler, J. L., and Katongole-Mbidde, E., 1996, Kaposi's sarcoma in childhood: An analysis of 100 cases from Uganda and relationship to HIV infection, *Int. J. Cancer* **65:**200–203.
120. Parravicini, C., Olsen, S. J., Capra, M. *et al.*, 1997, Risk of Kaposi's sarcoma-associated herpes virus transmission from donor allografts among Italian posttransplant Kaposi's sarcoma patients, *Blood* **90:**2826–2829.
121. Su, I.-J., Hsu, Y.-S., Chang, Y.-C., and Wang, I.-W., 1995, Herpesvirus-like DNA sequence in Kaposi's sarcoma from AIDS and non-AIDS patients in Taiwan, *Lancet* **345:**722–723.
122. Moore, P. S., and Change, Y., 1995, Detection of herpesvirus-like DNA sequences in Kaposi's sarcoma lesions from persons with and without HIV infection, *New Eng. J. Med.* **332:**1181–1185.
123. O'Neill, E., Henson, T. H., Ghorbani, A. J., Land, M. A., Webber, B. L., and Garcia, J. V., 1996, Herpes virus-like sequences are specifically found in Kaposi sarcoma lesions, *J. Clin. Pathol.* **49:**306–308.
124. Buonaguro, F. M., Tornesello, M. L. Beth-Giraldo, E., *et al.*, 1996, Herpesvirus-like DNA sequences detected in endemic, classic, iatrogenic and epidemic Kaposi's sarcoma (KS) biopsies, *Int. J. Cancer* **65:**25–28.
125. Cathomas, G., McGandy, C. E., Terracciano, L. M., Itin, P. H., De, R. G., and Gudat, F., 1996, Detection of herpesvirus-like DNA by nested PCR on archival skin biopsy specimens of various forms of Kaposi sarcoma, *J. Clin. Pathol.* **49**(8)**:**631–633.
126. Marchioli, C. C., Love, J. L., Abbott, L. Z., *et al.*, 1996, Prevalence of human herpesvirus 8 DNA sequences in several patient populations, *J. Clin. Microbiol.* **34**(10)**:**2635–2638.
127. Luppi, M., Barozzi, P., Maiorana, A., *et al.*, Frequency and distribution of herpesvirus-like DNA sequences (KSHV) in different stages of classic Kaposi's sarcoma and in normal tissues from an Italian population, *Int. J. Cancer* **66:**427–431.

128. Lebbé, C., Agbalika, F., de Cremaux, P., *et al.*, 1997, Detection of human herpesvirus 8 and human T-cell lymphotropic virus type 1 sequences in Kaposi sarcoma, *Arch. Dermatol.* **133**(1)**:**25–30.
129. Collandre, H., Ferris, S., Grau, O., Montagnier, L., and Blanchard, A., 1995, Kaposi's sarcoma and new herpesvirus, *Lancet* **345:**1043.
130. Humphrey, R. W., O'Brien, T. R., Newcomb, F. M., *et al.*, 1996, Kaposi's sarcoma (KS)-associated herpesvirus-like DNA sequences in peripheral blood mononuclear cells: Association with KS and persistence in patients receiving anti-herpesvirus drugs, *Blood* **88**(1)**:**297–301.

6

Immunobiology of Murine Gamma Herpesvirus-68

JAMES P. STEWART, EDWARD J. USHERWOOD, BERNADETTE DUTIA, and ANTHONY A. NASH

1. INTRODUCTION

The gamma herpesviruses are important because of their association with disease in humans and animals. Notable members of this subgroup are the Epstein–Barr virus (EBV) which is associated with infectious mononucleosis, Burkitt's lymphoma and nasopharyngeal carcinoma,[1] Kaposi's sarcoma-associated herpesvirus (KSHV; also called human herpesvirus type 8 [HHV-8]),[2] and agents associated with malignant catarrhal fever (MCF) in ruminants.[3] EBV is the best studied, and most knowledge of the biology of gamma herpesvirus has resulted from studies with EBV. Even so, the complete pattern of EBV infection in man has still to be fully delineated. Characterization of mice infected with murine gamma herpesvirus 68 (MHV-68) has shown that this is a model system for exploring the nature of gamma herpesvirus infection *in vivo,* the genesis of disease, and the host response, including prophylactic immune modulation.[4–7]

JAMES P. STEWART, EDWARD J. USHERWOOD, BERNADETTE DUTIA, ANTHONY A. NASH • Department of Veterinary Pathology, The University of Edinburgh, Summerhall, Edinburgh EH9 1QH, Scotland. *Present address for EJU:* Department of Immunology, St. Jude Children's Research Hospital, Memphis, Tennessee 38105.

Herpesviruses and Immunity, edited by Medveczky *et al.* Plenum Press, New York, 1998.

2. THE VIRUS

2.1. Aspects of Viral Infection

Murine gamma herpesvirus 68 was isolated from free-living murid rodents.[8] In contrast to most gamma herpesviruses, this virus forms a fully productive infection in conventional cell monolayer cultures.[9,10] Baby hamster kidney (BHK) cells are typically used to quantitate virus via a conventional plaque assay.[4] However, MHV-68 also infects and persists in cell lines of B lymphocyte origin.[6,11] The best example of this is the S11 line which is derived from a MHV-68-positive B lymphoma.[11] All of the cells in this culture are infected with MHV-68. The majority are latently infected with the genome in an episomal form. At any one time approximately 2% of S11 cells undergo spontaneous reactivation and become productively infected. The number of reactivating cells can be increased to around 10% by adding phorbol esters. The S11 line is clearly similar to EBV-positive B lymphoblastoid cell lines[1] and HHV-8-positive B lymphoma lines[12] and is an important tool for further defining the interaction of MHV-68 with B cells and identifying viral genes expressed during latency.

2.2. The Viral Genome

The MHV-68 genome, shown in Fig. 1, consists of 118kb of unique double-stranded DNA flanked by terminal repeat regions.[13] The complete genomic sequence has recently been determined independently by two separate groups, Virgin *et al.*[14] and A. Davison (unpublished results). There are some 80 open reading frames. A large number are clearly identifiable homolgues of other gamma herpesvirus genes. However, there are also a number of unique open reading frames, termed M1-14. Determining the sequence is a critical step in the study of this model system and leads the way to identifying genes associated with virus latency, pathogenicity and those genes which are immunological targets. Of interest, at the left hand end of the genome are eight genes homologous to mammalian tRNAs.[15] These genes are similar to the Epstein–Barr virus early RNA (EBER) genes in that they are expressed during viral latency and are therefore an extremely important tool. However, like the EBER genes, the biological significance of these unique genes is presently unclear. Of interest in an immunological context are two genes which may be involved in interfering with the immune response. These are homologues of complement regulatory protein *(crp)* and interleukin-8 receptor *(IL-8^R)*. Their location is shown by black arrows in Fig. 1. The *crp* sequences show particular homology with both decay-accelerating factor and membrane cofactor protein both of which regulate C3 convertase, suggesting that the crp protein regulates C3 function. This is consistent with the observation that a homologous protein encoded by herpesvirus saimiri inhibits

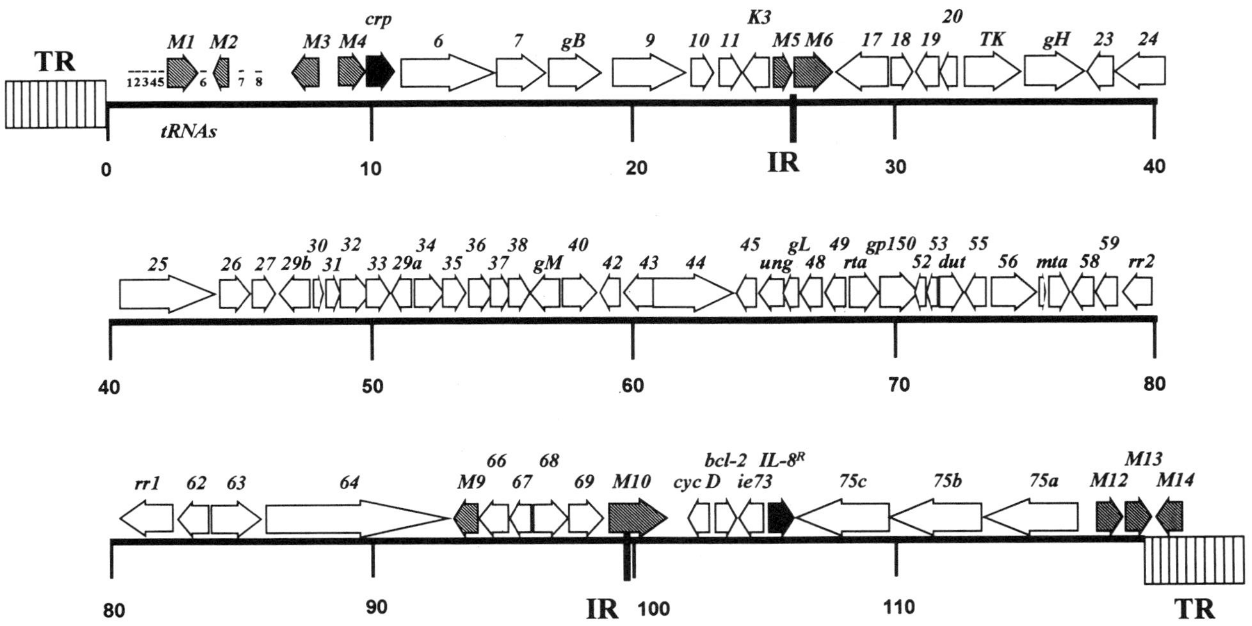

FIGURE 1. Genomic organization of MHV-68 after the sequence of Virgin *et al.*[14] The unique portion of the genome is represented by a single black line, and the coordinates are marked off in kilobase pairs. The terminal repeat (TR) elements bounding this are shown by open bars, and the position of the two internal repeats (IR) is marked by a solid line. The open reading frames are shown by arrows, and the direction of the arrow indicates the direction of transcription. Open reading frames with homologues in other gamma herpesviruses are shown by open arrows, and those which are unique to MHV-68 *(M1-14)* are shown by shaded arrows. Two genes which may be associated with immune modulation are highlighted by black arrows. The gene number (according to Virgin *et al.*[14]) is shown except where the function of the gene is known. These functions are abbreviated as follows: *crp,* complement regulatory protein; *gB,* glycoprotein B;[41] *K3,* homologue of the immediate-early gene of HHV-8; *TK,* thymidine kinase;[20] *gH,* glycoprotein H; *gM,* glycoprotein M; *ung,* uracil DNA glycosylase; *gL,* glycoprotein L; *rta,* R (immediate Early) transcriptional transactivator; *gp150,* glycoprotein homologous to EBV gp340;[42] *dut,* dUTPase; *mta,* M (immediate early) transcriptional transactivator. This gene is composed of two spliced exons;[43] *RR2,* small subunit ribonucleotide reductase; *RR1,* large subunit ribonucleotide reductase; *cyc D,* homologue of mammalian cyclin D; and *bcl-2,* homologue of mammalian bcl-2.

C3-mediated lysis.[16] The IL-8^R may have a function in binding chemokines because the homologous genes in herpesvirus saimiri and HHV-8 are functional receptors.[17,18] Thus, the IL-8^R could damp down an initial inflammatory response by 'mopping up' secreted chemokines.

2.3. Drug Sensitivity

MHV-68 productive replication is moderately sensitive to the antiherpes drug acyclovir (acycloguanosine), but other recently developed nucleoside analogs, such as 2′-deoxy-5-ethyl-beta-4′-thioruidine (4′-S-EtdU)[6,19] are much more effective. These drugs are effective only against productive replication and do not affect MHV-68 latency, making them an invaluable experimental tool for delineating latency-specific viral functions. A thymidine kinase has been identified,[20] and this is the most likely target for these drugs. Thus, the virus-encoded thymidine kinase phosphorylates the drug, resulting in termination of DNA synthesis.

3. INFECTION AND PATHOGENESIS

3.1. Acute Infection in the Lung

A summary of the pathogenesis of MHV-68 in inbred mice is shown in Fig. 2. The natural route if MHV-68 infection in murid rodents is not known. However, based on our knowledge of other gamma herpesvirus infections, it is likely that this involves close contact, probably respiratory transmission. Therefore we feel that in an experimental context, intra-nasal infection represents the most authentic route. Laboratory strains of mice inoculated intranasally with 4×10^5 p.f.u. of MHV-68 establish an initial productive infection in the lung, causing interstitial and peri bronchiolar pneumonia[4,21,22] involving alveolar epithelial cells and mononuclear cells surrounding airways and blood vessels.[4,23] The productive phase of infection lasts for around 10 days before clearance by the immune system.[24] Intense inflammatory infiltration accompanies primary infection (discussed later), and this persists up to 30 days post infection in the form of chronic granulomatous lesions.[23]

3.2. Persistent Infection in the Lung

In contrast to the spleen (see below), conventional biological analysis cannot detect either productive or latent virus in lungs later than 10 days postinfection.[4] Analysis of B cell-deficient (μMT transgenic) mice, however, using a sensitive PCR has shown that MHV-68 can persist in lungs indefinitely in the absence of

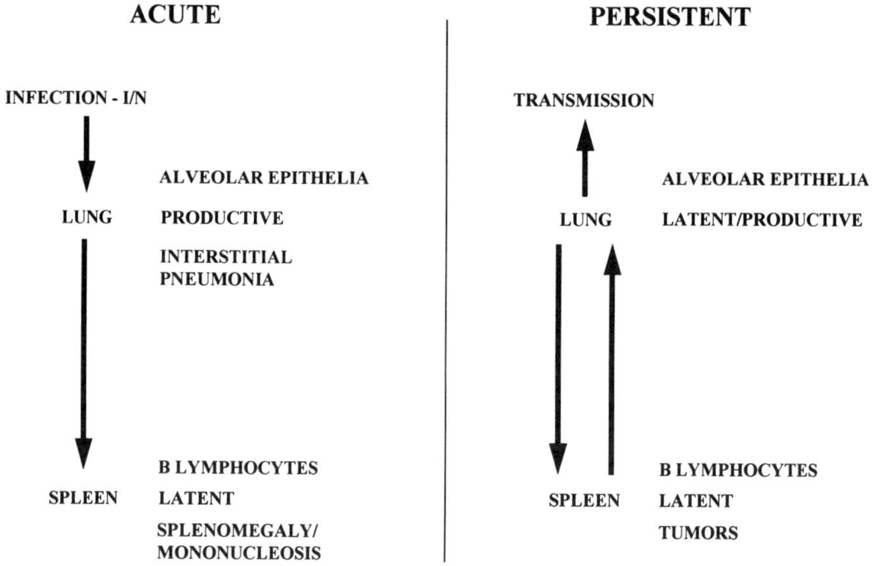

FIGURE 2. Pathways of MHV-68 infection and pathogenesis.

any evidence of infection of the spleen or other organs.[25] Both linear and circular forms of the genome are present which shows that both latent and productively infected cells are present (J. Stewart and E. Usherwood, unpublished results). We have also demonstrated latently infected alveolar epithelial cells by using *in situ* hybridization for virus tRNAs. Therefore epithelial tissue in the lung is a true independent site of gamma herpesvirus persistence and does not need to be constantly reseeded from circulating B cells. The identification of productive forms of genome in epithelial tissue suggests that it may also be a reservoir for transmission. Transgenic mice deficient in MHC class II have a greatly augmented chronic lung infection. The mice control the initial lung infection but infectious virus reappears by day 20 postinfection and is maintained for up to 90 days.[26] This pattern of biology is not unique among gamma herpesviruses and indeed EBV replication has been associated with chronic lung disease in humans.[27]

3.3. Latent Infection in the Spleen

Before resolving the productive infection in the lung, MHV-68 spreads to the spleen where the virus establishes latency in B lymphocytes.[4,5] Infectious virus is not normally seen in this organ. The biological criteria for detecting latent infection involves quantitating reactivating cells in an infective center assay. Using this technique, latently infected cells are detected in the first week post infection and reach peak levels during the second to third weeks (1 in $10^3/10^4$ splenocytes) before declining to relatively stable numbers of 1 in 10^6 spleen leukocytes (see Fig. 3).[4,5] Latently infected cells are also detected at a lower frequency in lymph nodes and bone marrow.[26] As with other herpesviruses, latency is maintained indefinitely in the host.

The virus-encoded tRNAs are a molecular marker of latency. Although these genes are expressed during a productive infection, they are also present in the cells of the germinal centers in spleens in the absence of any productive infection.[15] The number of cells latently infected exceeds that determined by the infectious center assay and follows the same time course, high numbers during the second to third weeks and declining thereafter. The biological significance of the tRNA expression is unknown, but there is a direct parallel with the expression of the EBV EBER RNAs of Epstein–Barr virus during latent B cell infections in humans.

3.4. Other Consequences of MHV-68 Infection

A productive viral infection occurs in tissues other than lung and spleen including the lever, heart, kidney, adrenal gland, and nervous system.[23] In these instances infection depends on various host factors, such as route of infection (introducing virus directly into the blood stream leads to a more disseminated infection); age of mouse on infection (young 3-week old mice) are more susceptible to viraemic spread); the immune status of the animal (immunosuppressed mice are more prone to a disseminated infection).

Infection of the nervous system is not usually associated with gamma herpesviruses. However, MHV-68 can infect neurons and meningeal cells (L.A. Terry, unpublished observations). Whether this occurs in the natural infection is unclear. However, ganglionic infection is observed naturally in HHV-8 infection of humans,[28] BHV-4 infected cattle,[29] and EHV-2 infected horses.[30]

A common feature of gamma herpesvirus infection is the genesis of lymphoproliferative disorders, such as infectious mononucleosis, malignant catarrhal fever, and lymphomas. The latter is seen in humans infected with EBV and HHV-8. MHV-68 has also been linked to lymphoma induction in mice. In 10% of Balb/C mice infected with MHV-68 for between 9 months and 2 years, lymphomas were identified of which 50% were scored as high-grade lymphomas.[7] These tumors were composed of T and B cells, and the B cells have a restricted immunoglobulin light chain restriction, suggesting a monoclonal origin.[7] Virus was associated with the tumors, as determined by *in situ* hybridization for viral DNA, and cell lines have been derived from these tumors. One cell line,

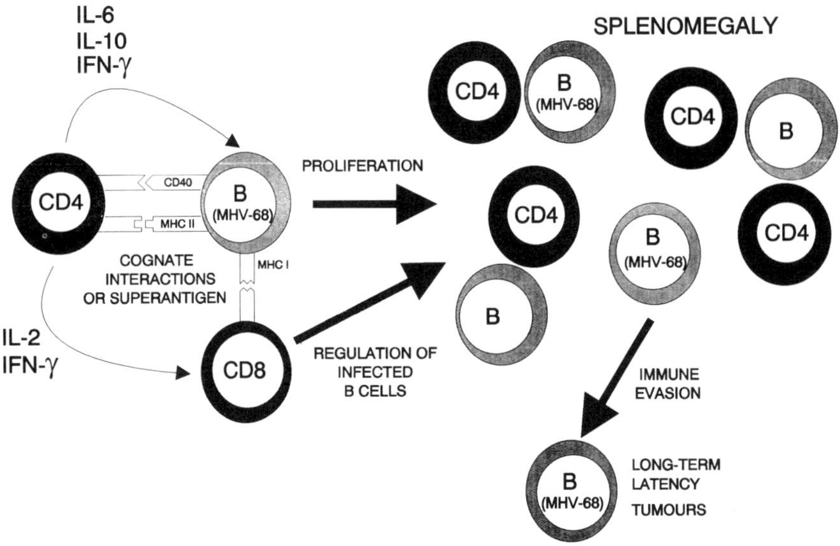

FIGURE 3. Mechanistic aspects of the genesis and resolution of splenomegaly. $CD4^+$ T cell–B cell interaction is necessary for splenomegaly and proliferation of infected B cells. This occurs via cell contact involving MHC class II and CD40 recognition or via cytokines. B cells also serve a key role in presenting viral antigen to $CD8^+$ T cells. We hypothesize that these cells limit the number of infected B cells, thereby reducing splenomegaly. Clearly, some latently infected B cells evade recognition and establish the pool of latent cells in the host.

termed S11, is B cell in origin and harbors the virus in a latent form. This cell line can be transplanted to nude mice where tumors develop.[11]

4. IMMUNOLOGICAL EVENTS DURING INFECTION

4.1. Influence of Adaptive Immune Response

4.1.1. Cellular Infiltration in the Lung

MHV-68 replication in the lung induces a typical inflammatory response consisting mainly of mononuclear phagocytes and lymphocytes.[4] Analysis of cellular infiltration by bronchoalveolar lavage revels an early appearance of monocytes, peaking around day 3 to 4. As the monocytes decline, $CD8^+$ T cells are detected peaking at day 8 to 10. Neither $CD4^+$ T cells nor B cells were detected in the lavage, although these were seen in lung sections (N.P. Sunil-Chandra and A.A. Nash, unpublished observations).

4.1.2. Influence of T Cells on Infection of the Lung

The appearance of large numbers of CD8+ T cells in the lung suggests these cells may play an important antiviral role in lung tissue. To investigate this, Balb/C mice depleted of CD8+, CD4+, or both CD4+ and CD8+ T cells were infected, and viral clearance from the lung was measured.[24] In the absence of CD8+ T cells, the infection progressed unchecked despite the presence of CD4+ T cells and antibody (see Table 1). In contrast, only a slight delay in virus clearance was noted in the absence of CD4+ cells. Depletion of both T cells subsets led to a severe infection and virus spread to infect other tissues (see Table 1).

The lack of CD4+ T cells in MHC class II-deficient mice leads to a deficit in immune surveillance by CD8+ T cells and a lack of anti viral antibody.[26] Although the infection is initially cleared from the lung in these mice, it later recurs and free virus is consistently detected thereafter at this site (see Table 1). Most infected MHC class II-deficient mice developed a wasting disease and died by day 133 post infection. This is most likely caused by deficit in T cell surveillance in these mice because there is an absence of immunological 'help' normally supplied by CD4+ T cells to CD8+ T cells.

4.1.3. The Genesis Of Splenomegaly

It seems likely that latently infected B cells in the spleen initiate splenomegaly because splenomegaly does not occur in the absence of such cells. This has been shown by the infection of transgenic μMT mice which are deficient in mature B cells.[25,31] When infected intranasally, these mice, have no latent infection in spleen cells as determined by the infective center assay or virus-specific PCR and do not display splenomegaly (Table 1).[25] After intraperitoneal inoculation of μMT mice, latency (in an undefined cell type) occurs but is still unaccompanied by splenomegaly.[31] This underlines the importance of the infected B cell for the genesis of splenomegaly and also possibly in trafficking of the virus from lung to spleen.

Then what factors are responsible for the increase in spleen cell number? During splenomegaly cell numbers double with an increase in both B cells and T cells.[32] Expansion of T cells is probably a response to virally infected B cells. Indeed, recent analysis of Vβ T cell receptor (TCR) usage during MHV-68 infection has revealed a disproportionate rise in Vβ4 T cells which are identified in the blood and in the spleen. The selective expansion of such T cells indicates a possible role for a virally encoded or virally induced superantigen[33] which would explain the rapid increase in CD4+ T cells.

As might be the case for EBV and herpesvirus saimiri, there is no evidence to suggest that MHV-68 gene products directly drive B cell proliferation by "transformation." Indeed, all attempts to "transform" or "immortalize" B cells in this

TABLE I
Immunologic Control of MHV-68 Infection in Mice Deficient of Lymphocyte Subsets and Cell-Surface Markers

Immunological parameter	Immune control of lung infection	Splenomegaly	Immune control of latent virus in spleen	Long-term latency	References
Untreated	Normal	Present	Normal	Normal	4,5,24
CD4-depleted and MHC Class II K.O.	Delayed	Absent	Normal	Normal	24,26,32
CD8-depleted and β-2 M K.O.	Absent	Present	Absent	nd[a]	24,31,32,39
CD4-plus CD8-depleted and nude	Absent	Present	Absent	nd[a]	24,39
B cell K.O.	Delayed	Absent	Normal	Absent/low[b]	25,31

[a]Not determined because mice do not survive long term.
[b]Latency in the spleen depends on the route of infection. Intranasal infection results in an absence of long-term latency[25] whereas intraperitoneal infection results in long-term latency.[31]

laboratory have been unsuccessful to date. We know, however, that the rapid expansion of infected B cells and splenomegaly depends on CD4+T cells because after depletion of $CD4^+$ T cells (but not $CD8^+$ T cells), no splenomegaly is observed (see Table 1).[32] B cell infection occurs in these mice, but the peak number of latently infected B cells in the spleen is reduced from 2×10^4, seen in normal mice, to 1×10^3. Similar findings have also been reported after infection of transgenic mice deficient in MHC class II with the virus.[26] Elevated levels of IL-6, IL-10, and IFN-γ are also detected during splenomegaly.[34] Therefore, taken together, it seems likely that T cell 'help' is responsible for driving the proliferation of both infected and uninfected B cells.

In conclusion, evidence suggests that infected B cells cause activation of $CD4^+$ T cells which in turn leads to B cell expansion, a proportion of which are MHV-68-infected. Then, splenomegaly occurs as a result of this positive feedback loop. A schematic diagram illustrating is is shown in Fig. 3).

4.1.4. *The Resolution of Splenomegaly*

If $CD4^+$ T cells influence the burst of infected B cells in the spleen, then $CD8^+$ T cells influence the decline (Table 1). Depletion of $CD8^+$ T cells at the time of infection leads to a slower decline in the number of spleen infective centers (E.J. Usherwood and A.A. Nash, unpublished), and infection of β_2-microglobulin-deficient mice, which lack $CD8^+$ T cells, leads to productive viral replication in the spleen, which cannot be cleared by the immune system.[31] If $CD8^+$T cells are depleted during splenomegaly, this has no effect on the fall in infective centers, suggesting they must exert their effects early in the infection (K. Robertson and A.A. Nash, unpublished observations). This suggests an important role for $CD8^+$ T cells in the regulating latent B cells, although other mechanisms, such as apoptosis of infected B cells,, cannot be ruled out. It also implies that some infected B cells are receptive to T cell recognition, whereas others evade this effector response and can persist for the lifetime of the host. There may be parallels with EBV latent B cell infection, in which type III latency (where nine virus proteins are expressed) is targeted by T cells, but type I latency (where only EBNA 1 is expressed) evades recognition.[35,36] Whether this is the case with MHV-68 awaits the discovery of MHV-68 latency-associated genes.

4.1.5. Mechanistic Aspects of $CD8^+$ T Cell Killing

The role of $CD8^+$ T cells in mediating control of MHV-68-infection is clearly critical, and conventional virus-specific CTLs have recently been found.[36a] The clearance of virus in transgenic mice, which have a lesion in the perforin gene, is identical to the clearance in wild-type mice, and antibodies to the Fas molecule failed to block the action of CTLs derived from infected mice.[37] Thus, either the perforin/granzyme and Fas-Fas ligand-based mechanisms are unessential in their own right for mediating the $CD8^+$ T cell control, or compensatory mechanisms are operating in these systems. Therefore the actual role of these effector mechanisms awaits further experimentation involving adoptive transfer of effector subsets.

Evidence suggests that cytokines might contribute to viral resistance mediated by $CD8^+$ T cells. Two groups have studied transgenic mice with a lesion in the interferon-γ (IFN-γ) gene. In both cases the transgenic mice cleared MHV-68 less efficiently than wild-type,[38,39] and in one set of experiments one third of the mice died in the first seventeen days of infection.[39] In addition, transgenic mice deficient in inducible nitric oxide synthase (iNOS) all died between 8 and 17 days post infection.[39] Therefore MHV-68-specific $CD8^+$ T cells may exert control by secreting IFN-γ which then induces NOS production in macrophages which in turn produce protective nitric oxide.

4.1.6. Role of the Interferon-γ Receptor

Studies have been performed using mice with a targeted deletion in the IFN-γ receptor α chain (IFN-$\gamma R^{-/-}$), which produces IFN-γ but cannot respond to it.[40] These mice control the lung infection efficiently. Spleens of IFN-γ $R^{-/-}$, however, exhibit profound pathological changes. By 14 days post infection, there is histological evidence of widespread destruction of splenic architecture in these mice. A marked decrease in the number of splenic B cells, $CD4^+$, and $CD8^+$ T cells occurs, accompanied by a 10-100-fold increase in the numbers of latently infected cells above those found in wild-type mice. The latent virus load, however, falls to wild-type levels by 21 days postinfection.

The precise mechanisms involved in the pathological and virological changes in IFN-γ $R^{-/-}$ are not understood. The differences in the spleen, however, are unaffected by antiviral drugs and are hence independent of virally mediated lysis of infected cells. Also, both $CD4^+$ and $CD8^+$ T cells are involved because depletion of either subset reverses the changes. At any rate, there is a fundamental difference after MHV-68 infection between mice deficient in IFN-γ and those deficient in the receptor. This may suggest that in the absence of the receptor IFN-γ either is interacting with a counterreceptor or that an unknown factor interacts with the receptor in intact mice.

4.1.7. Antiviral Antibody

The role of antibody in combating MHV-68 infection has not been directly assessed. Anti-MHV-68 antibody first appears in the serum toward the end of the first week of an intranasal infection.[39] The response gradually increases over the next few months although neutralizing antibodies do not appear until the second to third weeks of infection.[39] This delay may be attributed to infection of B cells in germinal centers and the events surrounding splenomegaly. However, antibodies play only a minor role in the immune response to MHV-68 because mice deficient in B cells (and hence antibody) clear the virus from the lung only slightly more slowly than intact mice.[25] The definition of the exact role of antibody awaits the passive infusion of immune sera into naive mice.

4.2. Influence of the Innate Immune Response

4.2.1. Role of Cells Involved in Innate Immunity

The role of either macrophages or cells with natural-killer (NK) activity during an MHV-68 infection has not been directly assessed. NK cells, however, are unlikely to play a role in direct lysis of infected cells because they kill via the perforin/granzyme system and perforin deficient mice show no obvious deficiency in clearing virus (see section 4.1.4). Both NK cells and macrophages, however, are important secretors of interferons, and these are dealt with below.

4.2.2. Role of Type I Interferons

The role of type I interferons in MHV-68 infection has been investigated by using gene knock out mice (B. Dutia and A.A. Nash, unpublished data). Infection of mice with a lesion in the α/β interferon receptor gene (IFN-α/β $R^{-/-}$) with a standard dose (4×10^5 p.f.u.) of MHV-68 results in 90% mortality. Wild-type mice show 100% recovery from the same dose of virus. Death occurs in the IFN-α/β R$^{-/-}$ mice 6–8 days after infection at which time the lung titer of virus is 100- to -1000-fold higher than that found in wild-type mice. However, IFN-α/β R$^{-/-}$ mice infected with a lower dose (4×10^3 p.f.u.) of MHV-68 have a 60% mortality rate, and the mice die at a later time. Interestingly, these mice develop similar lung virus (6 to 8 days post infection) to the mice infected with the higher dose of virus. Thus type I interferons play a crucial role in controlling the initial replication of the virus and in their absence the virus replicates to levels which overwhelm the adaptive immune system.

5. CONCLUSIONS

Study of gamma herpesvirus infections has been limited, to a large extent, to tissue culture models. Murine gamma herpesvirus 68 has enabled the detailed study of events in the gamma herpesvirus life cycle *in vivo* in its natural host. The ability to produce MHV-68 recombinants with targeted gene deletions means that gene function can be assessed in the animal. This has a bearing on genes involved in viral tropism, latency, and targets for the immune system. An additional benefit of working with a murid virus is the ability to analyze the relative importance of host components by using transgenic knockout technology. The impact of this in studying virus infections is already considerable, particularly in relation to studying components of the immune system involved in antiviral defense. We and others have begun studying the properties of MHV-68 infection in mice deficient in B cells, T cell subsets, cytokines or their receptors, and cell adhesion molecules. A major application of this model will be in the exploration of immunotherapeutic strategies. This will involve identifying viral proteins for use as vaccines or the prophylactic use of antibody and T cells. Important candidate vaccines against gamma herpesviruses, however, may also include latent gene products. The ability to target latent MHV-68 infections may provide important clues for treating these states of gamma herpesvirus infection in humans and domestic animals.

REFERENCES

1. Rickinson, A. B., and Kieff, E., 1996, Epstein–Barr virus, In: *Fields Virology* (B. N. Fields, D. M. Knipe, and P.M. Howley, eds.), Lippincott-Raven, New York, pp.2397–2446.
2. Chang, Y., Cesarman, E., Pessin, M. S., Lee, F., Culpepper, J., Knowles, D. M., and Moore, P. S., 1994, Indentification of herpesvirus-like DNA sequences in AIDS-associated Kaposi's sarcoma, *Science* **266:**1865–1869.

3. Roizman, B., Desrosiers, R. C., Fleckenstein, B., Lopez, C., Minson, A. C., and Studdert, M. J., 1992, The family of *Herpesviridae:* An update, *Arch. Virol.* **123:**425–449.
4. Sunil-Chandra, N.P., Efstathiou, S., Arno, J., and Nash, A. A., 1992, Virological and pathological features of mice infected with murine gammaherpesvirus 68, *J. Gen. Virol.* **73:**2347–2356.
5. Sunil-Chandra, N. P., Efstathiou, S., and Nash, A. A., 1992, Murine gammaherpesvirus 68 establishes a latent infection in mouse B lymphocytes *in vivo, J. Gen. Virol.* **73:**3275–3279.
6. Sunil-Chandra, N. P., Efstathiou, S., and Nash, A. A., 1993, Interactions of murine gammaherpesvirus 68 with B and T cell lines, *Virology* **193:**825–833.
7. Sunil-Chandra, N. P., Arno, J., Fazakerley, J., and Nash, A. A., 1994, Lymphoproliferative disease in mice infected with murine gammaherpesvirus 68, *Am. J. Pathol.* **145:**818–826.
8. Blaskovic, D., Stancekova, M., Svobodova, J., and Mistrikova, J., 1980, Isolation of five strains of herpesviruses from two species of free living small rodents, *Acta Virol.* **24:**468.
9. Ciampor, F., Stancekova, M., and Blaskovic, D., 1981, Electron microscopy of rabbit embryo fibroblasts infected with herpesvirus isolates from *Clethrionomys glareolus* and *Apodemus flavicollis, Acta Virol.* **25:**101–107.
10. Svobodova, J., Blaskovic, D., and Mistrikova, J., 1982, Growth characteristics of herpesviruses isolated from free living small rodents, *Acta Virol.* **26:**256–263.
11. Usherwood, E. J., Stewart, J. P., and Nash, A. A., 1996, Characterization of tumor cell lines derived from murine gammaherpesvirus 68-infected mice, *J. Virol.* **70:**6516–6518.
12. Moore, P. S., Gao, S.-J., Dominguez, G., Cesarman, E., Lungu, O., Knowles, D. M., Garber, R., Pellet, P. E., McGeoch, D. J., and Chang, Y., 1996, Primary characterization of a herpesvirus agent associated with Kaposi's sarcoma. *J. Virol.* **70:**549–558.
13. Efstathiou, S., Ho, Y. M., and Minson, A. C., 1990, Cloning and molecular characterisation of the murine herpesvirus 68 genome, *J. Gen. Virol.* **71:**1355–1364.
14. Virgin, H. W., Latreille, P., Wamsley, P., Hallsworth, K., Weck, K. A., Dal Canto, A. J., and Speck, S. H., 1997. Complete sequence and genomic analysis of murine gammaherpesvirus 68. *J. Virol.* **71:**5894–5904.
15. Bowden, R. J., Simas, J. P., Davis, A., and Efstathiou, S., 1997, Murine gammaherpesvirus 68 encodes tRNA-like sequences which are expressed during latency, *J. Gen. Virol.* **78:**1675–1687.
16. Ahuja, S. K., and Murphy, P. M., 1993. Molecular piracy of mammalian interleukin-8 receptor type B by herpesvirus saimiri. *J. Biol. Chem.* **268:**20691–20694.
17. Fodor, W. L., Rollins, S. A., Bianco-Caron, S., Rother, R. P., Guilmette, E. R., Buton, W. V., Albrecht, J.-C., Fleckenstein, B., and Squinto, S. P., 1995. The complement control protein of herpesvirus saimiri regulates serum complement by inhibiting C3 convertase activity, *J. Virol.* **69:**3889–3892.
18. Arvanitakis, L., Geras-Raaka, E., Varma, A., Gershengorn, M. C., and Cesarman, E., 1997. Human herpesvirus KSHV encodes a constitutively active G-protein-coupled receptor linked to cell proliferation, *Nature.* **385:**347–350.
19. Sunil-Chandra, N. P., Efstathiou, S., and Nash, A. A., 1994, The effect of acyclovir on the acute and latent murine gammaherpesvirus-68 infection of mice, *Antiviral Chem. Chemother.* **5:**290–296.
20. Pepper, S. deV., Stewart, J. P., Arrand, J. R., and Mackett, M, 1996, Murine gammaherpesvirus-68 encodes homologues of thymidine kinase and glycoprotein H: Sequence, expression and characterization of pyrimidine kinase activity, *Virology* **219:**475–479.
21. Blaskovic, D., Stanekova, D., and Rajcani, J., 1984, Experimental pathogenesis of murine herpesvirus in new-born mice, *Acta Virol.* **28:**225–231.
22. Rajcani, J., Blaskovic, D., Svododova, J., Ciampor, F., Huckova, D., and Stanekova, D., 1985, Pathogenesis of acute and persistent murine herpesvirus infection in mice, *Acta Virol.* **29:**51–60.
23. Sunil-Chandra, N. P., 1991, Studies on the pathogenesis of a murine gammaherpesvirus (MHV-68), Ph.D. Thesis, University of Cambridge.

24. Ehtisham, S., Sunil-Chandra, N. P., and Nash, A. A., 1993, Pathogenesis of murine gammaherpesvirus infection in mice deficient in CD4 and CD8 T cells, *J. Virol.* **67:**5247–5252.
25. Usherwood, E. J., Stewart, J. P., Robertson, K., Allen, D. J., and Nash, A. A., 1996, Absence of splenic latency in murine gammaherpesvirus 68-infected B-cell deficient mice, *J. Gen. Virol.* **7:**2819–2825.
26. Cardin, R. D., Brooks, J. W., Sarawar, S. R., and Doherty, P. C., 1996, Progressive loss of CD8$^+$ T cell-mediated control of a γ-herpesvirus in the absence of CD4$^+$ T cells, *J. Exp. Med.* **184:**863–871.
27. Egan, J. J., Stewart, J. P., Hasleton, P. S., Arrand, J. R., Carroll, K., and Woodcock, A., 1995, Epstein–Barr virus replication within pulmonary epithelial cells in cryptogenic fibrosing alveolitis, *Thorax* **50:**1234–1239.
28. Corbellino, M., Parravicini, C., Aubin, J. T., and Berti, E., 1996. Kaposi's sarcoma and herpesvirus-like DNA sequences in sensory ganglia. *N. Engl. J. Med.* **334:**1341–1342.
29. Castrucci, G., Frigeri, F., Ferrari, M., Di Luca, D., and Traldi, V., 1991, A study of some biologic properties of bovid herpesvirus-4, *Comp. Immunol. Microbiol. Infect. Dis.* **14:**1341–1342.
30. Rizvi, S. M., Slater, J. D., Wolfinger, U., Borchers, K., Field, H. J., and Slade, A. J., 1997, Detection and distribution of equine herpesvirus 2 DNA in the central and peripheral nervous system of ponies, *J. Gen. Virol.* **78:**1115–1118.
31. Weck, K. E., Barkon, M. L., Yoo, L. I., Speck, S. H., and Virgin, H. W., 1996, Mature B cells are required for acute splenic infection but not for establishment of latency by murine gammaherpesvirus 68, *J. Virol.* **70:**6775–6780.
32. Usherwood, E. J., Ross, A. J., Allen, D. J., and Nash, A. A., 1996, Murine gammaherpesvirus-induced splenomegaly: A critical role for CD4 T cells, *J. Gen. Virol.* **77:**627–630.
33. Tripp, R. A., Hamilton-Easton, A. M., Cardin, R. C., Nguyen, P., Behm, F., Woodland, D. L., Doherty, P. C., and Blackman, M. A., 1997, Pathogenesis of an infectious mononucleosis-like disease induced by a murine γ-herpesvirus: role for a viral superantigen? *J. Exp. Med.* **185:**1641–1650.
34. Sarawar, S. R., Cardin, R. D., Brooks, J. W., Mehrpooya, M., Tripp, R. A., and Doherty, P. C., 1996, Cytokine production in the immune response to murine gammaherpesvirus 68, *J. Virol.* **70:**3264–3268.
35. Rooney, C. M., Rickinson, A. B., Moss, D. J., Lenoir, G. M., and Epstein, M. A., 1984, Paired Epstein–Barr virus-carrying lymphoma and lymphoblastoid cell lines from Burkitt's lymphoma patients: Comparative sensitivity to non-specific and to allo-specific cytotoxic responses *in vitro*, *Int. J. Cancer* **34:**339–348.
36. Rooney, C. M., Rowe, M., Wallace, L. E., and Rickinson, A. B., 1985, Epstein–Barr virus-positive Burkitt's lymphoma cells not recognised by virus-specific T cell surveillance, *Nature* **317:**629–631.
36a. Stevenson, P. G., and Doherty, P. C., 1988, Kinetic analysis of the specific host response to a murine gammaherpesvirus, *J. Virol.* **72:**943–949.
37. Usherwood, E. J., Brooks, J. W., Sarawar, S. R., Cardin, R. C., Young, W. D., Allen, D. J., Doherty, P. C., and Nash, A. A., 1997, Immunological control of murine gammaherpesvirus infection is independent of perforin, *J. Gen. Virol.* **78:**2025–2030.
38. Sarawar, S. R., Cardin, R. C., Brooks, J. W., Mehrpooya, M., Hamilton-Easton, A-M., Mo, X. Y., and Doherty, P. C., 1997, Gamma interferon is not essential for recovery from acute infection with murine gammaherpesvirus 68. *J. Virol.* **71:**3916–3921.
39. Kulkarni, A. B., Holmes, K. L., Fredrickson, T. N., Hartley, J. W., and Morse, H. C., 1997, Characteristics of a murine gammaherpesvirus infection in immunocompromised mice *in vivo*, **11:**281–292.
40. Dutia, B. M., Clarke, C. J., Allen, D. J., and Nash, A. A., 1997, Pathological changes in the spleens of gamma interferon receptor-deficient mice infected with murine gammaherpesvirus: A role for CD8 T cells, *J. Virol.* **71:**4278–4283.

41. Stewart, J. P., Janjua, N. J., Sunil-Chandra, N. P., Nash, A. A., and Arrand, J. R., 1994, Characterisation of murine gammaherpesvirus 68 glycoprotein B homolog: Similarity to Epstein–Barr virus gB (gp110), *J. Virol.* **68:**6496–6504.
42. Stewart, J. P., Janjua, N. J., Pepper, S. deV., Bennion, G., Mackett, M., Allen, T., Arrand, J. R., and Nash, A. A., 1996, Identification and characterization of murine gammaherpesvirus 68 (MHV-68) gp150: A virion membrane glycoprotein, *J. Virol.* **70:**3528–3535.
43. Mackett, M., Stewart, J. P., Pepper, S., Chee, M., Efstathiou, S., Nash, A. A., and Arrand, J. R., 1997, Genetic content and preliminary transcriptional analysis of a representative region of murine gammaherpesvirus 68, *J. Gen. Virol.* **78:**1425–1433.

7

EBV and B Cell Lymphomas

GEORGE KLEIN

The renaissance of viral oncology in the 1950s and 60s resulted from the discovery that a large number of retroviruses cause a variety of malignancies in birds, mice, cats, and other vertebrates. Many of them caused leukemias or lymphomas under experimental conditions, and a few could also be proven to do so under natural conditions. In contrast, DNA tumor viruses do not cause lymphomas or leukemias in experimental animals, as a rule. Marek's disease virus (MDV), a member of the herpesvirus family that can cause epizootic lymphomas in the chicken, is the most notable exception.

The subsequent extensive search for leukemogenic human retroviruses has been disappointing. Only one such virus was found, HTLV-1.[1] A member of the lentivirus group like HIV, HTLV-1 can cause adult T cell leukemia (ATL) and can immortalize normal T cells *in vitro*. Such lines are not immediately tumorigenic, however. Frank malignant transformation requires additional changes that have not been defined. It is believed that HTLV-1 creates a preleukemia condition, probably by inducing the potential target cells to divide and thereby expanding the target cell population at risk.[1] The likelihood of ultimate cytogenetic change increases with the number of cell divisions. Chromosomal aberrations in ATL often involve chromosome 7, but no single, consistent cytogenetic aberration has been identified.

Epstein–Barr virus (EBV), a human lymphotropic herpesvirus, is presently the best known viral contributor to the development of human lymphomas. It is a

---

GEORGE KLEIN • Microbiology and Tumor Biology Center (MTC), Karolinska Institutet, S-171 77 Stockholm, Sweden.

Herpesviruses and Immunity, edited by Medveczky *et al.* Plenum Press, New York, 1998.

highly powerful transforming agent for B lymphocytes that it converts into immortalized cell lines in vitro.[2] Morphologically, the transformed cells correspond to activated immunoblasts. They secrete immunoglobulins and a variety of cytokines, similarly to mitogen or antigen-activated immunoblasts.

The virus transforms normal resting B cells into immunoblasts *in vivo*, just as it does *in vitro*. This can be most readily observed in infectious mononucleisos (IM), a self limiting lymphorproliferative disease that can be regarded as a pathological form of primary EBV infection. In most young children and in about half of the primary infections in adolescents and in adults, there are no disease symptoms at all. The other half come down with mononucleosis, where dividing virus-carrying B blasts can be detected in the blood and in the lymphoid organs.[3,4] This proliferation and/or the reactions it provokes are responsible for the disease symptoms. The reasons for the age-related difference between silent and pathogenic infection are not known.

Only a minority of the atypical cells in the blood during IM are EBV-carrying B cells. The majority are T cells and other immune effectors that kill the virally infected B blasts with high efficiency. During the convalescent phase of mononucleosis, the EBV-positive immunoblasts disappear but the virus persists, probably largely, if not exclusively, in the small resting B cell fraction.[4,5] The reasons for the rejection of the virus-carrying immunoblasts and the persistence of the resting cells may be sought in the cell-phenotype-dependent defferences in viral expression. Infected immunoblasts express nine virally encoded proteins. Six of them are nuclear antigens, designated as EBNA1-6, and three are membrane antigens (LMP1, 2A, and 2B). (According to an alternative nomenclature, EBNA3, 4, and 6 are designated EBNA3a, b, c, and EBNA5 is called EBNA-LP.) Resting B cells express only EBNA1 and do not generate cytotoxic T cells (CTLs)[4,6–8] The difference results from the differential usage of promoters and splice programs. Because of its exclusive expression of EBNA1, the virus-harboring, resting B cell apparently remains unrecognizable to the immune system. The virus persists through life without causing any disease in the vast majority of all infected individuals.

These and other features of EBV biology make the virus and its normal and neoplastic host cells highly interesting for both molecular and biological studies. The interaction of EBV with one of its main host cells, the B lymphocyte, is best known. The virus has adapted to the B cell at several levels. It uses a B cell specific surface moiety, CD21, also known as CR2, a complement (C3d) receptor, as its receptor.[9,10] Normally, CD21 is involved in the activation of B cells by antigen-antibody-complement complexes. The virus is a polyclonal B cell activator in itself and its attachment to CD21 may only provide it with a particularly favorable point of entry that facilitates activation of the host cell.

The virally encoded growth-transformation-associated EBNA and LMP proteins participate in this activation in a cascadelike fashion that has not been clarified in detail. It is clear, however, that the expression of EBNA2 is an essential prerequisite for activation. LMP1, the main membrane antigen, prevents apoptotic death, a frequent fate of B lymphocytes, probably because it activates the apoptosis-antagonizing bcl-2 gene.[11] The virally activated immunoblasts produce a variety of cytokines, including B cell growth factors, and express corresponding receptors, such as CD23.[12] An autocrine loop is generated that may stimulate the proliferation of the virally infected B cells *in vivo,* as it certainly does *in vitro.* In consequence, the number of virally infected cells may reach a sufficiently high level, before the onset of the rejection response, to secure the persistence of the virus and spreading it to new hosts.

Eight of the nine virally encoded proteins expressed in the transformed immunoblasts are competent to induce a rejection response. The predominant rejection target is determined by the HLA constitution of the host.[7] In immunocompetent persons, the response is very powerful. It clears away the virally infected blasts within a few days. Persons with congenital or acquired immunodeficiencies may succumb to progressive lymphoproliferative disease, however.

In spite of the efficient immune response, the virus persists in resting B cells. It probably does so by switching to the "EBNA1 only" program. EBV-carrying resting B cells do not expand and are probably long-lived. It may be surmised that their activation and subsequent immunoblastic proliferation are prevented by the switch-on of the highly immunogenic EBNAs. It is not known how the latently infected cells that express only EBNA1 manage to maintain themselves at an apparently constant level throughout life.[13]

How did the virtually watertight surveillence that prevents the proliferation of the virally transformed blasts evolve? This must be seen in relation to the fact that closely related viruses are present in all Old World, but not in New World primates. In immunologically naive New World monkeys, the virus can cause fatal lymphoproliferative disease, quite as it does in immunodefective humans.[2] It may be surmised that humans and other Old World primates have reached the present state of an essentially nonpathogenic coexistence with this potentially very dangerous virus family only after having been selected for efficient recognition of the virus-carrying immunoblasts. Our HLA class I allotype spectrum is obviously highly competent to deal with the problem. Analyses performed at different laboratories have shown that different responders target their cytotoxic T cells against peptides derived from eight of nine virally encoded, transformation-associated proteins, with the notable exception of EBNA1 (for a review see Ref.[7]). The precise choice of the target depends on the immunodominant HLA class I restriction element. The total MHC equipment of our species, combined with the eight virally encoded target proteins, apparently provides full protection against unlimited proliferation of the virally transformed immunoblasts in virtually all normal immunocompetent individuals.

1. EBV-ASSOCIATED PROLIFERATIVE DISEASES

1.1. EBV and Human Malignancy

It follows from what has been said about the stable, nonpathogenic equilibrium between EBV and the human host that EBV-induced or EBV-associated disease arises only as a biological accident. The nature of this accident is fairly well known for some but not for all EBV-associated neoplastic diseases.

1.1.1. Infectious Mononucleosis

This disease is caused by primary viral infection above the age of about 10, as already mentioned. The reasons for the age-related difference in viral pathogenicity is not yet known. It may be noted, however, that the symptom-free, early childhood infection is the rule in all developing countries and in low socioeconomic groups throughout the world.[14] The virus is believed to spread largely, if not exclusively, through the saliva. Postponement of the regular early childhood infection by good hygienic conditions thus is one "accident" that may disrupt the normal "rejection-geared" immune response that prevents proliferative disease in young children and also in half of adolescents and adults. The IM syndrome may be viewed as a delayed and, initially at least, quite disorganized rejection reaction that permits a considerable degree of immunoblast proliferation, perhaps because of the remarkably strong T cell suppression that precedes it.[15] Nevertherless, the rejection reaction is unerringly efficient, unless the host is immunocompromised. This is the next accident.

1.1.2. Lymphoproliferative Disease in Immunodefectives

Congenital, iatrogenic (posttransplant) and infectious (AIDS) immunideficiencies may be accompanied by the fatal proliferation of EBV-transformed immunoblasts.[16,17] It starts as polyclonal growth, as in mononucleosis, but it fails to regress. In the course of time, it may turn into oligo- and later monoclonal growth.[18,19] The cells maintain their immunoblastic phenotype in the course of this progression. They resemble the lymphoblastoid cell lines *in vitro* and express the full set of EBNAs and LMPs accordingly.[20,21] Apparently, the cells proliferate because the host response has broken down. Initially at least, no cellular change is required over and above the EBV-induced transformation.

At least 90% of the posttransplant lymphoproliferative diseases (PT-LPD) carry EBV.[22] On the basis of their immunoglobulin rearrangement patterns, the lesions are subdivided into polymorphic (oligo- or polyclonal) and monomorphic (monoclonal). The polymorphic group is often associated with polyclonal EBV infection. As discussed in a separate section below, part of the HIV-infected group but not other groups, consists of monoclonal lymphomas that have a rearranged c-myc gene, caused by an Ig/myc translocation that corresponds to sporadic Burkitt lymphoma in its details. About 70% of this group are EBV-negative. For conceptual and clinical reasons it is important to distinguish between lymphomas with and without myc-rearrangement in the HIV-infected group.

The occurrence and clinical features of PT-LPD have been summarized by Nalesnik *et al.*[23] Its EBV aspects have been reviewed by Ho *et al.*[24] In the Pittsburgh-Denver organ transplant series, the total detection rate was 1.7%. Kidney recipients had the lowest (1.0%), and heart/lung recipients the highest (4.6%) frequency of PT-LPD. Bone marrow transplant cases were not included. The main clinical categories included a mononucleosis-like syndrome, gastrointestinal/abdominal disease, and solid organ disease. In cyclosporin A-(CSA) treated patients, the median time of onset was 4.4 months after transplantation, regardless of tumor clonality. The histological spectrum ranged from polymorphic to monomorphic. Polymorphic lesions could be either clonal or nonclonal, whereas monomorphic lesions were clonal. Clonal and nonclonal lesions were about equally frequent. All nonclonal and about half of the clonal lesions responded to reduced immunosuppression and supportive surgery.

Patients who received other forms of immunosuppression developed lymphoproliferative disease after significantly longer intervals.[16]

Patients with preferentially localized symptomology show head, neck, gastrointestinal, and other organ involvement. The allografted organ is also frequently affected. The most famous case of multiple PT-LPD presentation is the "bubble boy" who had concurrent clonal and nonclonal tumors, including more than 20 lesions with polymorphic lymphoid tumor histology.[25] Some of the phenotypically polymorphic proliferations were genotypically oligoclonal, and one had a monoclonal component. This is consistent with a progression series where monoclonal tumors evolve from a polyclonal background. It is noteworthy in this context that clonal B cell infiltrates are also observed in fatal mononucleosis,[23] although the proliferation associated with mononucleosis is generally polyclonal. As Nalesnik points out, it is often difficult to clearly separate malignant from nonmalignant PT-LPD. Many lesions that would be self-limiting in an immunocompetent host may act malignantly in immunosuppressed hosts. Monoclonal tumors capable of invasion and distant metastasis are considered malignant, regardless whether they regress or not. The concepts of malignancy and progressive disease must be obviously dissociated in this system. This is not surprising in view of the highly immunogenic nature of EBV-transformed cells, the potentiality of tumor progression by genetic changes in the EBV-generated polyclonal lymphoproliferative lesion, and the counteracting influence of the host response that may, depending on the EBV-expression phenotype of the tumor cell, act more or less efficiently. The three independent variables, progression of the tumor by genetic changes, the EBV expression phenotype that strongly influences the potential immunogenicity of the cells, and the pharmacologically modulated host response can obviously produce a large variety of clinical phenotypes.

The precarious balance between progression and regression and the potential strength of the immune response, unparallelled in other tumor systems, is well brought out by the early experience that kidney transplant recipients, who developed PT-LPD, may reject their immunoblastomas, even if they are of recipient origin, if immunosuppression is lifted. Following rejection of the kidney and return to dialysis, a second kidney may be grafted. Renewed immunosuppression does not lead to recurrence of the immunoblastoma.

Hickey *et al* has studied a large amount of data that is relevant.[26] Renal transplant cases who lost their allografts during the management of a previous PT-LPD were drawn from a total of 1226 transplant cases managed at the University of Pittsburgh. All had been treated with CSA. Fourteen patients (1.1%) developed PT-LPD during a 7-year period. Following the reduction of immunosuppression, the immunoblastomas regressed. Six of the 14 patients have retained their allografts. Three were retransplanted. Two died. All retransplanted patients were treated with CSA without adverse effects. Two of the three second transplants were still funcitoning 18–26 months after retransplantation.

In this study, the simple expedient of stopping or reducing immunosupression in CSA treated and prednisone-treated patients has led to prompt and apparently permanent resolution of the lymphoroliferative disease, provided that infection and other complications could be controlled. Monoclonal and polyclonal immunoblastomas regressed equally easily.

Special risk factors that promote the development of PT-LPD in bone marrow transplant recipients include T cell depletion of the graft, HLA mismatching between donor and recipient, and treatment of acute GVHD with T cell antibodies.[27,28]

Spectacular results were achieved by infusing T cells from the bone marrow donor in the recipient that developed donor type LPD. Papadopoulos *et al*,[29] infused donor leucocytes in five bone-marrow transplant recipients who developed EBV-carrying LPD after allogeneic BMT. Unirradiated donor leucocytes were infused with doses calculated to provide approximately 10^6 $CD3^+$ T cells per kilogram body weight. Each of the four immunoblastoma specimens that could be evaluated were of donor cell origin. EBV-DNA was detected by PCR in all five samples. In all five patients, complete, clinically and histologically documented regression occurred within 8–21 days after infusion. Clinical remissions were achieved 14–30 days after the infusions and were sustained without further therapy in the three surviving patients for 10, 16, and 16 months.[30]

1.2. Burkitt's Lymphoma

In contrast to the immunoblastomas that arise in immunodefective hosts, high endemic and sporadic Burkitt lymphoma (BL) develop in immunocompetent individuals. Its progressive growth is not caused by the breakdown of immune surveillance but by the escape of the BL cell from immune rejection. It is not driven by the same growth-transformation-associated EBV encoded proteins as the immunoblast, although this must be said with the reservation that EBNA1, the only virally encoded protein expressed by BL cells, induces B cell lymphomas in transgenic mice.[31] The main driving force of the BL cell is undoubtedly provided by the c-myc gene that is activated by its translocation to one of the three Ig loci.[32] The chromosomal translocation accident generates a permanent stimulus for cell proliferation because of the constitutive activation of c-myc. Moreover, it positions the BL cell in a phenotypic window where it is prone to escape rejection because of the down-regulation of the potentially immunogenic EBNA2-6 and LMP proteins, low expression of cellular adhesion molecules, and a relative deficiency of MHC class I antigen expression.

Only 97% of the highly endemic BLs prevalent in the rain forest areas of Africa and New Guinea and not more than 20% of the sporadic BLs that occur at lower frequency all over the world carry EBV (for a review, see Ref.[33]). One of the three alternative forms of the If/myc translocation, 8;14 (myc/IgH), 2;8 (kappa/myc), and 8;22 (myc/lambda), is regularly present in all BLs, whether EBV-positive or negative (for a review, see Ref.[32]) The subordination of c-myc to one of the constitutively active immunoglobulin regions interferes with the normal regulability of the gene. As a result, the cells are prevented from leaving the cycling compartment.

Thus, the translocation, rather than EBV, must be considered the main rate-limiting event in the development of BL. EBV may increase the probability of this event by expanding the target cell population at risk, prolonging the life span of the target cells, and increasing the number of their divisions. The risk of genetic accidents is proportional to the number of consecutive mitoses.

1.2.1. The BL Cell Phenotype

In 1975 Nilsson and Pontén already found that BL-derived cell lines differ from EBV-transformed LCLs of nonneoplastic origin in several respects.[34] The ultrastructure and surface glycoprotein pattern of the LCLs resembled mitogen-transformed immunoblasts, whereas BL cells were more like resting B cells. Later, surface marker analysis confirmed and further emphasized these differences (for review see Ref.[34]). BL cells express CD10 (CALLA) and CD77 (BLA), but they do not express the numerous "activation markers" and adhesion molecules associated with mitogen-induced blast transformation.[35] LCLs showed the opposite pattern. The difference in the expression of the adhesion molecules explains why LCLs grow as large clusters, whereas BLs proliferate as single-cell suspensions.

The phenotype of the BL tumor cells *in vivo* is faithfully reproduced within the freshly established cell lines, as a rule. They are designated type I or group I

BL lines. In the course of serial propagation *in vitro,* most EBV-carrying but not EBV-negative BL lines "drift" phenotypically, i.e., convert spontaneously into more "LCL-like" (type III) cultures.[36] Their drift is reflected by the appearance of activation markers and adhesion molecules and the gradual disappearance of the BL type I associated CD10 and CD77 markers. Concurrently, the EBV expression of the cells changes dramatically. Type I BL cells express EBNA1 only. Type II/III BLs express all six EBNAs and the LMP1-2 membrane antigens.

The differential EBNA expression of the BL cells can be related to the alternative use of promoters and splicing programs. In EBV-transformed normal B cells and in BL type II/III cells, one of two alternative promoters (designated as Wp and Cp, respectively), is used to generate a giant, approximately 85-kb long message, out of which all six EBNAs are spliced.[37] The W/C promoter system is inactive in type I BL cells and in all other EBV-carrying cells that do not have an immunoblastic phenotype. An alternative promoter, located in the Q-region, is used. It generates a short message that encodes only EBNA1.[38] Promoter usage and the corresponding antigen expression can be shifted experimentally in both directions.[39] BL type I cells that express only EBNA1 can switch on the W/C program either in connection with the phenotypic drift mentioned above or upon exposure to 5-azacytidine.[40] The latter finding indicates that DNA methylation is involved in down-regulating the missing proteins, as also shown by more direct experimental evidence.[41,42]

W/C usage can be switched to the Q-program by fusing LCLs with non-B cells, whereupon the expression of EBNA2-6 is eclipsed, and the exclusive EBNA1 program is switched on.[43]

We and others have suggested that the differential usage of alternative viral programs in immunoblasts and in BL type I cells reflects the viral life cycle in corresponding normal cells. Although the lymphoproliferative diseases of the immunodefectives are caused by progressive growth of virally transformed immunoblasts, the type I BL cell represents the neoplastic counterpart of latently infected resting B-cells. The latter provide one, if not the only reservoir of the persisting virus in normal seropositive individuals, as already discussed. We have shown that such cells express only EBNA1 and LMP2.[4] Our group has also shown[8] that the long glycine-alanine repeat of EBNA1 prevents appropriate processing of the protein, a prerequisite for the transport of the derived peptides to the MHC class I molecules on the surface. This is probably one reason for the nonrejectability of the BL cell in immunocompetent hosts. There are also two other reasons: the allele-specific HLA class I antigen suppression, characteristic of BL type I cells,[44] and the deficient expression of adhesion molecules.[35] Together with the down-regulation of the immunogenic EBV proteins, these phenotypic traits may all stem from the fact that the Ig/myc translocation "fixes" the cellular phenotype in a state where it resembles resting, rather than dividing cells. It is a peculiar coincidence that the two most important features of the EBV-carrying

BL cell, the presence of the latent virus in a down-regulated state, where it expresses only EBNA1, and the proliferative driving force of the illegitimate Ig/myc translocation have the resting B cell as their common target. The former is due to the fact that the resting B cells serve as the main, if not the only reservoir of latently persisting virus, whereas the Ig/myc translocation generates monoclonal neoplastic growth by preventing the precursor of the resting (virgin or memory) B cell from leaving the cycling compartment.

1.2.2. *Etiological Considerations*

Among the two, potentially important etiological factors so far considered, EBV and the Ig/myc translocation, the former serves obviously in a contributory capacity, particularly in the highly endemic forms of the lymphoma, whereas the latter is an essential, rate-limiting facctor, as indicated by its virtually 100% association with the tumor.

Among the environmental factors that can be related to the highly enedemic form, chronic hyper- or holoendemic malaria occupies a dominant position, for geographical and climatic reasons.[45] It may act by stimulating chronic cell proliferation within the target cell population at risk. Prolonged stimulation of cell division is also involved in the development of the rodent Ig/myc translocation-carrying tumors.[32] Chronic local granuloma formation and high genetic susceptibility are the two most prominent requirements for inducing mouse plasmacytoma. Hypersentitization to helminths in the ileocoecal region may play a similar role in the induction of Louvain rat immunocytoma. Chronic proliferation involves a long series of cell divisions in the tumor precursor cells. The probability of genetic accidents increases with the number of divisions. The development of autonomously growing tumors is promoted when the translocation accident happens to juxtapose the coding exons of c-myc and immunoglobulin sequences, as already discussed. Additional genetic changes are probably required before the tumor clone reaches full autonomy.

The "molecular anatomy" of the Ig/myc translocation shows certain differences between endemic and sporadic BL (for a review, see Magrath[33]). In the highly endemic, largely EBV-associated form, chromosome 8 breaks usually upstreams of the intact c-myc gene. In the largely EBV-negative sporadic tumors, the break occurs usually within the first intron, the first exon, or immediately upstream of the gene. There are also certain differences in the immunoglobulin sequences involved. In the sporadic form, a switch region within the IgH locus, usually Smu, is the most common receiving site. The breaks are more variable in the highly endemic form. They are higher upstream, as a rule, in a J or a V region. This implies that the distance between the myc and the Ig sequences is often greater in the EBV-carrying than in the EBV-negative form. Conceivably, activating and activated sequences may have to be closer to each other in the absence of EBV. This would be understandable if EBV contributed positively to the constitutive activation of myc. The myc breakpoint difference between endemic and sporadic tumors implies, furthermore, that the gene is transcribed from its own promoters in the highly endemic tumors whereas in the sporadic tumors transcription starts at a cryptic promoter within the first intron. These differences between the highly endemic and the sporadic tumors with regard to breakpoint location, together with several differences in primary localization, bone marrow involvement, and other clinical features, suggest that the precursor cells of the two tumor forms may represent different stages of maturity, perhaps because different etiological factors are involved.[33]

The accidental nature of the translocation is confirmed by the fact that the c-myc-carrying chromosome can break at many different points, upstream or within the c-myc gene in the typical (IgH-myc) translocation, or downstream of the gene in the variant translocations that involve the light-chain genes. No breakpoint cluster, fragile sites, or hot spots are involved. This indicates that the break occurs at random. The immunoglobulin locus-carrying chromosomes break at more specific points, corrsponding to sites that serve as targets for the recombinases involved in physiological Ig-rearrangement.

Although the Ig/myc juxtaposition is thus a common molecular feature of all BLs, irrespective of geographical origin or EBV-carrying status, it is probably not the only change. Evidence from immunoglobulin-enhancer c-myc-construct-carrying transgenic mice[46] and examination of preneoplastic tissues during the development of mouse plasmacytoma[47] concur in suggesting that constitutive activation of c-myc by its juxtaposition to immunoglobulin suquences is not sufficient to cause a lymphoma by itself. Secondary changes may include point mutations in the c-myc locus itself, affecting some probably critical phosphorylation sites.[48,49] Additional oncogenic activation and/or suppressor gene losses may also be involved. Mutation of p53 may be one such event.[51,52] Further changes, including a deletion of 6q and mutation of K-ras have been found in HIV-infected NHL lymphoma patients [22] (see section 1.3).

Highly endemic BLs usually present as well-circumscribed solid tumors, wherease sporadic BLs may be more generalized, involving lymph nodes and bone marrow. African BLs appear often as jaw tumors around the age of dental development, gonadal tumors in prepubertal children, and long bone tumors in adolescents. Bilateral breast tumors were found in a lactating woman.[53] This indicates that local growth factors may play a promoting role.

1.2.3. Immunological Features

BL cells are relatively or completely resistant to CTL-mediated killing, in comparison with LCLs derived from the same patient.[54] The phenotypic features responsible for this have been already discussed. They are also less able to process antigen,[55,56] and less effective in stimulating allogeneic T cells.[57] The reduced expression of adhesion molecules[58] is consistent with the resemblance between BL cells and resting B cells. This may also contribute to CTL resistance, together with the down-regulation of certain HLA class I antigens.[59]

These features of the EBV-carrying BL cell provide an adequate explanation for its ability to grow in immunocompetent individuals, in contrast to the EBV-carrying immunoblast. The fact that EBV-carrying BLs are more localized than their EBV-negative counterparts and more curable by chemotherapy, nevertheless, indicate that the specific immune response of the patient can affect the former more than the latter.

1.3. Non-Hodgkin's Lymphoma (NHL) in HIV-Infected Patients

AIDS-related NHL is 60 times more frequent than NHL in the general population.[60] Seventy percent of the tumors are high-grade NHL. They are almost exclusively B cell-derived. They belong to a variety of types, such as small noncleaved cell lymphoma (SNCCL), large cell immunoblastic plasmacytoid lymphoma (LC-IBPL), and large noncleaved cell lymphoma (LNCCL), as reviewed by Ballerini *et al.*[61] In the data reviewed, 42% carried EBV, 29% had c-myc rearrangement, 15% had a ras-mutation, and 37% had lost or carried mutated p53. The frequency of each of these changes varies with the type of AIDS-NHL. All LC-IBPLs carried EBV and were similar to, if not identical with, the immunoblastic lymphomas seen in congenital and iatrogenic (posttransplant) T cell deficiencies. It may be surmised that the immunoblasts proliferate as a result of immune breakdown in all these cases. Only about 30–40% of the SNCCLs carry EBV, however. Virtually all of them carry an Ig/myc translocation. Unlike the sharp association of the Ig/myc translocation with both highly endemic and sporadic BL, the translocations carried by the HIV-associated lymphomas are present in a variety of histological types. Many of them would not be normally classified as BL by a pathologist, but the presence of the Ig/myc translocation indicates that they all originate from a similar oncogenic activation. Conceivably, they may diversify phenotypically as a result of the T cell immunodeficiency.

According to the review of Gaidano and Della Favera,[22] EBV is present in 30–40% of the HIV-associated SNCCL and LNCCL and in 100% of systemic LC-IBPL and CNS-NHL.

Thus, EBV-carrying and/or c-myc-rearranged B-cells can undergo oligoclonal expansion that may give rise to monclonal neoplasia in the context of AIDS.[62] A positive correlation has been found between the appearance of detectable EBV-carrying B cell clones in patients with HIV-associated, persistent generalized lymphadenopathy (PGL) and the later development of EBV-carrying NHL.[63] The clonal expansion was preceded by EBV infection, as shown by EBV-termini analysis.[61,64]

EBV infection is a universal feature of systemic AIDS-LC-IBPL, as already mentioned. A subgroup in this category originates in the central nervous system and was found to carry EBV in 100% of the cases.[65,66] This is reminiscent of our early finding, showing that EBV-transformed LCLs of normal origin do not grow subcutanously in nude mice, in contrast to BL cells, but grow progressively in the nude mouse brain.[67,68]

The EBV-carrying status of 128 AIDS-related lymphomas (ARLs) has been also surveyed by Hamilition-Dutoit *et al.*[69] Presence of EBV in the tumor cells was detected by *in situ* staining for the virally encoded small RNA, EBER1. EBV was present in 85 of the 128 lymphomas (66%), but the frequency of virus-carrying tumors differed according to the histological type. EBER1 expression was detected in all 11 HIV postivie Hodgkin's disease cases (100%), in 15 of 16 NHLs originating in the central nervous system (94%), in 46 to 60 immunoblastomas (77%), but only in 12 of 35 (34%) of the Burkitt type (SNCL) lymphomas. One of six diffuse, noncleaved, large-cell lymphomas also carried the virus. EBV-carrying immunoblastoma patients showed more severe immunosuppression, as indicated by the level of $CD4^+$ cells, than patients with BLs.[70]

The authors concluded that all AIDS-related HD cases, virtually all primary CNS-derived ARLs and most immunoblastomas carry EBV, but only a minority of BLs and monomorphic centroblastic lymphomas are EBV-positive. Similar conclusions were reached by other authors; see Hamilton-Dutoit (1993) for references.[69,71] It is noteworthy that primary CNS lymphomas that arise in immunocompetent, HIV-negative persons only infrequently carry EBV.[72] The presence of EBV in virtually all tumor cells in the EBV positive ARLs indicated that EBV infection precedes clonal expansion. This suggests that it plays a pathogenic role in EBV-positive tumors.

Hamiton–Dutoit *et al* have studied the expression of EBV-genes in relation to the tumor cell phenotype in AIDS-associated NHLs.[73] Forty-nine cases were examined by Southern blotting, in situ hybridization, and immunophenotyping. The choice of the viral program by the EBV-carrying cells was assessed by meauring the expression of EBNA2 and LMP1. As already discussed in the section on EBV biology, immunoblastic cells express both proteins within their "latency III" program. BLs express neither (latency I), whereas cells with a non-B phenotype express LMP1 but not EBNA2 (latency II). All three patterns were found in the 22 EBV-positive, immunoblast-rich, large-cell lymphomas. EBNA2 and LMP1 were expressed in nine lymphomas, corresponding to the type III latency of the *in vitro* LCLs. They carried activation markers, as expected. Extranodal lymphomas belonged to this category. A BL-like, $EBER^+$, $EBNA2^-$, $LMP1^-$ type I latency was found in six lymphomas. The seven remaining lymphomas were EBER1 positive, EBNA2 negative, but LMP-1 positive, as in latency II. The latter category is particularly interesting because latency II was previously seen only in EBV-carrying tumors of non-B cell origin, such as Hodgkin's disease, T cell lymphomas, and nasopharyngeal carcinomas.

Thus it appears that the ARLs belong to at least three major, pathogenetically distinct groups. The EBV-carrying immunoblastomas are phenotypically similar to the proliferative lesions that arise in organ transplant recipients. They arise because of the deficiency of the immune surveillance mechanisms that keep EBV-infected B cells under control in normal EBV-seropositive persons but can evolve towards oligo-and monoclonality by further cellular changes. The BLs and other Ig/myc-translocation-carrying lymphomas correspond phenotypically to the sporadic, largely EBV-negative BLs. Only one-third of them carry EBV. The latter express only EBNA1, according to the type I latency model, characteristic of highly endemic BLs (for a review, see Ref.[33]). There is also a group of Hodgkins lymphomas. They carry EBV in 100% of cases, unlike their non-AIDS-associated counterparts that are only 50% EBV-positive (see section 1.4).

The LCL-like type III latency is a general characteristic of all immunoblastomas, also observed in transplant recipients, as already mentioned. It is also found in the EBV-induced lymphoproliferative lesions in tamarins[74] and in LCL lines or EBV-carrying normal B cells that grow progressively in SCID mice.[75]

In contrast to the immunoblastomas that are believed to result from immunosuppression and deficient surveillance, the myc/Ig translocation-carrying Burkitt or Burkitt-like tumors in HIV-infected patients have a more complex pathogenesis. It is basically comparable to what has been discussed above for highly endemic Burkitt lymphomas with only some minor differences. It is noteworthy that BL and other Ig/myc-translocation-carrying lymphomas are not observed in transplant recipients or in congenital T cell deficiencies. Chronic stimulation of cell division in the B cell population at risk must be an essential requirement for facilitating the translocation. This may be provided by HIV itself, and/or by the many AIDS-associated opportunistic infections. HIV-associated lymphadenopathy is known to involve chronic stimulation of the B cell population, leading to high proliferative activity and hypergammaglobulinemia. This may predispose the cell to translocations and other genetic accidents. Activated cells may also act by increased cytokine production, providing autocrine and paracrine stimulation for the target cells at risk. T cells, monocytes, bone marrow, and stromal cells are known to respond to HIV infection by increased cytokine production. The B cells of HIV-infected persons secrete TNF alpha and IL6 without *in vitro* stimulation. They may promote the expansion of oligoclonal B cell populations.

1.4. Hodgkin's Disease (HD)

The idea that EBV may play an important role in the genesis of BL and also NPC came originally from serological studies. The interpretation of the frequently elevated EBV antibody titers in HD patients remained ambiguous, in contrast to the viral association of high EBV titers with African BL and NPC of any geographic origin that were exclusively found in EBV-seropositive persons.

The small EBV-seronegative subpopulation was uniquely absent in BL and NPC, in contrast to all other groups, and most patients had high EBV titers. Detection of viral genomes in the tumor cells of BL and NPC has confirmed the postulated association (for a review, see Ref.[76]).

EBV seropositivity in HD was not equally regular. HD patients were more frequently seropositive than matched controls, and their mean antibody titers were higher, but a substantial minority of the patients was seronegative. Therefore, the titer elevations were written off as probable secondary consequences of the HD-associated immune defects. It was known that the antiviral (EA and VCA) antibody titers increase in T cell deficiencies.

The detection of EBV at the cellular level by immunohistological techniques brought the possible role of EBV in HD to the forefront of interest after a dormancy of two decades (for a review, see Ref.[77]). Improved techniques of *in situ* hybridization and the use of probes for the highly transcribed EBV-encoded small RNA (EBERs) cells opened the field for detection and identification of virally infected cells in complex tissues.

It was found that mixed cellularity (MC) and nodular sclerosis (NS) were the most frequent EBV-positive forms of HD, in that order.[77] Lymphocyte predominance (LP) is largely EBV-negative in line with earlier serological findings.[78]

The detection of EBV DNA in HD tissues has provoked much discussion. Are the viral genomes harbored by the neoplastic cells? If so, are all of them positive in a given tumor or only some? Are they also present in reactive normal cells?

Localization of EBV-DNA to Reed-Sternberg cells was shown by a large number of authors.[79–87] The episomal EBV in HD is clonal, as indicated by the terminal repeat hybridization method. This supports the idea that the virus may play an etiological role in the EBV-positive HDs.[86–92]

The variable appearance and the complexity of HD, its approximately 50% overall association with the virus, and the differences in the positivity of the morphologic subtypes create problems. The immunohistochemical distinction between EBV-carrying HD cells and normal lymphocytes, using a highly sensitive, single-stranded EBER antisense probe, was an important improvement.[77] In contrast to the uniform EBER expression of all neoplatic cells in the EBV-harboring HD cases, only a small number of normal B cells were positive. They were also present in EBV-negative HD lesions. These distinctions have brought the possible role of the virus in the EBV-positive HD lesions into much sharper focus but that does not mean that the relationship has been explained.

Mutatis mutandis, the partial EBV positivity of the HLs is not entirely unlike the much better known relationship between EBV and BL. The regular association of EBV with highly endemic BL indicates that the virus must play some role in the etiology of the tumor, perhaps by expanding and/or immortalizing a subset of Ig/myc translocation-prone B cells, as already discussed. In the absence of any information on the steps involved in generating HD, many questions remain enigmatic. If the neoplastic HL cells belong to the B cells series, why does their viral antigen expression (EBNA2- LMP+)[93] resemble the epithelial cells of NPC, rather than BL cells (EBNA1 only) or the EBV-carrying immunoblastomas that arise in immunodefectives (EBNA1-6 and LMP)?

Critical studies excluded the possible presence of EBNA2 in HL cells.[83,93–97]

The overall frequency of EBV-positive HD and the subtype distribution (more MC than NS) of Chinese EBV-positive HD was similar to what was found in developing countries.[98] Immunodeficiency-associated HD cases are rare.[80,82,89,91,99–101]

Is there any relationship between infectious mononucleosis (IM) and HL? Such a relationship has been postulated on the basis of certain epidemiologic similarities, the preferential apperance of both diseases in high socioeconomic groups, and the documented appearance of HD in the wake of IM. Cohort studies have shown 2–4 times increased risk of HD after IM, and the greatest risk is within 3 years after IM.[102,103]

Does the delayed EBV infection of hygienically protected subpopulations increase the risk of HD similarly to the way it increases the likelihood of clinically manifest mononucleosis, or is this relatively slight epidemiological similarity of the two diseases mainly coincidental, perhaps related to the EBV-like socioepidemiology of some unknown infectious agent in HD? Comparative epidemiological studies on EBV-carrying and EBV-negative HD cases would be of great interest in this context.

Because the EBV-encoded LMP1 protein is potentially immunogenic and is regularly expressed by EBV-carrying cases of HD, the question arises how such cells escape immune surveillance. The findings of Knecht *et al.*[104] may give a clue. In five cases of EBV-carrying HD, deletions were found near the 3′ end of the LMP1 gene. Similar deletions were seen in nasopharyngeal carcinoma. Conceivably, these deletions may impair the immunogenicity of the LMP1 protein.

1.5. EBV and T Cell Lymphomas

The unexpected association of EBV with some T cell lymphomas was initiated by Jones *et al.*[105] who found EBV-DNA by Southern blotting and by *in situ* hybridization in three cases. Subsequently, EBV genomes were detected in a certain proportion of human postthymic T cell malignancies and related lymphoproliferative disorders.[73,88,98,106–111] Surprisingly, the virus was more frequently associated with T than with B cell lymphomas in immunocompetent patients with NHL. The EBV-carrying neoplastic T cells belonged to a range of immunophenotypically different cell types, indicating that functionally distinct subpopulations of neoplastic T cells are involved. T cell lymphomas arising in the upper aerodigestive tract are particularly frequent carriers of EBV. They include sinonasal PTL, lethal midline granuloma, lymphomatoid granulomatosis, and angiocentric lymphoproliferative lesions.[112] According to Pallesen,[113] EBV is often particularly associated with AILP, pleomorphic medium and large cell type, and Lennert's lymphoma. See also a review by Ambinder and Mann.[114]

De Bruin *et al.*[115] showed that the association of EBV with T cell lymphomas differs, depending on site of presentation. Nasal T cell tumors are invariably EBV-positive, and those of the skin predominantly negative. Sato *et al.*[116] found that ATL leukemia/lymphoma carried EBV DNA in 21 cases according to PCR. EBER hybridization confirmed this in 16 cases. De Bruin *et al.*[117] studied 46 nodal T cell lymphomas for EBV-DNA by PCR, EBER 1 and 2, and LMP1 by *in situ* hybridization and immunohistochemistry. Twenty-one of 45 had EBV DNA by PCR. Eight of 15 cases showed clusters of EBER-positive cells. Therefore these lymphomas were considered strongly EBV-associated.

Kumar *et al.*[118] reported an EBV associated T cell immunoblastic lymphoma that occurred in a renal transplant recipient after seven years. Tein *et al.*[119] found chromosomal anomalies in EBV-carrying T cell lymphomas, particularly affecting chromosome 7, as in other T cell lymphomas.

Su *et al.*[111] found that 10 of 35 subcutaneous T cell lymphomas were EBV-associated. Three distinct clinical pathological subgroups could be recognized. The most consistent EBV association was seen with angiocentric T cell lymphoma or lymphomatoid granulomatosis. EBV-associated T cell lymphomas were quite resistant to therapy and had poor prognoses.

Monoclonal proliferation of cytotoxic T-cells containing the EBV-genome was also seen in a young child with sporadic fatal infectious mononucleosis.[120]

The pattern of latent gene expression in EBV-carrying T cell lymphomas corresponded to latency II with EBNA1 and LMP1, but without EBNA2-6.

Most of the EBV-positive T lymphomas contained monoclonal EBV-episomes, indicating that EBV infection has occcurred before the clonal expansion of the lymphoma. This is consistent with the possibility that EBV plays a direct or indirect pathogenic role.

T-cell lymphomas are rare in both AIDS patients and in organ recipients. There is no strong evidence to suggest that defects of EBV immunity are involved in their genesis.

2. CONCLUSIONS

EBV is a highly transforming virus. For human B cells, its transforming (immortalizing) ability is superior to any other known transforming virus. In spite of this, it lives in a virtually nonpathogenic equilibrium with a normal human host.

Following primary infection, the virus transforms a fraction of the B cell population into proliferating immunoblasts. Similarly to their *in vitro* transformed counterparts, they express nine viral proteins. Several of them are required to stimulate B cell proliferation. Eight of the nine proteins are potentially immunogenic and can generate powerful cytotoxic T cells (CTLs). They are targeted against peptides derived from one or several of these proteins. The choice of the target depends on the HLA class I consitution of the host. Primary infection usually goes unnoticed in young children. If infection is delayed to adolescence or to adulthood by modern hygienic conditions, it leads to mononucleosis in about half of the cases. Rejection may fail in immunodeficient hosts whereupon the proliferation of B blasts may take a lethal course. EBV-carrying blasts may also grow progressively in immunodeficient states of congenital (e.g., XLP), iatrogenic (e.g., organ transplant recipients), and infectious (HIV) origin. Immunoblastic disease usually starts as a polyclonal proliferation but may progress to oligo- and then to monoclonal disease by as yet undefined events.

The origin of EBV-carrying Burkitt lymphoma (BL) is fundamentally different. It occurs in immunocompetent individuals and is monoclonal from its inception. BL cells do not resemble immunoblasts but are more like resting B cells in ultrastructructure and marker equipment. They invariably carry a chromosomal translocation that has juxtaposed the c-myc prontooncogene to one of the three immunoglobulin loci. Although the precise relationship and the relative timing of EBV infection and chromosomal translocation has not been defined in relation to the developmental history of the BL precursor cells, both the translocation and the EBV latency can be seen in relation to a scenario where the CTL-mediated rejection of the virally transformed immunoblasts is followed by silent persistence of the virus in the resting B cell fraction. There, it expresses only EBNA1, required for viral episome maintenance, but no other EBNAs nor LMP1.[4] EBNA1 is apparently unable to induce a CTL response. As long as the cell maintains a resting phenotype, akin to a virgin or memory B cell, it is not recognized by the host rejection response. Following blast transformation, it expresses the highly immunogenic EBNA2-6 and LMP proteins and is promptly rejected. Such periodic activation may explain the maintenance of a sensitized state toward the growth-transformation-associated proteins in healthy seropositive persons.

It appears likely that most of the highly endemic BLs originate in EBV-carrying cells where a chromosomal translocation has occurred during the immunoglobulin rearrangement process, accidentally juxtaposing c-myc to one of the three Ig loci. This leads to the deregulated, constitutive expression of c-myc that prevents the cell from leaving the cycling compartment. A phenotypically resting B cell that is not resting has a number of properties that prevent immune elimination. In addition to the previously mentioned eclipse of EBNA2-6 and LMP, adhesion molecules and certain MHC class I antigens are also down-regulated.

In addition, BL cells process antigen more poorly than immunoblasts. All of this contributes to the "cellular escape" of BL.

This scenario and the added fact that some of the highly endemic and the majority of the sporadic BLs are EBV-negative, raises questions about the way in which EBV may contribute to the etiology of the lymphoma. The fact that EBV-carrying B cells represent only a small minority of the total B cell population even in high EBV antibody titered persons who live in the endemic region and the 97% EBV positivity of the African BLs must mean, *ipso facto,* that the probability of the lymphomatous transformation is greater in the EBV-carrying than in the EBV-negative B cell. This implies that the virus contributes to the etiology of the highly endemic African form. The same argument does not apply equally forcefully to the HIV-associated BLs that carry EBV in about 30% of cases and exhibit the molecular geometry of the sporadic, rather than the highly endemic BL-associated Ig/myc translocations (for a review, see Ref.[33]). This suggests that different cofactors act on different B cell subcompartments. Chronic holo- or hyperendemic malaria is the most likely cofactor in the highly endemic African form whereas HIV itself and/or the opportunistic infections associated with it may act in AIDS-associated lymphomas.

By what mechanism could the EBV carrier status increase the probability of BL development? It could act directly or indirectly. In an indirect scenario, it would expand the target cell population at risk by extending the life span of the infected lymphocyte and increasing the number of cell divisions before its elimination from the B cell pool. Alternatively, it could contribute directly to the oncogenic process. This effect would have to be mediated by EBNA1, the only virally encoded protein expressed in the BL cell. This possibility is supported by the finding that EBNA1 transgenic mice develop B cell lymphomas in high frequency.[31]

Little can be said about the involvement of EBV in HD and T cell lymphomas. The fact that the neoplastic cell clones carry EBV that has infected the cells on a single occasion, as indicated by the terminal repeat hybridization test, is in line with the assumption that the viral infection preceded the appearance of the neoplastic clone and that therefore the virus is somehow involved in causing the tumor. Because the interaction of the virus with the normal precursor cells of these tumors has not been defined, it is premature to speculate about the nature of this involvement.

REFERENCES

1. Hunsmann, G., and Hinuma, Y., 1985, Human adult T-cell leukemia virus and its association with disease, *Adv. Viral Oncol.* **5:**147.
2. Miller, G., 1980, Biology of Epstein–Barr virus, in *Viral Oncology,* (G. Klein, ed.), Raven Press, New York, pp.713–738.

3. Klein, G., Svedmyr, E., Jondal, M., and Persson, P. O., 1976, EBV-determined nuclear antigen (EBNA)-positive cells in the peripheral blood of infectious mononucleosis patients, *Int. J. Cancer* **17:**21–26.
4. Chen, F., Zou, J. Z., di Renzo, L., Winberg, G., Hu, L. F., Klein, E., Klein, G., and Ernberg, I., 1995, A subpopulation of normal B cells latently infected with Epstein–Barr virus resembles Burkitt lymphoma cells in expressing EBNA-1 but not EBNA-2 or LMP1, *J. Virol.* **69:**3752–3758.
5. Lewin, N., Aman, P., Masucci, M. G., Klein, E., Klein, G., Oberg, B., Strander, H., Henle, W., and Henle, G., 1987, Characterization of EBV-carrying B-cell populations in healthy seropositive individuals with regard to density, release of transforming virus and spontaneous outgrowth, *Int. J. Cancer* **39:**472–476.
6. Klein, G., 1994, Epstein–Barr virus strategy in normal and neoplastic B cells [Review], *Cell* **77:**791–793.
7. Masucci, M. G., Ernberg, I., 1994, Epstein–Barr virus: Adaptation to a life within the immune system [Review], *Trends Microbiol.* **2:**125–130.
8. Levitskaya, J., Coram, M., Levitsky, V., Imreh, S., Steigerwald-Mullen, P. M., Klein, G., Kurilla, M. G., and Masucci, M. G., 1995, Inhibition of antigen processing by the internal repeat region of the Epstein–Barr virus nuclear antigen-1, *Nature* **375:**685–688.
9. Marin, D. R., Marlowe, R. L., and Ahearn, J. M., 1994, Determination of the role for CD21 during Epstein–Barr virus infection of B-lymphoblastoid cells, *J. Virol.* **68:**4716–4726.
10. Kondo, N., Inoue, R., and Orii, T., 1992, Various responses of lymphocytes to Epstein–Barr virus in patients with common variable immunodeficiency, *J. Clin. Lab. Immunol.* **37:**183–189.
11. Henderson, S., Rowe, M., Gregory, C., Croom-Carter, D., Wang, F., Longnecker, R., Kieff, E., and Rickinson, A., 1991, Induction of bcl-2 expression by Epstein–Barr virus latent membrane protein 1 protects infected B cells from programmed cell death, *Cell* **65:**1107–1115.
12. Gordon, J., and Cairns, J. A., 1991, Autocrine regulation of normal and malignant B lymphocytes [Review], *Adv. Cancer Res.* **56:**313–334.
13. Decker, L. L., Klaman, L. D., and Thorley-Lawson, D. A., 1996, Detection of the latent form of Epstein–Barr virus DNA in the peripheral blood of healthy individuals, *J. Virol.* **70:**3286–3289.
14. Henle, G., and Henle, W., 1979, The virus as the etiologic agent of infectious mononucleosis, in *The Epstein–Barr Virus*, (M. Epstein, and B. Achong, eds.) Springer-Verlag, Berlin, pp. 297–320.
15. Svedmyr, E., Ernberg, I., Seeley, J., Weiland, O., Masucci, G., Tskuda, K., Szigeti, R., Masucci, M. G., Blomgren, H., Berthold, W., Henle, W., and Klein, G., 1984, Virologic, immunologic, and clinical observations on a patient during the incubation, acute, and convalescent phases of infectious mononucleosis, *Clin. Immunol. Immunopathol.* **30:**437–450.
16. Penn, I., 1987, Cancers following cyclosporine therapy, *Transplantation* **43:**32–35.
17. Rickinson, A. B., Murray, R. J., Brooks, J., Griffin, H., Moss, D. J., and Masucci, M. G., 1992, T cell recognition of Epstein–Barr virus associated lymphomas [Review], *Cancer Surv.* **13:**53–80.
18. Cleary, M. L., and Sklar, J., 1984, Lymphoproliferative disorders in cardiac transplant recipients are multiclonal lymphomas, *Lancet* **2:**489–493.
19. Kaplan, M. A., Ferry, J. A., Harris, N. L., and Jacobson, J. O., 1994, Clonal analysis of posttransplant lymphoproliferative disorders, using both episomal Epstein–Barr virus and immunoglobulin genes as markers, *Am. J. Clin. Pathol.* **101:**590–596.
20. Gratama, J. W., Zutter, M. M., Minarovits, J., Oosterveer, M. A., Thomas, E. D., Klein, G., and Ernberg, I., 1991, Expression of Epstein–Barr virus-encoded growth-transformation-associated proteins in lymphoproliferations of bone-marrow transplant recipients, *Int. J. Cancer* **47:**188–192.
21. Young, L., Alfieri, C., Hennessy, K., Evans, H., O'Hara, C., Anderson, K. C., Ritz, J., Shapiro, R. S., Rickinson, A., Kieff, E., *et al.*, 1989, Expression of Epstein–Barr virus transformation-

associated genes in tissues of patients with EBV lymphoproliferative disease, *N. Engl. J. Med.* **321:**1080–1085.

22. Gaidano, G., and and Dalla-Favera, R., 1995, Molecular pathogenesis of AIDS-related lymphomas, *Adv. Cancer Res.* **67:**113–153.
23. Nalesnik, M. A., Jaffe, R., Starzl, T. E., Demetris, A. J., Porter, K., Burnham, J. A., Makowa, L., Ho, M., and Locker, J., 1988, The pathology of posttransplant lymphoproliferative disorders occurring in the setting of cyclosporine A-prednisone immunosuppression, *Am. J. Pathol.* **133:**173–192.
24. Ho, M., Jaffe, R., Miller, G., Breinig, M. K., Dummer, J. S., Makoeka, L., Atchison, R. W., Karrer, F., Nalesnik, M. A., and Starzl, T. E., 1988, The frequency of Epstein–Barr virus infection and associated lymphoproliferative syndrome after transplantation and its manifestations in children, *Transplantation* **45:**719–727.
25. Shearer, W. T., Ritz, J., Finegold, M. J., Guerra, I. C., Rosenblatt, H. M., Lewis, D. E., Pollack, M. S., Taber, L. H., Sumaya, C. V., Grument, F. C., *et al.,* 1985, Epstein–Barr virus-associated B-cell proliferations of diverse clonal origins after bone marrow transplantation in a 12-year-old patient with severe combined immunodeficiency, *N. Engl. Med.* **312:**1151–1159.
26. Hickey, D., Nalesnik, M., and Vivas, C., 1990, Renal retransplantation in patients who lost their allografts during management of previous posttransplant lymphoproliferative disease. *Clin. Transpl.* **4:**187.
27. Shapiro, R. S., McClain, K., Frizzera, G., Gajl-Peczalska, K. J., Kersey, J. H., Blazar, B. R., Arthur, D. C., Patton, D. F., Greenberg, J. S., Burke, B. *et al.,* 1988, Epstein–Barr virus associated B cell lymphoproliferative disorders following bone marrow transplantation, *Blood* **71:**1234–1243.
28. Zutter, M. M., Marin, P. J., Sale, G. E., Shulman, H. M., Fisher, L., Thomas, E. D., and Durnam, D. M., 1988, Epstein–Barr virus lymphoproliferation after bone marrow transplantation, *Blood* **72:**520–529.
29. Papadopoulos, E. B., Ladanyi, M., Emanuel, D., Mackinnon, S., Boulad, F., Carabasi, M. H., Castro-Malaspina, H., Childs, B. H., Gillio, A. P., Small, T. N., *et al.,* 1994, Infusions of donor leukocytes to treat Epstein–Barr virus-associated lymphoproliferative disorders after allogeneic bone marrow transplantation [see comments], *N. Engl. J. Med.* **330:**1185–1191.
30. Lucas, K. G., Small, T. N., Heller, G., Dupoint, B., and O'Reilly, R. J., 1996, The development of cellular immunity to Epstein–Barr virus after allogeneic bone marrow transplantation, *Blood* **87:**2594–2603.
31. Wilson, J. B., Bell, J. L., and Levine, A. J., 1996, Expression of Epstein–Barr virus nuclear antigen-1 induces B cell neoplasis in transgenic mice, *EMBO J.* **15:**3117–3126.
32. Klein, G., 1989, Multiple phenotypic consequences of the Ig/Myc translocation in B cell-derived tumors [Review], *Genes, Chromosomes & Cancer* **1:**3–8.
33. Magrath, I., 1990, The pathogenesis of Burkitt's lymphoma [Review], *Adv. Cancer Res.* **55:**133–270.
34. Nilsson, K., and Ponten, J., 1975, Classification and biological nature of established human hematopoietic cell lines [Review], *Int. J. Cancer* **15:**321–341.
35. Gregory, C., 1992, Epstein–Barr virus and B cell survival, *Med. Virol.* **2:**205–220.
36. Rowe, M., Rowe, D. T., Gregory, C. D., Young, L. S., Farrell, P. J., Rupani, H., and Rickinson, A. B., 1987, Differences in B cell growth phenotype reflect novel patterns of Epstein–Barr virus latent gene expression in Burkitt's lymphoma cells *EMBRO J.* **6:**2743–2751.
37. Woisetschlaeger, M., Strominger, J. L., and Speck, S. H., 1989, Mutually exclusive use of viral promoters in Epstein–Barr virus latently infected lymphocytes, *Proc. Natl. Acad. Sci. USA* **86:**6498–6502.
38. Schafer, B. C., Strominger, J. L., and Speck, S. H., 1995, Redefining the Epstein–Barr virus-

encoded nuclear antigen EBNA-1 gene promoter and transcription initiation site in group I Burkitt lymphoma cell lines, *Proc. Natl. Acad. Sci. USA* **92:**10565–10569.

39. Hu, L.-F., Chen, F., Altiok, E., Winberg, G., Klein, G., and Ernberg, I., 1996, Cell phenotype dependent alternative splicing of EBNA mRNAs in EBV-carrying cells, *J. Gen. Virol.*, in press.
40. Masucci, M. G., Contreras-Salazar, B., Ragnar, E., Falk, K., Minarovits, J., Ernberg, L., and Klein, G., 1989, 5-Azacytidine up regulates the expression of Epstein–Barr virus nuclear antigen 2 (EBNA-2) through EBNA-6 and latent membrane protein in the Burkitt's lymphoma line rael. *J. Virol.* **63:**3135–3141.
41. Hu, L. F., Minarovits, J., Cao, S. L., Contreras-Salazar, B., Rymo, L., Falk, K., Klein, G., and Ernberg, I., 1991, Variable expression of latent membrane protein in nasopharyngeal carcinoma can be related to methylation status of the Epstein–Barr virus BNLF-1 5'-flanking region, *J. Virol.* **65:**1558–1567.
42. Minarovits, J., Minarovits-Kormuta, S., Ehlin-Henriksson B., Falk, K., Klein, G., and Ernberg, I., 1991, Host cell phenotype-dependent methylation patterns of Epstein–Barr virus DNA, *J. Gen. Virol.* **72:**1591–1599.
43. Contreras-Brodin, B. A., Anvret, M., Imreh, S., Altiok, E., Klein, G., and Masucci, M. G., 1991, B cell phenotype-dependent expression of the Epstein–Barr virus nuclear antigens EBNA-2 to EBNA-6: Studies with somatic cell hybrids, *J. Gen. Virol.* **72:**3025–3033.
44. Andersson, M. L., Stam, N. J., Klein, G., Ploegh, H. L., and Masucci, M. G., 1991, Aberrant expression of HLA class-I antigens in Burkitt lymphoma cells, *Int. J. Cancer* **47:**544–550.
45. Burkitt, D. P., 1969, Etiology of Burkitt's lymphoma—an alternative hypothesis to a vectored virus [Review], *J. Natl. Cancer Inst.* **42:**19–28.
46. Adams, J. M., Harris, A. W., Pinkert, C. A., Corcoran, L. M., Alexander, W. S., Cory, S., Palimiter, R. D., and Brinster, R. L., 1985, The c-myc oncogene driven by immunoglobulin enhancers induces lymphiod malignancy in transgenic mice, *Nature* **318:**533–538.
47. Muller, J. R., Janz, S., Goedert, J. J., Potter, M., and Rabkin, C. S., 1995, Persistence of immunoglobulin heavy chain/c-myc recombination-positive lymphocyte clones in the blood of human immunodeficiency virus-infected homosexual men, *Proc. Natl. Acad. Sci. USA* **92:**6577–6581.
48. Axelson, H., Henriksson, M., Wang, Y., Magnusson, K. P., and Klein, G., 1995, The amino-terminal phosphorylation sites of C-MYC are frequently mutated in Burkitt's lymphoma lines but not in mouse plasmacytomas and rat immunocytomas, *Eur. J. Cancer* **31A:**2099–2104.
49. Henriksson, M., Bakardjiev, A., Klein, G., and Luscher, B., 1993, Phosphorylation sites mapping in the N-terminal domain of c-myc modulate its transforming potential, *Oncogene* **8:**3199–3209.
50. Farrell, P. J., Allan, G. J., Shanahan, F., Vousden, K. H., and Crook, T., 1991, p53 is frequently mutated in Burkitt's lymphoma cell lines, *EMBO J.* **10:**2879–2887.
51. Gaidano, G., and Dalla-Favera, R., 1996, Molecular biology of lymphoid neoplasms, in *The Molecular Basis of Cancer,* (J. Mendelsohn, P. Howley, M. Israel, and L. Liotta, eds.), in press, W.B. Saunders, Philadelphia.
52. Wiman, K. G., Magnusson, K. P., Ramqvist, T., and Klein, G., 1991, Mutant p53 detected in a majority of Burkitt lymphoma cell lines by monoclonal antibody PAb240, *Oncogene* **6:**1633–1639.
53. Burkitt, D. P., and Kyalwazi, S. K., 1967, Spontaneous remission of African lymphoma, *Br. J. Cancer* **21:**14–16.
54. Rooney, C. M., Edwards, C. F., Lenoir, G. M., Rupani, H., and Rickinson, A. B., 1986, Differential activation of cytotoxic responses by Burkitt's lymphoma (BL)-cell lines: Relationship to the BL-cell surface phenotype. *Cell. Immunol.* **102:**99–112.
55. de Campos-Lima, P. O., Torsteinsdottir, S., Cuomo, L. Klein, G., Sulitzeanu, D., and Masucci,

M. G., 1993, Antigen processing and presentation by EBV-carrying cell lines: Cell-phenotype dependence and influence of the EBV-encoded LMP1, *Int. J. Cancer* **53:**856–862.

56. Rowe, M., Khanna, R., Jacob, C. A., Argaet, V., Kelly, A., Powis, S., Belich, M., Croom-Carter, D., Lee, S., Burrows, S. R., and *et al.*, 1995, Restoration of endogenous antigen processing in Burkitt's lymphoma cells by Epstein–Barr virus latent membrane protein-1: Coordinate up-regulation of peptide transporters and HLA-class I antigen expression, *Eur. J. Immunol.* **25:**1374–1384.
57. Avila-Carino, J., Torsteinsdottir, S., Ehlin-Henriksson, B., Lenoir, G., Klein, G., Klein, E., and Masucci, M. G., 1987, Paired Epstein–Barr virus (EBV)-negative and EBV-converted Burkitt lymphoma lines: Stimulatory capacity in allogeneic mixed lymphocyte cultures, *Int. J. Cancer* **40:**691—697.
58. Gregory, C. D., Murray, R. J., Edwards, C. F., and Rickinson, A. B., 1988, Downregulation of cell adhesion molecules LFA-3 and ICAM-1 in Epstein–Barr virus-positive Burkitt's lymphoma underlies tumor cell escape from virus-specific T cell surveillance, *J. Exp. Med.* **167:**1811–1824.
59. Masucci, M. G., Torsteindottir, S., Colombani, J., Brautbar, C., Klein, E., and Klein, G., 1987, Down-regulation of class I HLA antigens and of the Epstein–Barr virus-encoded latent membrane protein in Burkitt lymphoma lines, *Proc. Natl. Acad. Sci. USA* **84:**4567–4571.
60. Beral, V., Peterman, T., Berkelman, R., and Jaffe, H., 1991, AIDS-associated non-Hodgkin lymphoma [see comments], *Lancet* **337:**805–809.
61. Ballerini, P., Gaidano, G., Gong., J. Z., Tassi, V., Saglio, G., Knowles, D. M., and Dalla-Favera, R., 1993, Multiple genetic lesions in acquired immunodeficiency syndrome-related non-Hodgkin's lymphoma, *Blood* **81:**166–176.
62. Pelicci, P. G., Knowles, D. M., Arlin, Z. A., Wieczorek, R., Luciw, P., Dina, D., Basilico, C., and Dalla-Favera, R., 1986, Multiple monoclonal B cell expansions and c-myc oncogene rearrangements in acquired immune deficiency syndrome-related lymphoproliferative disorders. Implications for lymphomagenesis, *J. Exp. Med.* **164:**2049–2060.
63. Shibata, D., Weiss, L. M., Nathwani, B. N., Brynes, R. K., and Levine, A. M., 1991, Epstein–Barr virus in benign lymph node biopsies from individuals infected with the human immunodeficiency virus is associated with concurrent or subsequent development of non-Hodgkin's lymphoms, *Blood* **77:**1527–1533.
64. Neri, A., Barriga, F., Inghirami, G., Knowles, D. M., Neequaye, J., Magrath, I. T., and Dalla-Favera, R., 1991, Epstein–Barr virus infection precedes clonal expansion in Burkitt's and acquired immunodefiency syndrome-associated lymphoam [see comments], *Blood* **77:**1092–1092.
65. Karp, J. E., and Border, S., 1991, Acquired immunodeficiency syndrome and non-Hodgkin's lymphomas [Review], *Cancer Res.* **51:**4743–4756.
66. MacMahon, E. M., Glass, J. D., Hayward, S. D., Mann, R. B., Becker, P. S., Charache, P., McArthur, J. C., and Ambinder, R. F., 1991, Epstein–Barr virus in AIDS-related primary central nervous system lymphoma, *Lancet* **338:**969–973.
67. Giovanella, B., Nilsson, K., Zech, L., Yim, O., Klein, G., and Stehlin, J. S., 1979, Growth of diploid, Epstein–Barr virus-carrying human lymphoblastoid cell lines heterotransplanted into nude mice under immunologically privileged conditions, *Int. J. Cancer* **24:**103–113.
68. Nilsson, K., Gionvaella, B. C., Stehlin, J. S., and Klein, G., 1977, Tumorigenicity of human hematopoietic cell lines in athymic nude mice, *Int. J. Cancer* **19:**337–344.
69. Hamilton-Dutiot, S. J., Raphael, M., Audouin, J., Diebold, J., Lisse, I., Pedersen, C., Oksenhendler, E., Marelle, L., and Pallesen, G., 1993, *In situ* demonstration of Epstein-Barr virus small RNAs (EBER 1) in acquired immunodeficiency syndrome-related lymphomas: Correlation with tumor morphology and primary site, *Blood* **82:**619–624.
70. Pedersen, C., Gerstoft, J., Lundgren, J. D., Skinhoj, P., Bottzauw, J., Geisler, C., Hamilton-Dutiot, S. J., Thorsen, S., Lisse, I., Ralfkiaer, E., *et al.,* 1991, HIV-associated lymphoma:

Histopathology and association with Epstein–Barr virus genome related to clinical, immunological and prognostic features, *Eur. J. Cancer* **27:**1416–1423.

71. Hamilton-Dutoit, S. J., Rea, D., Rapheal, M., Sandvej, K., Delecluse, H. J., Gisselbrecht, C., Marelle, L., van Krieken, H. J., and Pallesen, G., 1993, Epstein–Barr virus-latent gene expression and tumor cell phenotype in acquired immunodeficiency syndrome-related non-Hodgkin's lymphoma. Correlation of lymphoma phenotype with three distinct patterns of viral latency, *Am. J. Pathol.* **143:**1072–1085.
72. Geddes, J. F., Bhattacharjee, M. B., Savage, K., Scaravilli, F., and McLaughlin, J. E., 1992, Primary cerebral lymphoma: A study of 47 cases probed for Epstein–Barr virus genome, *J. Clin. Pathol.* **45:**587–590.
73. Hamilton-Dutoit, S. J., and Pallesen, G., 1992, A survey of Epstein–Barr virus gene expression in sporadic non-Hodgkin's lymphomas. Detection of Epstein–Barr virus in a subset of peripheral T-cell lymphomas, *Am. J. Pathol.* **140:**1315–1325.
74. Young, L. S., Finerty, S., Brooks, L., Scullion, F., Rickinson, A. B., and Morgan, A. J., 1989, Epstein–Barr virus gene expression in malignant lymphomas induced by experimental virus infection of cottontop tamarins, *J. Virol.* **63:**1967–1974.
75. Rowe, M., Young, L. S., Crocker, J., Stokes, H., Henderson, S., and Rickinson, A. B., 1991, Epstein–Barr virus (EBV)-associated lymphoproliferative disease in the SCID mouse model: Implications for the pathogenesis of EBV-positive lymphomas in man, *J. Exp. Med.* **173:**147–158.
76. Klein, G., 1973, The Epstein–Barr virus, in *The Herpesviruses,* (A. Kaplan, ed.), Academic Press, New York, pp. 521–555.
77. Herbst, H., Steinbrecher, E., Niedobitek, G., Young, L. S., Brooks, L., Muller-Lantzsch, N., and Stein, H., 1992, Distribution and phenotype of Epstein–Barr virus-harboring cells in Hodgkin's disease, *Blood* **80:**484–491.
78. Johansson, B., Klein, G., Henle, W., and Henle, G., 1970, Epstein–Barr virus (EBV)-associated antibody patterns in malignant lymphoma and leukemia. I. Hodgkin's disease, *Int. J. Cancer* **6:**450–462.
79. Brousset, P., Chittal, S., Schlaifer, D., Icart, J., Payen, C., Rigal-Huguet, F., Voigt, J. J., and Delsol, G., 1991, Detection of Epstein–Barr virus messenger RNA in Reed–Sternberg cells of Hodgkin's disease by *in situ* hybridization with biotinylated probes on specially processed modified acetone methyl benzoate xylene (ModAMeX) sections [see comments], *Blood* **77:**1781–1786.
80. Coates, P. J., Slavin, G., and D'Ardenne, A.J., 1991, Persistence of Epstein–Barr virus in Reed–Sternberg cells throughout the course of Hodgkin's disease, *J. Pathol.* **164:**291–297.
81. Guarner, J., del Rio, C., Hendrix, L., and Unger, E. R., 1990, Composite Hodgkin's and non-Hodgkin's lymphoma in a patient with acquired immune deficiency syndrome. *In situ* demonstration of Epstein–Barr virus, *Cancer* **66:**796–800.
82. Herbst, H., Niedobitek, G., Kneba, M., Hummel, M., Finn, T., Anagnostopoulos, I., Bergholz, M., Krieger, G., and Stein, H., 1990, High incidence of Epstein–Barr virus genomes in Hodgkin's disease, *Am. J. Pathol.* **137:**13–18.
83. Neidobitek, G., Deacon, E. M., Young, L. S. Herbst, H., Hamilton-Dutoit, S. J., and Pallesen, G., 1991, Epstein–Barr virus gene expression in Hodgkin's disease [letter; comment], *Blood* **78:**1628–1630.
84. Uccini, S., Morardo, F., Stoppacciaro, A., Gradilone, A., Agliano, A. M., Faggioni, A., Manzari, V., Vago, L., Costanzi, G., Ruco, L. P., *et al.,* 1990, High frequency of Epstein–Barr virus genome detection in Hodgkin's disease of HIV-positive patients, *Int. J. Cancer* **46:**581–585.
85. Uhara, H., Sato, Y., Mukai, K., Akao, I., Matsuno, Y., Furuya, S., Hoshikawa, T., Shimosato, Y., and Saida, T., 1990, Detection of Epstein–Barr virus DNA in Reed–Sternberg cells of

Hodgkin's disease using the polymerase chain reaction and *in situ* hybridization, *Jpn. J. Cancer Res.* **81:**272–278.

86. Weiss, L. M., Movahed, L. A., Warnke, R. A., and Sklar, J., 1989, Detection of Epstein–Barr viral genomes in Reed–Sternberg cells of Hodgkin's disease, *N. Engl. J. Med.* **320:**502–506.
87. Weiss, L. M., Strickler, J. G., Warnke, R. A., Purtilo, D. T., and Sklar, J., 1987, Epstein–Barr viral DNA in tissues of Hodgkin's disease, *Am. J. Pathol.* **129:**86–91.
88. Anagnostopoulos, I., Hummel, M., Finn, T., Tiemann, M., Korbjuhn, P., Dimmler, C., Gatter, K., Dallenbach, F., Parwaresch, M. R., and Stein, H., 1992, Heterogeneous Epstein–Barr virus infection patterns in peripheral T-cell lymphoma of angioimmunoblastic lymphadenopathy type, *Blood* **80:**1804–1812.
89. Boiocchi, M., Carbone, A., De Re, V., and Dolcetti, R., 1989, Is the Epstein–Barr virus involved in Hodgkin's disease?, *Tumori* **75:**345–350.
90. Gledhill, S., Gallagher, A., Jones, D. B., Krajewski, A. S., Alexander, F. E., Klee, E., Wright, D. H., O'Brien, C., Onions, D. E., and Jarrett, R. F., 1991, Viral involvement in Hodgkin's disease: Detection of clonal type A Epstein–Barr virus genomes in tumour samples, *Br. J. Cancer* **64:**227–232.
91. Jarrett, R. F., Gallagher, A., Jones, D. B., Alexander, F. E., Krajewski, A. S., Kelsey, A., Adams, J., Angus, B., Gledhill, S., Wright, D. H. *et al.*, 1991, Detection of Epstein–Barr virus genomes in Hodgkin's disease: relation to age [see comments], *J. Clin. Pathol.* **44:**844–848.
92. Staal, S. P., Ambinder, R., Beschorner, W. E., Hayward, G. S., and Mann, R., 1989, A survey of Epstein–Barr virus DNA in lymphoid tissue. Frequent detection in Hodgkin's disease [see comments], *Am. J. Clin. Pathol.* **91:**1–5.
93. Pallesen, G., Hamilton-Dutoit, S. J., Rowe, M., and Young, L. S., 1991, Expression of Epstein–Barr virus latent gene products in tumour cells of Hodgkin's disease [see comments], *Lancet* **337:**320–322.
94. Deacon, E. M., Pallesen, G., Niedobitek, G., Crocker, J., Brooks, L., Rickinson, A. B., and Young, L. S., 1993, Epstein–Barr virus and Hodgkin's disease: Transcriptional analysis of virus latency in the malignant cells, *J. Exp. Med.* **177:**339–349.
95. Delsol, G., Brousset, P., Chittal, S., and Rigal-Huguet, F., 1992, Correlation of the expression of Epstein–Barr virus latent membrane protein and *in situ* hybridization with biotinylated BamHI-W probes in Hodgkin's disease, *Am. J. Pathol.* **140:**247–253.
96. Herbst, H., Dallenbach, F., Hummel, M., Niedobitek, G., Pileri, S., Muller-Lantzsch, N., and Stein, H., 1991, Epstein–Barr virus latent membrane protein expression in Hodgkin and Reed–Sternberg cells, *Proc. Natl. Acad. Sci. USA* **88:**4766–4770.
97. Poppema, S., van Imhoff, G., Torensma, R., and Smit, J., 1985, Lymphadenopathy morphologically consistent with Hodgkin's disease associated with Epstein–Barr virus infection, *Am. J. Clin. Pathol.* **84:**385–390.
98. Zhou, X. G., Hamilton-Dutoit, S. J., Yan, Q. H., and Pallesen, G., 1993, The association between Epstein–Barr virus and Chinese Hodgkin's disease, *Int. J. Cancer* **55:**359–363.
99. Libetta, C. M., Pringle, J. H., Angel, C. A., Craft, A. W., Malcolm, A. J., and Lauder, I., 1990, Demonstration of Epstein–Barr viral DNA in formalin-fixed, paraffin-embedded samples of Hodgkin's disease [see comments], *J. Pathol.* **161:**255–260.
100. Vestlev, P. M., Pallesen, G., Sandvej, K., Hamilton-Dutoit, S. J., and Bendtzen, S. M., 1992, Prognosis of Hodgkin's disease is not influenced by Epstein–Barr virus latent membrane protein [letter], *Int. J. Cancer* **50:**670–671.
101. Weinreb, M., Day, P. J., Murray, P. G., Raafat, F., Crocker, J., Parkes, S. E., Coad, N. A., Jones, J. T., and Mann, J. R., 1992, Epstein–Barr virus (EBV) and Hodgkin's disease in children: Incidence of EBV latent membrane protein in malignant cells, *J. Pathol.* **168:**365–369.
102. Munoz, N., Davidson, R. J., Witthoff, B., Ericsson, J. E., and De-The, G., 1978, Infectious mononucleosis and Hodgkin's disease, *Int. J. Cancer* **22:**10–13.

103. Rosdahl, N., Larsen, S. O., and Clemmesen, J., 1974, Hodgkin's disease in patients with previous infectious mononucleosis: 30 years' experience, *Br. Med. J.* **2:**253–256.
104. Knecht, H., Bachmann, E., Brousset, P., Sandvej, K., Nadal, D., Bachmann, F., Odermatt, B. F., Delsol, G., and Pallesen, G., 1993, Deletions within the LMP1 oncogene of Epstein–Barr virus are clustered in Hodgkin's disease and identical to those observed in nasopharyngeal carcinoma, *Blood* **82:**2937–2942.
105. Jones, J. F., Shurin, S., Abramowsky, C., Tubbs, R. R., Sciotto, C. G., Wahl, R., Sands, J., Gottman, D., Katz, B. Z., and Sklar, J., 1988, T-cell lymphomas containing Epstein–Barr viral DNA in patients with chronic Epstein–Barr virus infections, *N. Engl. J. Med.* **318:**733–741.
106. Chan, J. K., Ng, C. S., Lau, W. H., and Lo, S. T., 1987, Most nasal/nasopharyngeal lymphomas are peripheral T-cell neoplasms, *Am. J. Surg. Pathol.* **11:**418–429.
107. Chott, A., Rappersberger, K., Schlossarek, W., and Radaszkiewicz, T., 1988, Peripheral T cell lymphoma presenting primarily as lethal midline granuloma, *Hum. Pathol.* **19:**1093–1101.
108. Hastrup, N., Hamilton-Dutoit, S. Ralfkiaer, E., and Pallesen, G., 1991, Peripheral T-cell lymphomas: An evaluation of reproducibility of the updated Kiel classification, *Histopathology* **18:**99–105.
109. Ohshima, K., Kikuchi, M., Eguchi, F., Masuda, Y., Sumiyoshi, Y., Mohtai, H., Takeshita, M., and Kimura, N., 1990, Analysis of Epstein–Barr viral genomes in lymphoid malignancy using Southern blotting, polymerase chain reaction and *in situ* hybridization. *Virchows Arch. B Cell Pathol.* **59:**383–390.
110. Ott, G., Ott, M. M., Feller, A. C., Seidl, S., and Muller-Hermelink, H. K., 1992, Prevalence of Epstein–Barr virus DNA in different T-cell lymphoma entities in a European population, *Int. J. Cancer* **51:**562–567.
111. Su, I. J., Tsai, T. F., Cheng, A. L., and Chen, C. C., 1993, Cutaneous manifestations of Epstein–Barr virus-associated T-cell lymphoma, *J. Am. Acad. Dermatol.* **29:**685–692.
112. Minarovits, J., Hu, L. F., Imai, S., Harabuchi, Y., Kataura, A., Minarovits-Kormuta, S., Osato, T., and Klein, G., 1994, Clonality, expression and methylation patterns of the Epstein–Barr virus genomes in lethal midline granulomas classified as peripheral angiocentric T cell lymphomas, *J. Gen. Virol.* **75:**77–84.
113. Pallesen, G., Hamilton-Dutoit, S. J., and Zhou, X., 1993, The association of Epstein–Barr virus (EBV) with T cell lymphoproliferations and Hodgkin's disease: Two new developments in the EBV field [Review], *Adv. Cancer Res.* **62:**179–239.
114. Ambinder, R. F., and Mann, R. B., 1994, Detection and characterization of Epstein–Barr virus in clinical specimens [Review], *Am. J. Pathol.* **145:**239–252.
115. de Bruin, P. C., Jiwa, M., Oudejans, J. J., van der Valk, P., van Heerde, P., Sabourin, J. C., Csanaky, G., Gaulard, P., Noorduyn, A. L., Willemze, R. *et al.*, 1994, Presence of Epstein–Barr virus in extranodal T-cell lymphomas: Differences in relation to site, *Blood* **83:**1612–1618.
116. Sato, E., Tokunaga, M., Hasui, K., Kitajima, S., and Nomoto, M., 1990, Pathoepidemiological features of adult T-cell lymphoma/leukemia in an endemic area: Kagoshima, Japan, *Cancer Detection Prev.* **14:**423–429.
117. De Bruin, P. C., Jiwa, N. M., Van der Valk, P., Van Heerde, P., Gordijn, R., Ossenkoppele, G. J., Walboomers, J. M., and Meijer, C. J., 1993, Detection of Epstein–Barr virus nucleic acid sequences and protein in nodal T-cell lymphomas: Relation between latent membrane protein-1 positivity and clinical course, *Histopathology* **23:**509–518.
118. Kumar, S., Kumar, D., Kingma, D. W., and Jaffe, E. S., 1993, Epstein–Barr virus-associated T-cell lymphoma in a renal transplant patient [see comments], *Am. J. Surg. Pathol.* **17:**1046–1053.
119. Tien, H. F., Su, I. J., Chuang, S. M., Lee, F. Y., Liu, M. C., Tsai, T. F., Lin, K. H., and Chen,

R. L., 1993, Cytogenetic characterization of Epstein–Barr virus-associated T-cell malignancies. *Cancer Genet. Cytogenet.* **69:**25–30.

120. Mori, M., Kurozumi, H., Akagi, K., Tanaka, Y., Imai, S., and Osato, T., 1992, Monoclonal proliferation of T cells containing Epstein–Barr virus in fatal mononucleosis [letter], *N. Engl. J. Med.* **327:**58.

8

Immune Responses to Epstein–Barr Viral Infection

TAKESHI SAIRENJI and TAKESHI KURATA

1. INTRODUCTION

The Epstein–Barr virus (EBV) is the cause of infectious mononucleosis (IM), and there is substantial evidence linking it with two types of human cancer, Burkitt's lymphoma (BL) and nasopharyngeal carcinoma (NPC).[1–3] EBV has also been linked to a variety of other lymphoid- and epithelial-derived proliferative diseases. The former include X-linked recessive lymphoproliferative syndrome, atypical B cell lymphoproliferations and non-Hodgkin's lymphoma of B-cell origin in individuals with primary or secondary immunodeficiency, and Hodgkin's disease, T cell lymphomas, and leukemia. The latter include thymic carcinoma, oral hairy leukoplakia, and gastric carcinoma.[4] Other categories of EBV-related diseases, chronic active EBV infection, and autoimmune diseases have been described. The association of EBV has been demonstrated by the detection of EBV genomes and/or antigens in the tissues or cells and by the humoral and cellular immune responses to EBV antigens in patients.[1–3]

TAKESHI SAIRENJI • Department of Biosignaling, School of Life Science, Faculty of Medicine, Tottori University, Yonago 683; and Department of Pathology, National Institute of Infectious Diseases, Tokyo 162, Japan. TAKESHI KURATA • Department of Pathology, National Institute of Infectious Diseases, Tokyo 162, Japan.

Herpesviruses and Immunity, edited by Medveczky *et al.* Plenum Press, New York, 1998.

2. EBV-SPECIFIC ANTIGENS

EBV antigens are classified as latent antigens, EBV-nuclear antigens (EBNA),[5] and latent membrane proteins (LMP)[6] which are expressed in viral latency and by activated antigens, early antigens (EA),[7] viral capsid antigens (VCA),[8] and membrane antigens (MA)[9] expressed in the viral productive cycle.

The EBV was found in the cultured cells from biopsy specimens of BL.[10] Henle *et al.* used the indirect immunofluorescence technique to detect patients' antibodies to the virus and found positive cells in a small proportion of cultured cells from BL.[8] The antigen was later recognized as VCA. The MA was determined on the cell surface of BL cells from sera of patients with BL in the search for tumor-specific immune reactions in BL patients.[9] The EA was detected in superinfected EBV-nonproducer cell lines.[7] The EBNA was found in EBV-seropositive human serum by anticomplement immunofluorescence techniques.[5] The LMP was found by the nucleotide sequence of an mRNA transcribed in a latent growth-transforming virus.[6] The EBV genome in virion is a linear double-stranded DNA, 172 kbp long, and is maintained as a plasmid molecule in EBV-infected cells. The entire genome of the B95-8 strain of EBV has been cloned and sequenced, and the regions encoding for specific antigens have been mapped.[2]

2.1. EBNA

Six of these proteins constitute a family of EBNAs, 1, 2, 3A, 3B, 3C, and leader protein (LP) derived from a common transcription unit under the control of two alternatively used promoters, Wp and Cp.[2,3] The gene encoding for EBNA1 is mapped to BKRF1.[11] EBNA1 is involved in maintaining plasmid viral DNA in latently infected cells[12] and activates viral DNA replication from a cis-acting viral DNA sequence, oriP.[13] EBNA1 is a 60-to 85-kDa polymorphic polypeptide containing a variable number of glycine-alanine repeats.[14]

EBNA-2 is mapped to BYRF2 and is essential for the immortalization of lymphocytes by EBV[2]. EBNA2 is detected with a molecular weight of 81–85kDa. P3HR-1 and Daudi EBV, which have large deletions encompassing EBNA2, cannot immortalize B cells.[15] EBNA2 is the first gene expressed in virally infected primary B cells together with EBNA-LP and serves as a master switch in viral and cellular genes involved in transformation by EBV.[16] EBNA2 is a pleiotropic activator of viral and cellular genes and is targeted to DNA by interacting with RBP-Jk,[17] a ubiquitously expressed and evolutionarily conserved gene essential in *Drosophila melanogaster* embryonic development.

The genes encoding for EBNA3 are mapped to different three open reading frames on the Bam HI E fragment of EBV-DNA. The molecular weights for EBNA3A, 3B, and 3C are 136, 142, and 147kd, respectively.[18] EBNA3A, 3B, and 3C are important in EBV transformation of primary human B lymphocytes.[19]

EBNA1, but not the rest of the EBNAs, is located on chromosomes at metaphase.[20] EBNA gene expression in several EBV-associated tumors is restricted, in contrast to lymphoblastoid cell lines (LCL).[3] EBNA1 is expressed in all EBV genome positive cells, but the other EBNAs are not. The expression of EBNA1 alone, known as type I latency, is characteristic of group I BL, whereas expression of LMP in addition to EBNA1 (type II latency) is characteristic of undifferentiated NPC and EBV-positive Hodgkin's and T cell lymphomas.[21] The full sets of EBNA and LMP are expressed in LCL referred as type III latency.

With the exception of EBNA1, each of the EBNAs contains epitopes recognized by MHC-restricted virus-specific cytotoxic T lymphocytes (CTL) in EBV-positive individuals.[22] Current evidence is consistent with the possibility that EBNA1 may not be able to provoke CTL response and that the persistence of the virus in the B cell fraction depends on the down-regulation of all latent viral proteins. For the CTL response, it has been reported that the internal repeat region of EBNA1 inhibits antigen processing.[23]

2.2. LMP

The LMP was defined from the nucleotide sequence encoded by BNLF1, while attempting to identify the lymphocyte-determined membrane antigen (LYDMA) which is the target antigen of CTL.[6] LMP is a 62-kDa membrane protein containing a transmembrane and cytoplasmic membrane protein with a cytoplasmic amino terminus of 23 residues, six membrane-spanning segments, and a cytoplasmic carboxy terminus of 200 residues.[2] LMP has tumorigenic effects in rodent fibroblast cell lines[24] and contributes to the immortalizing activity in primary human B lymphocytes. LMP1 interacts with tumor necrosis factor receptor-associated factor 1 (TRAF1).[25] The interaction of TRAF1 with LMP1 is evidence that this protein has a central role as an effector of cell growth or death signaling pathways. LMP2A interacts with LMP1 and members of the src tyrosine kinase family in the plasma membrane. Recently it has been reported that prevalent EBV strains with a LMP deletion associate with T cell lymphoma and NPC.[26]

2.3. Induction of Lytic Cycle by Signaling from the Cell Surface

The switch to a virus lytic cycle can be induced by various stimuli.[7,27,28] We demonstrated that second-messenger pathways regulate the activation of EBV in membrane immunoglobulin (Ig)-crosslinked Akata cells.[29–33] The agonists and antagonists of certain second-messenger pathways mimic or inhibit the anti-Ig stimulus that activates EBV in the cells.[29,33] A positive signal for EBV activation was mediated by calcium and calmodulin. A diacylglycerol analog diC_8 or TPA, which acts on protein kinase C, synergized with the calcium signal but alone did not activate EBV. The requirement of tyrosine kinase activation for the anti-Ig-mediated EBV activation was also demonstrated by the tyrosine kinase inhibitor genistein or herbimycin.[30,32,33]

2.4. EA

EA is the name of an antigen complex expressed in the early phase of the EBV lytic cycle and detected by the immunofluorescence technique. The EA was originally demonstrated by the differential appearance of EA and VCA in EBV-nonproducer cells superinfected with P3HR-1 EBV.[7] It was also demonstrated in nonproducer cells by treatment with certain drugs.[27,28]

The EA appears before the cycle of viral replication begins in productively infected cells.[7,34,35] Therefore, EA expression does not require viral DNA replication. Two distinct patterns of immunofluorescence for EA were demonstrated, showing the existence of at least two EA components.[36] The diffuse component (D) of the antigen is found mainly in the nucleus. The restricted component (R) is found in the cytoplasm. Immunoprecipitation analyses revealed about 15 peptides as EA components. mAb 9240 recognizes a major EA-D component which was mapped to BMRF1[37] and associated with EBV DNA polymerase.[38] The activity of EBV/DNA polymerase depends on the functional interaction between the catalytic subunit (BALF5 protein) and the accessory subunit (BMRF1 protein).[38] Other mAbs recognize EA-R components.[41–43] An EA-R component mapped to BORF2 appeared to be a subunit of an EBV-encoded ribonucleotide reductase.[41] Another EA-R mAb recognizes BHRF1 protein which is homologous to *bcl-2*.[42,43] We reported two unique EA components. One is associated with intermediate filaments of the cytoskeleton,[40] and the other was not a classical EA-D or -R but is associated with viral DNA replication.[44] The sequence of viral gene expression during the lytic cycle has been partially elucidated. Immediate-early genes (BZLF1, BRLF1) are expressed first and then early genes (BMLF1, BHRF1, BMRF1, BORF2) in viral replication.[31] Sequential appearance of BZLF1 protein (ZEBRA) of the BHRF1 (EA-R), BORF2 (EA-R), and BMRF1 (EA-D) components and viral DNA was observed.[35,45]

2.5. VCA

VCA appears following the replication of viral DNA and this protein constitutes the virion. VCA is present in both the nucleus and the cytoplasm. The virion is composed of at least seven proteins. The major VCA is encoded by the BCLF1 open reading frame[2] and is expressed as a 150–160-kDa non-glycosylated protein in producer cell lines.[46] A protein of 36 kDa encoded by BGLF2 has been described.[47] Other 18-kDa and 40-kDa VCA proteins have been mapped on BFRF3 and BdRF1, respectively.[48] Progeny virus particles are assembled in the cell nucleus, after which some of them pass into the cell cytoplasm and are released through the cell membrane.[49]

2.6. MA

MA is a general term which applies to EBV-induced membrane proteins, including, at least, those of molecular weights 350kDa(gp 350) and 220kDa(gp 220) from an alternately spliced mRNA or BLLF1, 110kDa(gp110), and 85kDa(gp85).[50–52] gp350/220 on the enveloped virus binds to CD21, and the virus enters into B lymphocytes by receptor-mediated endocytosis.[53] gp 85, the product of BXLF2, is the gH homologue of herpes simplex virus and has been implicated in the penetration of the virus into B cells.[54] gp85 localizes to the cytoplasm and nuclear rim rather than to the plasma membrane and is important in fusion between the virus and the cell membrane. A fourth MA molecule, gp 110, resides principally in the nuclear membranes and endoplasmic reticulum of productively infected cells.[55]

The components of MA play a central role in many of mechanisms for attack on EBV-infected cells. MA determinants could participate in neutralization of viral infectivity.[56] gp350 and gp85 can be targets for virus-neutralizing antibodies.[52,55] The protein backbone of gp 350/220 contains neutralizing epitopes.[50,51] gp85 also plays a role in virus neutralization in the presence of complement.[57] MA is a target for ADCC killing.[58] gp 350 is a target for the ADCC[59] and T cell-mediated responses.[60] gp350 is a strong candidate for a subunit vaccine. The cottontop tamarin has proved to be an invaluable model for designing and evaluating EBV vaccine candidates. Recombinant viruses, subunit vaccines, and features of novel vaccine preparations and adjuvants have been evaluated in this model.[60–62]

3. ANTIBODY RESPONSE TO EBV INFECTION

3.1. Immunofluorescence Test

The course of infection was monitored during the patient's immune response to IM.[1] In primary infection the first antibody to appear after infection is immunoglobulin M (IgM), which is specific for VCA. The production of IgM stops after the acute phase of IM and is replaced by the production of IgG. In most patients with IM, transient IgG antibodies to the EA also develop by the time of the onset of illness. In contrast, antibodies to EBNA develop late in the course of the disease. The IgG antibodies to VCA and EBNA, but not to EA, persist for life. Titers of the IgM antibodies to MA were higher in the earlier stages of IM,[63] a pattern similar to that for IgM antibodies to VCA.[1] The titer of IgG antibodies arose later than those to VCA and EA.[64] Patients with EBV-associated diseases generally have high antibody titers to VCA and EA. Antibodies to EA are good indicators of the diagnoses and prognoses in BL, NPC, and other EBV-associated diseases because the antibody to EA is rarely present with a few exceptions in the sera of healthy individuals carrying EBV.[1] IgA antibodies against EA and VCA are mainly restricted to NPC and are used as diagnostic markers.[65] The lack of antibodies to EBNA1 is seen in some patients with chronic active EBV infection.[66]

3.2. Enzyme-Linked Immunosorbent Assay (ELISA)

ELISA has been employed to detect antibodies to EBV antigens in crude, partially purified extracts of chemically induced cells and in purified EBV-specific proteins.[67] By using recombinant DNA technology, defined EBV-specific antigens can be produced in *E. coli* or in a baculovirus. ELISA has been used to assess antibodies to EBNA1,[68] EBNA2,[69] ZEBRA,[70] EA-D (BMRF1),[67,71] MA gp 350[72] and VCA p150 (BcLF1), and p125 (BALF4). The results correlate well with the results of the standard immunofluorescence technique. The ELISA provides the means for specific and sensitive serodiagnosis. In standard rapid diagnosis, several EBV antigens, which are diagnostically useful, were combined in an ELISA.[73]

3.3. Antibody Responses to EBV-Specific Enzymes

Sera from NPC patients had the capacity to neutralize the activity of several EBV specific enzymes: DNase,[74] thymidine kinase,[75] and DNA polymerase.[76] The frequency and levels of antibodies to DNase were elevated in patients with NPC.[74,77] An increase in anti-DNase titers is a marker for the early detection of NPC and is associated with a higher risk of local recurrence or metastatic relapse. Neutralizing antibodies decrease during prolonged periods of remission.

4. ANTIBODY-MEDIATED IMMUNE MECHANISMS

4.1. Neutralizing Antibodies

Neutralizing antibodies are measured by the prevention of either superinfection of nonproducer cells or transformation of B lymphocytes by EBV. The P3HR-1 virus is used for the superinfection, and its neutralization becomes evident from the prevention of EA synthesis or death in the superinfected cells.[78] For the second type of test, transforming virus is used, and the criteria for neutralization of the virus are prevention of the establishment of LCLs.[79,80]

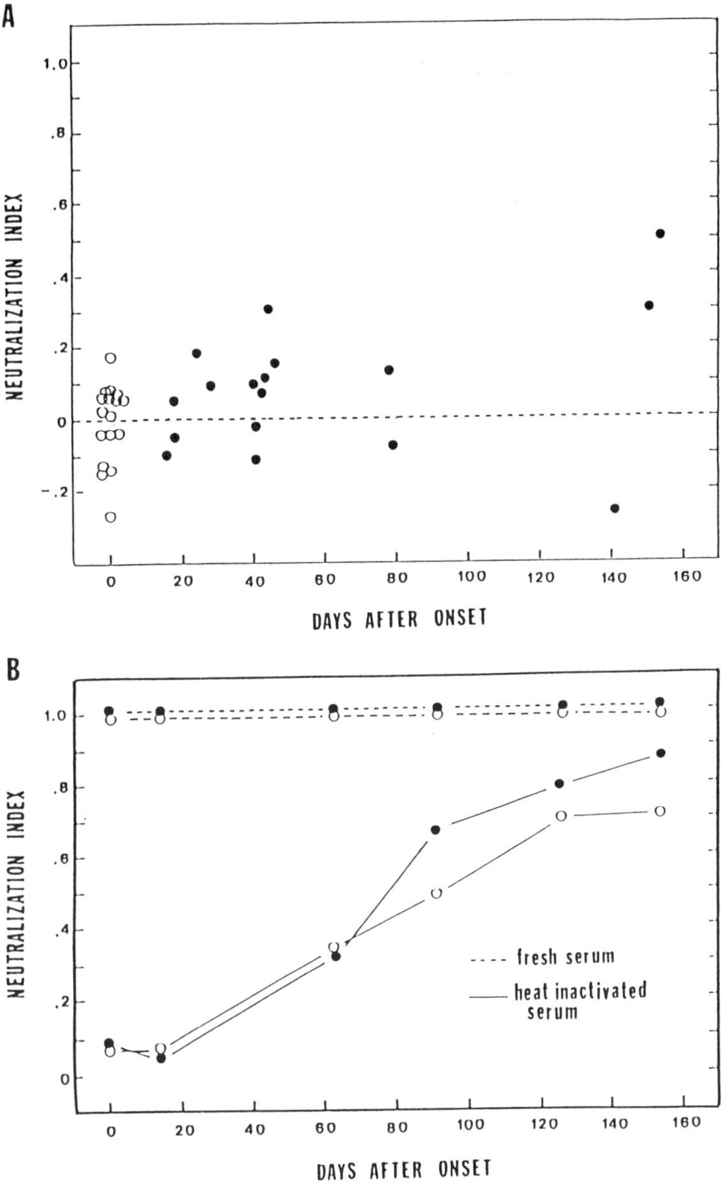

FIGURE 1. Neutralizing antibody in sera of patients with IM. (a) Neutralizing antibody was assayed at a 1:10 dilution of sera obtained at onset (○) and 15–154 days later (●). Serum was heated at 56 °C for 30 min to inactivate complement before the neutralization test. (b) The titers of neutralizing antibody in fresh and heat-inactivated sera were followed in a patient with IM. A neutralization index was calculated according to the formula described previously.[78]

We have tested neutralizing antibodies in IM and demonstrated the presence of EBV-specific complement(C′)-dependent neutralizing antibody in the sera of patients with IM (Fig. 1).[75] The first serum obtained after the onset of illness and that obtained later from the patients were examined for neutralization of P3HR-1 EBV. Without C′, the viral neutralization indices of sera obtained soon after the onset of disease were all < 0.2, and those of sera obtained later were < 0.5. The addition of C′ to those sera yielded neutralization indices of > 0.95. C′-dependent neutralizing antibodies were also demonstrated in sera from EBV-seropositive healthy adults. Two boys with XLP and fatal IM had no significant C′-dependent neutralizing antibody. C′-dependent neutralizing antibody may be relevant to immune disease against EBV early in the course of IM before the development of C′-independent neutralizing antibody.

4.2. Modulation of EBV Production with Anti-EBV Antibodies

To understand how the immune system controls EBV infection, we have studied the effects of EBV antibodies on modulating the expression of EBV genome functions. We found that the expression of EA in P3HR-1 EBV-superinfected Raji cells decreased during culturing of the EBV-seropositive cells, but not with EBV-seronegative, human serum.[81] The effect of human serum with EBV antibodies was also tested on virus production in P3HR-1 cells.[64] Cell culturing with EBV-seropositive sera reduced both production of infectious virus and the amount of virus DNA in the cells and in the supernatants. The inhibitory effect of seropositive serum was reversed upon switching the cells to a medium with seronegative serum. In serial sera of an acute IM patient, EBV DNA-reducing activity rose in parallel with antibodies against MA. We hypothesized that EBV production is modulated by EBV antibodies.

4.3. Inhibition of EBV Release by mAbs to gp350/220

We found that an anti-gp350/220 mAb had no neutralizing activity, but inhibited the release of virus from P3HR-1 cells.[52] The inhibition of EBV release was probed with the other mAbs, 72A1 and 2L10.[82] IB6 recognized MA on P3HR-1 cells but did not recognize (or barely recognized) a determinant on B95-8 cells. 2L10 and 72A1 mAbs reacted with B95-8 and P3HR-1 cells. IB6 and 2L10 neutralized neither P3HR-1 nor B95-8 virus, but 72A1 neutralized both viruses. IB6 and 72A1 inhibited P3HR-1 virus release, and 72A1, but not IB6, inhibited release of EBV from B95-8 cells. These results have been interpreted to reflect the roles of gp350/220 epitopes in virus neutralization and release. This inhibition of EBV release might be part of an immune defense against EBV infection.

5. EBV-INDUCED AUTOIMMUNE RESPONSES

Paul and Bunnell first reported the development of heterophilic antibodies, directed against antigens on the surface of sheep erythrocytes, in the sera of IM patients. Shope and Miller suggested that a heterophilic antigen of the Paul–Bunnell type was present on EBV-transformed lymphoblasts because they observed the development of the heterophilic antibody in the sera of squirrel monkeys inoculated with autologous EBV-transformed cells.[83] We reported evidence for the presence of antigen(s) on the surface of EBV genome-carrying lymphoid cells in culture, which is reactive with the Paul–Bunnell antibody in the serum of IM patients.[84] IM patients also make rheumatoid factor and a variety of autoantibodies to cytoskeletal components, such as vimentin and keratin.[85] We demonstrated an immune response to intermediate filament-associated EA-R which was recognized by mAbs produced by EBV-transformed B cells derived from IM patients.[40] Such autoantibodies could react with an induced viral protein, which associates with the cytoskeleton, or with a virally-induced modification of the cytoskeleton. It has been also reported that autoantibodies in IM are specific for the glycine-alanine repeat region of EBNA 1.[86] Many data indicated EBV reactivation in some patients with auto-immune diseases, as evidenced by elevated antibodies to EBV, increased viral DNA in their saliva, and an increased number of circulating B cells containing EBV in the blood.[87,88] During EBV infection several autoantibodies can transiently arise because of infection and proliferation of specific B cell clones. Autoimmune diseases, such as systemic lupus erythematosus, Sjögrens' syndrome, Grave's disease, and hepatitis, have been described following EBV infection.

Several models have been proposed to account for the etiology and pathogenesis of autoimmune diseases. One of these, the molecular mimicry model, states that an infecting agent, such as a bacterium or virus might be involved in the onset or pathology of the disease.[89] A role for the EBV in initiating Type 1 diabetes mellitus has been proposed because EBV BOLF I has an 11 amino acid identity with HLA-DQw8 β. We tested the relating homology between the BOLF1 molecule and the HLA-DQw8 β chain in recent onset type 1 diabetes mellitus. However, no simple relationship was observed.[90]

6. CONCLUSION

The EBV has been studied mostly because of its association with human cancer and lymphoproliferative diseases, although EBV is a ubiquitous virus which does not cause any disease in the majority of infected individuals. Many studies analyzed the structures and functions of EBV-induced proteins (Table 1).

TABLE I
Characteristics of EBV Antigens in EBV-Infected Cells

Antigens		ORF[a]	MW[b]	Localization[c]	Major functions
EBNA	1	BKRF 1	65–80	N	Maintainence of EBV genome
	2	BYRF 1	81–85		Cell transformation
	3A	BLRF1/BERF1	136		Cell transformation
	3B	BERF2a/BERF2b	142		
	3C	BERF3/BERF4	147		Transactivation
	LP	BWRF	22–70		
LMP	1	BNLF1	62	M, C	Cell transformation
	2A	BARF2/BNRF			Cell signaling
	2B	BNRF			Cell signaling
EA-D		BRLF1	93–96	N, C	Transactivation
		BMLF1	60		Transactivation
		BMRF 1	45–54		DNA replication
		BZLF 1	38		Transactivation
		BALF 2	130–138		DNA replication
EA-R		BHRF 1	17	C	Homology to bcl 2
		BGLF 5	52–60		EBV-DNase
		BOLF2	85		EBV-ribonucleotide rednctase
VCA		BcLF 1	150–160	N, C	Viral capsid
		BALF 4	110–125		
		BdRF 1	41		
		BGLF 1	36		
		BFRF 2	21		
		BFRF3	18–21		
		BCRF1	19		Homology to IL-10(viral IL-10)
MA		BLLF 1	350/220	M, C, N	Viral envelope
			110–125		
		BXLF2	85		

[a]ORF, open reading frame.
[b]MW, molecular weight (−kD).
[c]N, nucleus: C, cytoplasm; M, membrane.

The latent antigens, EBNA and LMP have been the focus of studies of EBV-induced cell transformation and oncogenicity which revealed their associations with the functions of B cells. It is noteworthy that the expression of EBV genes is regulated differently in normal cells and neoplastic cells, which is reflected in differences in cell growth and CTL responses.

The EA components function during the lytic cycle. The immediate early BZLF1 gene encodes ZEBRA protein, which transactivates other immediate early and early genes, and it can disrupt the latency of EBV. The early genes encode the early proteins which associate with viral DNA replication. The EBV DNA polymerase, catalytic subunits, ribonucleotide reductase, and DNase have been characterized. After viral DNA replication, viral proteins VCA and MA are synthesized, and the assembled virions are released from the cells.

The EBV reactivation plays a central role in the pathogenesis of EBV, however, we do not know how EBV reactivation is associated with it. It has been reported that the EBV genome encodes a homologue of IL-10 (BCRF1, viral (v) IL-10), which shares many of the cellular cytokine's biological activities.[91] The v-IL-10 is induced during EBV activation. There is evidence that vIL-10 inhibits CTL to EBV-infected cells and enhances the growth of EBV-transformed cells. Some cytokines produced upon EBV activation may promote EBV-associated diseases.

Antibodies to the virus envelope neutralize viral infectivity, kill lytically infected cells by ADCC, and protect against reinfections by EBV. However, latently infected cells would not be killed by these antibodies. Immune defenses against infected cells are thought to be mediated by CTL. EBV infection is also associated with several autoimmune diseases. It has been hypothesized that antigen mimicry is responsible for the mechanisms of autoantibody production.

Antibody response is a critical tool for diagnosing virally associated diseases. Its utility is clear for IM, the typical primary EBV infection. However, most of the EBV-associated diseases are chronic infections seen in EBV-associated tumors and chronically active EBV infection. Serodiagnosis should be used during the early period before the development of diseases. The difficulty is that the anti-EBV antibody is positive in almost all adults, and levels are relatively high. What is the best marker for early diagnosis and the prognosis for the various EBV associated diseases? The development of an ideal diagnostic system is still to be developed.

Acknowledgments. We thank Drs. Herman Friedman, Peter Medveczky, and Mauro Bendinelli for the opportunity of writing this paper and Ms. Sayuri Matsumoto for typing this paper.

REFERENCES

1. Henle, W., Henle, G., and Lennette, E. T., 1979, The Epstein–Barr virus, *Sci. Am.* **241:**48–59.
2. Kieff, E., 1996, Epstein–Barr virus and its replication, in *Fields Virology*, 3rd ed. (B. N. Fields, D. M. Knipe, and P. M. Howley, eds.), Lippincott-Raven, Philadelphia, pp. 2343–2396.
3. Rickinson, A. B., and Kieff, E., 1996, Epstein–Barr virus, in *Fields Virology*, 3rd ed. (B. N. Fields, D. M. Knipe, and P. M. Howley, eds.), Lippincott-Raven, Philadelphia, pp. 2397–2446.
4. Tokunaga, M., Land, C. E., Uemura, Y., Tokudome, T., Tanaka, S., and Sato, E. 1993, Epstein–Barr virus gastric carcinoma, *Am. J. Pathol.* **143:**1250–1254.
5. Reedman, B., and Klein, G., 1973, Cellular localization of an Epstein–Barr virus (EBV)-associated complement-fixing antigen in producer and non-producer lymphoblastoid cell lines, *Int. J. Cancer* **11:**499–520.

6. Fennewald, S., van Santen, V., and Kieff, E., 1984, The nucleotide sequence of a messenger RNA transcribed in latent growth transforming virus infection indicates that it may encode a membrane protein, *J. Virol.* **51:**411–419.
7. Henle, W., Henle, G., Zajac, B. A., Pearson, G., Waubke, R., and Scriba, M., 1970, Differential reactivity of human serums with early antigens induced by Epstein–Barr virus, *Science* **169:**188–190.
8. Henle, G., and Henle, W., 1966, Immunofluorescence in cells derived from Burkitt's lymphoma, *J. Bacteriol.* **91:**1248–1256.
9. Klein, G., Clifford, P., Klein, E., and Stjernsward, J., 1966, Search for tumor specific immune reactions in Burkitt's lymphoma patients by the membrane immunofluorescence reaction, *Proc. Natl. Acad. Sci. USA* **55:**1628–1635.
10. Epstein, M. A., Achong, B. G., and Barr, Y. M., 1964, Virus particles in cultured lymphoblasts from Burkitt's lymphoma, *Lancet* **I:**702–703.
11. Summers, W. P., Grogan, E. A., Sheed, D., Robert, M., Liu, C. R., and Miller, G., 1982, Stable expression in mouse cells of nuclear neoantigen after transfer of a 3.4 megadalton cloned fragment of Epstein–Barr virus DNA, *Proc. Natl. Acad. Sci. USA* **79:**5688–5692.
12. Yates, J., Warren, N., Reisman, D., and Sugden, B., 1984, A cis-acting element from the Epstein–Barr viral genome that permits stable replication of recombinant plasmids in latently infected cells, *Proc. Natl. Acad. Sci. USA* **81:**3806–3810.
13. Yates, J. L., Warren, N., and Sugden, B., 1985, Stable replication of plasmids derived from Epstein–Barr virus in various mammalian cells, *Nature* **313:**812–815.
14. Strnad, B. C., Schuster, T. C., Hopkins, R. F., III, Neubeauer, R. H., and Rabin, H., 1981, Identification of an Epstein–Barr virus nuclear antigen by fluoroimmunoelectrophoresis and radioimmunoelectrophoresis, *J. Virol.* **38:**996–1004.
15. Jones, M., Foster, L., Sheedy, T., and Griffin, B. E., 1984, The EB virus genome in Daudi Burkitt's lymphoma cells has a deletion similar to that observed in a nontransforming strain (P3HR1) of the virus, *EMBO J.* **3:**813–821.
16. Kempkes, B., Zimber-Strobl, U., Eissner, G., Pawlita, M., Falk, M., Hammerschmidt, W., and Bornkamm, G. W., 1996, Epstein–Barr virus nuclear antigen 2 (EBNA-2) - estrogen receptor fusion proteins complement the EBNA 2-deficient Epstein–Barr virus strain P3HR-1 in transformation of primary B cells but suppress growth of human B cell lymphoma lines, *J. Gen. Virol.* **77:**227–237.
17. Henkel, T., Ling, P. D., Hayward, S. D., and Peterson, M. G., 1994, Mediation of Epstein–Barr virus EBNA2 transactivation by recombination signal-binding protein J kappa, *Science* **265:**92–95.
18. Shimizu, N., Yamaki, M., Sakuma, S., Ono, Y., and Takada, K., 1988, Three Epstein–Barr virus-determined nuclear antigens induced by the BamHI E region of Epstein–Barr virus DNA, *Int. J. Cancer* **41:**744–751.
19. Robertson, E. S., Lin, J., and Kieff, E., 1996, The amino-terminal domains of Epstein–Barr virus nuclear proteins 3A, 3B, and 3C interact with RBPJk, *J. Virol.* **70:**3068–3074.
20. Petti, L., Sample, C., and Kieff, E., 1990, Subnuclear localization and phosphorylation of Epstein–Barr virus latent infection nuclear proteins, *Virology* **176:**563–574.
21. Chen, C-L., Sadler, R. H., Walling, D. M., Su, I-J., Hsieh, H-C., and Raab-Traub, N., 1993, Epstein–Barr virus (EBV) gene expression in EBV-positive peripheral T cell lymphomas, *J. Virol.* **67:**6303–6308.
22. Klein, G., 1994, Epstein–Barr virus strategy in normal and neoplastic B cells, *Cell* **77:**791–793.
23. Levitskaya, J., Coram, M., Levitsy, V., Imreh, S., Stelgerwald-Mullen, P. M., Klein, G., Kurilla, M. G., and Masucci, M. G., 1995, Inhibition of antigen processing by the internal repeat region of the Epstein–Barr virus nuclear antigen-I, *Nature* **375:**685–688.

24. Wang, D., Liebowitz, D., and Kieff, E., 1985, An Epstein–Barr virus membrane protein expressed in immortalized lymphocytes transforms established rodent cells, *Cell* **43:**831–840.
25. Mosialos, G., Birkenbach, M., Van Arsdale, T., Ware, C., Yalamanchili, R., and Kieff, E., 1995, The Epstein–Barr virus transforming protein LMP1 engages signaling proteins for the tumor necrosis factor receptor family, *Cell* **80:**389–399.
26. Chang, Y-S., Su, I-J., Shu, P-J., Chung, C-H., Ng, C. K., Wu, S-J., and Lui, S-T., 1995, Detection of an Epstein–Barr virus variant in T-cell lymphoma tissues identical to the distinct strains in nasopharyngeal carcinoma in the Taiwanese population, *Int. J. Cancer* **62:**673–677.
27. Gerber, P., 1972, Activation of Epstein–Barr virus by 5-bromodeoxyuridine in "virus-free" human cells, *Proc. Natl. Acad. Sci. USA* **69:**83–85.
28. zur Hausen, H., O'Neill, F. J., and Freese, U. K., 1978, Persisting oncogenic herpesvirus induced by the tumor promoter TPA, *Nature* **272:**373–375.
29. Daibata, M., Humphreys, R. E., Takada, K., and Sairenji, T., 1990, Activation of latent EBV via anti-IgG-triggered, second messenger pathways in the Burkitt's lymphoma cell line Akata, *J. Immunol.* **144:**4788–4793.
30. Daibata, M., Mellinghoff, I., Takagi, S., Humphreys, R. E., and Sairenji, T., 1991, Effect of genistein, a tyrosine kinase inhibitor, on latent EBV activation induced by cross-linkage of membrane IgG in Akata B cells, *J. Immunol.* **147:**292–297.
31. Mellinghoff, I., Daibata, M., Humphreys, R. E., Mulder, C., Takada, K., and Sairenji, T., 1991, Early events in Epstein–Barr virus genome expression after activation: Regulation by second messengers of B cell activation, *Virology* **185:**922–928.
32. Takagi, S., Daibata, M., Last, T. J., Humphreys, R. E., Parker, D. C., and Sairenji, T., 1991, Intracellular localization of tyrosine kinase substrates beneath cross-linked surface immunoglobulins in B cells, *J. Exp. Med.* **174:**381–388.
33. Daibata, M., Speck, S. H., Mulder, C., and Sairenji, T., 1994, Regulation of the BZLF1 promoter of Epstein–Barr virus by second messengers and anti-immunoglobulin-treated B cells, *Virology* **198:**446–454.
34. Sairenji, T., Hinuma, Y., Sekizawa, T., Yoshida, M., 1978, Appearance of early and late components of Epstein–Barr virus-associated membrane antigen in Daudi cells superinfected with P3HR-1 virus, *J. Gen. Virol.* **38:**111–120.
35. Takagi, S., Takada, K., and Sairenji, T., 1991, Formation of intranuclear replication compartments of Epstein–Barr virus with redistribution of BZLF1 and BMRF1 gene products, *Virology* **185:**309–315.
36. Henle, G., Henle, W., and Klein, G., 1971, Demonstration of two distinct components in the early antigen complex of Epstein–Barr virus-infected cells, *Int. J. Cancer* **8:**272–282.
37. Pearson, G. R., Vroman, B., Chase, B., Sculley, T., Hummel, M., and Kieff, E., 1983, Identification of polypeptide components of the Epstein–Barr virus early antigen complex with monoclonal antibodies, *J. Virol.* **47:**193–201.
38. Tsurumi, T., Kobayashi, A., Tamai, K., Yamada, H., Daikoku, A. T., Yamashita, Y., and Nishiyama, Y., 1996, Epstein–Barr virus single-stranded DNA-binding protein; Purification, characterization, and action on DNA synthesis by the viral DNA polymerase, *Virology* **222:**352–364.
39. Luka, J., Miller, G., Jornvall, H., and Pearson, G. R., 1986, Characterization of the restricted component of Epstein–Barr virus early antigens as a cytoplasmic filamentous protein, *J. Virol.* **58:**748–756.
40. Sairenji, T., Nguyen, Q. V., Woda, B., and Humphreys, R. E., 1987, Immune response to intermediate filament-associated, Epstein–Barr virus-induced early antigen, *J. Immunol.* **138:**2645–2652.
41. Goldschmidts, W. L., Ginsburg, M., and Pearson, G. R., 1989, Neutralization of Epstein–Barr

virus-induced ribonucleotide reductase with antibody to the major restricted early antigen polypeptide, *Virology* **170:**330–333.
42. Pearson, G., Luka, J., Petti, L., Sample, J., Brikenback, M., Braun, D., and Kieff, E., 1987, Identification of an Epstein–Barr virus early gene encoding for a second component of the restricted early antigen complex, *Virology* **160:**151–161.
43. Marchini, A., Tomkinson, B., Cohen, J., and Kieff, E., 1991, BHRF1, the Epstein–Barr virus gene with homology to Bcl-2, is dispensable for B-lymphocyte transformation and virus replication, *J. Virol.* **65:**5991–6000.
44. Freemer, C. S., Bertoni, G., Takagi, S., and Sairenji, T., 1993, A novel antigen associated with Epstein–Barr virus productive cycle, *Virology* **194:**387–392.
45. Daibata, M., and Sairenji, T., 1993, Epstein–Barr virus (EBV) replication and expressions of EA-D (BMRF1 gene product), virus-specific deoxyribonuclease, and DNA polymerase in EBV-activated Akata cells, *Virology* **196:**900–904.
46. Vroman, B., Luka, J., Rodriquez, M., and Pearson, G. R., 1985, Characterization of a major protein with a molecular weight of 160,000 associated with the viral capsid of Epstein–Barr virus, *J. Virol.* **53:**107–113.
47. Chen, M. R., Hsu, T. Y., Lin, S. W., Chen, J. Y., and Yang, C. S., 1991, Cloning and characterization of cDNA clones corresponding to transcripts from the BamHI G region of the Epstein–Barr virus genome and expression of BGLF2, *J. Gen. Virol.* **72:**3047–3055.
48. van Grunsven, W. M. J., van Heerde, E. C., de Haard, H. J. W., Spaan, W. M. J., and Middeldorp, J. M., 1993, Gene mapping and expression of two immunodominant Epstein–Barr virus capsid proteins, *J. Virol.* **67:**3980–3916.
49. Gong, M., and Kieff, E., 1990, Intracellular trafficking of two major Epstein–Barr virus glycoproteins, gp350/220 and gp110, *J. Virol.* **64:**1507–1516.
50. Hummel, M., Thorley-Lawson, D., and Kieff, E., 1984, An Epstein–Barr virus DNA fragment encodes messages for the two major envelope glycoproteins (gp350/330 and gp220/200), *J. Virol.* **49:**413–417.
51. Beisel, C. J., Tanner, T., Matsuo, T. T., Thorley-Lawson, D., Kezdy, F., and Kieff, E., 1985, Two major outer envelope glycoproteins of Epstein–Barr virus are encoded by the same gene, *J. Virol.* **54:**665–674.
52. Sairenji, T., Reisert, P. S., Spiro, R. C., Connolly, T., and Humphreys, R., 1985, Inhibition of Epstein–Barr virus (EBV) release from the P3HR-1 Burkitt's lymphoma cell line by a monoclonal antibody against a 200,000-dalton EBV membrane antigen, *J. Exp. Med.* **161:**1097–1111.
53. Tanner, J. J., Weis, D., Fearon, D., Whang, Y., and Kieff, E., 1987, Epstein–Barr virus gp350/220 binding to the B lymphocyte C3d receptor mediates adsorption, capping, and endocytosis, *Cell* **50:**203–213.
54. Miller, N., and Hutt-Fletcher, L. M., 1988, A monoclonal antibody to glycoprotein gp85 inhibits fusion but not attachment of Epstein–Barr virus, *J. Virol.* **62:**2366–2372.
55. Gong, M., Ooka, T., Matsuo, T., and Kieff, E., 1987, The Epstein–Barr virus glycoprotein gene homologous to HSVgB, *J. Virol.* **61:**499–508.
56. Miller, G., Niederman, J. C., and Stitt, D. A., 1972, Infectious mononucleosis: Appearance of neutralizing antibody to Epstein–Barr virus measured by inhibition of formation of lymphoblastoid cell lines, *J. Infect. Dis.* **125:**403–406.
57. Strnad, B. C., Schuster, T., Klein, R., Hopkins, R. F. III, Witmer, T., Neubauer, R. H., and Rabin, H., 1982, Production and characterization of monoclonal antibodies against the Epstein–Barr virus membrane antigen, *J. Virol.* **41:**258–264.
58. Takaki, K., Harada, M., Sairenji, T., and Hinuma, Y., 1980, Identification of target antigen for antibody-dependent cellular cytotoxicity on cells carrying Epstein–Barr virus genome, *J. Immunol.* **125:**2112–2117.
59. Khyatti, M., Patel, P. C., Stefanescu, I., and Menezes, J., 1991, Epstein–Barr virus (EBV)

glycoprotein gp350 expressed on transfected cells resistant to natural killer cell activity serves as a target antigen for EBV-specific antibody-dependent cellular cytotoxicity, *J. Virol.* **65:**996–1001.

60. Wallace, L. E., Wright, J., Ulaeto, D. O., Morgan, A. J., and Rickinson, A. B., 1991, Identification of two T-cell epitopes on the candidate Epstein–Barr virus vaccine glycoprotein gp340 recognized by CD4+ T cell clones, *J. Virol.* **65:**3821–3828.
61. Pither, R. J., Zhang, C. X., Shiels, C., Tarlton, J., Finerty, S., and Mogan, A. J., 1992, Mapping of B-cell epitopes on the polypeptide chain of the Epstein–Barr virus major envelope glycoprotein and candidate vaccine molecule gp340, *J. Virol.* **66:**1246–1251.
62. Mackett, M., Cox, C., Pepper, S. de V., Lee, J. F., Naylor, B. A., Wedderburn, N., and Arrand, J. R., 1996, Immunization of common marmosets with vaccinia virus expressing Epstein–Barr virus (EBV) gp340 and challenge with EBV, *J. Med. Virol.* **50:**263–271.
63. Harada, M., Sairenji, T., Takaki, K., and Hinuma, Y., 1980, IgM antibodies to Epstein–Barr virus-associated membrane antigen in sera of infectious mononucleosis patients, *Microbiol. Immunol.* **24:**123–132.
64. Sairenji, T., Bertoni, G., Medveczky, M. M., Medveczky, P. G., and Humphreys, R. E., 1991, Inhibition of Epstein–Barr virus production in P3HR-1 cells by Epstein–Barr virus-seropositive human serum, *Intervirology* **32:**37–51.
65. Zeng, Y., Zhang, L. G., Wu, Y. C., and Huang, Y. S., Huang, N. G., 1985, Prospective studies on nasopharyngeal carcinoma in Epstein–Barr virus - Ig A/VAC-antibody-positive persons in Wuzhou City, China, *Int. J. Cancer* **36:**545–547.
66. Henle, W., Henle, G., Andersson, J., Ernberg, I., Klein, G., Horwitz, C. A., Marklund, G., Rymo, L., Wellinder, C., and Straus, S. E., 1987, Antibody responses to Epstein–Barr virus-determined nuclear antigen (EBNA)-1 and EBNA2 in acute and chronic Epstein–Barr virus infection. *Proc. Natl. Acad. Sci. USA* **84:**570–574.
67. Luka, J., Chase, R. C., and Pearson, G. R., 1984, A sensitive enzyme-linked immunosorbent assay (ELISHA) against the major EBV-associated antigens. I. Correlation between ELISHA and immunofluorescence titers using purified antigens, *J. Immunol. Methods* **67:**145–156.
68. Inoue, N., Kuranari, J., Harada, S., Nakajima, H., Ohbayashi, M., Nakamura, Y., Miyasaka, N., Ezawa, K., Ban, F., and Yanagi, K., 1992, Use of enzyme-linked immunosorbent assays with chimeric fusion proteins to titrate antibodies against Epstein–Barr virus nuclear antigen 1, *J. Clin. Microbiol.* **30:**1442–1448.
69. Geertsen, R., Espander-Jansson, A., Dobec, M., Price, P., Wunderli, W., and Ryno, L., 1994, Development of a recombinant enzyme-linked immunosorbent assay for detection of antibodies against Epstein–Barr virus nuclear antigens 2A and 2B, *J. Clin. Microbiol.* **32:**112–120.
70. Joab, I., Triki, H., Martin, J. de S., Perricaudet, M., and Nicolas, J. C., 1991, Detection of anti-Epstein–Barr virus transactivator (ZEBRA) antibodies in sera from patients with human immunodeficiency virus, *J. Infect. Dis.* **163:**53–56.
71. Gorgievski-Hrishoho, M., Hinderer, W., Nebel-Schickel, H., Horn, J., Vornhagen, R., Sonneborn, H-H., Wolf, H., and Siegl, G., 1990, Serodiagnosis of infectious mononucleosis by using recombinant Epstein–Barr virus antigens and enzyme-linked immunosorbent assay technology, *J. Clin. Microbiol.* **28:**2305–2311.
72. Durda, P. J., Sullivan, M., Kieff, E., Pearson, G., and Rabin, H., 1993, An enzyme-linked immunosorbent assay for the measurement of human IgA antibody responses to Epstein–Barr virus membrane antigen, *Intervirology* **36:**11–19.
73. Farber, I., Wultzler, P., Wohlrabe, P., Wolf, H., Hinderer, W., and Sonneborn, H-H., 1993, Serological diagnosis of infectious mononucleosis using three anti-Epstein–Barr virus recombinant ELISAs, *J. Virol. Methods* **42:**301–308.
74. Cheng, Y. C., Chen, J. Y., Glaser, R., and Henle, W., 1980, Frequency and levels of antibodies to Epstein–Barr virus specific DNase are elevated in patients with nasopharyngeal carcinoma, *Proc. Natl. Acad. Sci. USA* **77:**6162–6165.

75. De Turenne-Tessier, M., Ooka, T., Calender, A., De Thé, G., and Daillie, J., 1989, Relationship between nasopharyngeal carcinoma and high antibody titers to Epstein–Barr virus specific thymidine kinase, *Int. J. Cancer* **43:**45–48.
76. Lin, L. S., Ro, L. H., Lo, M. S., Huang, W. L., Ma, J., Chang, T. H., Shu, C. H., Chow, K. C., Liu, W. T., Chen, K. Y., and Yang, H. L., 1995, Expression of the Epstein–Barr virus DNA polymerase in *Escherichia coli* for use as antigen for the diagnosis of nasopharyngeal carcinoma, *J. Med. Virol.* **45:**99–105.
77. Chen, J. Y., Chen, C. J., Liu, M. Y., Cho, S. M., Hsu, M. M., Lynn, T. C., Shieh, T., Tu, S. M., Lee, H. H., Kuo, S. L., Lai, M. Y., Hsieh, C. Y., Hu, C. P., and Yang, C. S., 1987, Antibodies to Epstein–Barr virus specific DNase in patients with nasopharyngeal carcinoma and control groups, *J. Med. Virol.* **23:**11–21.
78. Sairenji, T., Sullivan, J. L., and Humphreys, R. E., 1984, Complement-dependent, Epstein–Barr virus-neutralizing antibody appearing early in the sera of patients with infectious mononucleosis, *J. Infect. Dis.* **149:**763–768.
79. Sairenji, T., Katsuki, T., and Hinuma, Y., 1976, Cell killing by Epstein–Barr virus: Analysis by colony inhibition procedure, *Int. J. Cancer* **17:**389–395.
80. Sairenji, T., and Hinuma, Y., 1980, Re-evaluation of a transforming strain of Epstein–Barr virus from the Burkitt lymphoma cell line, Jijoye, *Int. J. Cancer* **26:**337–342.
81. Sairenji, T., Reisert, P. S., Spiro, R. C., and Humphreys, R. E., 1996, Decreased expression of early antigens in P3HR-1-EBV superinfected Raji cells cultured in EBV-seropositive human sera, *Yonago Acta Medica* **39:**99–107.
82. Sairenji, T., Bertoni, G., Medveczky, M. M., Medveczky, P. G., Nguyen, Q. V., Humphreys, R. E., 1988, Inhibition of Epstein–Barr virus (EBV) release from P3HR-1 and B95-8 cell lines by monoclonal antibodies to EBV membrane antigen gp350/220, *J. Virol.* **62:**2614–2621.
83. Shope, T., and Miller, G., 1973, Epstein–Barr virus: Heterophile responses in squirrel monkeys inoculated with virus-transformed autologous leukocytes, *J. Exp. Med.* **137:**140–147.
84. Maeda, M., Sairenji, T., and Hinuma, Y., 1979, Reactivity of Paul–Bunnell type-heterophile antibody in sera from infectious mononucleosis patients with the surface of lymphoid cells carrying Epstein–Barr virus genomes, *Microbiol. Immunol.* **23:**1189–1197.
85. Linder, E., Kurki, P., and Andersson, L. C., 1979, Autoantibody to "intermediate filaments" in infectious mononucleosis, *Clin. Immunol. Immunopathol.* **14:**411.
86. Rhodes, G., Rumpold, H., Kurki, P., Patrick, K. M., Carson, D. A., Vaughan, J. H., 1987, Autoantibodies in infectious mononucleosis have specificity for the glycine-alanine repeating region of the Epstein–Barr virus nuclear antigen, *J. Exp. Med.* **165:**1026–1040.
87. Sculley, T. B., Walker, P. J., Moss, D. J., and Pope, J. H., 1984, Identification of multiple Epstein–Barr virus-induced nuclear antigens with sera from patients with rheumatoid arthritis, *J. Virol.* **52:**88–93.
88. Tosato, G., Steinberg, A. D., Yarchoan, R., Heliman, C. A., Pike, S. E., DeSeau, V., and Blaese, R. M., 1984, Abnormally elevated frequency of Epstein–Barr virus-infected B cells in the blood of patients with rheumatoid arthritis, *J. Clin. Invest.* **73:**1789–1795.
89. Oldstone, M. B. A., 1987, Molecular mimicry and autoimmune disease, *Cell* **50:**819–820.
90. Sairenji, T., Daibata, M., Sorli, C. H., Qvistbäck, H., Humphreys, R. E., Ludvigsson, J., Palmer, J., Landin-Olsson, M., Sundkvist, G., Michelsen, B., Lernmark, A., and Dyrberg, T., 1991, Relating homology between the Epstein–Barr virus BOLF1 molecule and HLA-DQw8 B chain to recent onset type I (insulin-dependent) diabetes mellitus, *Diabetologica* **34:**33–39.
91. Moore, K. W., Vieira, P., Fiorentino, D. F., Trounstine, M. L., Khan, T. A., and Mosmann, T. R., 1990, Homology of cytokine synthesis inhibitory factor (IL-10) to the Epstein–Barr virus gene BCRF1, *Science* **248:**1230–1234.

9

EBV Persistence *in Vivo*

Invading and Avoiding the Immune Response

DAVID A. THORLEY-LAWSON

1. INTRODUCTION

The human herpesvirus Epstein–Barr virus (EBV), like other herpesviruses, sets up a lifelong, persistent infection in the host. To achieve this, it invades the immune response by infecting normal B lymphocytes and persisting within them as a latent infection. At the same time the virus avoids the immune response because it persists despite aggressive humoral and cellular immunity directed against the virus and virally infected cells. The characteristics of the antibody responses to the virus[1] and the cytotoxic T cell (CTL) response to virus infected cells[2] have been described in detail in many reviews and are not discussed here.

EBV has long been known to be trophic for B lymphocytes. The virus has also been found, however, in a much wider range of cell types, usually in association with tumors (for a general review and specific references see Rickinson[3]). These include epithelial cells in the form of nasopharyngeal carcinoma (NPC) and gastric carcinoma, nasal lymphomas originating from T lymphocytes and natural killer (NK) cells and the precursor cells of Hodgkin's lymphoma (HD). Earlier claims of finding EBV in association with normal epithelial cells[4] led to extensive speculation that epithelial cells play a role in the normal biology of EBV (see, for

---

DAVID A. THORLEY-LAWSON • Department of Pathology, Tufts University School of Medicine, Boston, Massachusetts 02111.

Herpesviruses and Immunity, edited by Medveczky *et al.* Plenum Press, New York, 1998.

example, Rickinson[3,5]). However, recent studies with highly sensitive techniques have failed to reproduce these findings[5a] and have shown that all of the infected cells in mucosal epithelial tissue are B lymphocytes.[6] Therefore, it is reasonable to conclude that the normal pathogenesis of EBV occurs entirely within the B lymphoid compartment. Access to other cell types is fortuitous and plays no role in the viral life cycle but is a significant risk factor in the subsequent development of neoplastic diseases.

Therefore, it is the goal of this chapter to describe the known properties of EBV in the context that it is a "stealth" virus that has efficiently and effectively invaded the B lymphocyte arm of the immune response. Specifically, the site of immunologic memory is a perfect site for viral persistence because immunologic memory, like EBV, is for life. The form of this chapter is to follow the virus from the moment of entry into the host, through initial spread in the body, to the subsequent establishment of long-term persistence, and finally reactivation and release for reinfection. The goal is to take many diverse properties of the virus and show how they can be best understood in terms of a model where the virus uses B lymphocyte biology for its own ends. Where evidence is lacking, speculations are offered that are consistent with the model and hopefully stimulate experimentation.

2. VIRAL ENTRY AND THE ESTABLISHMENT OF INFECTION

2.1. Viral Entry

It is generally accepted that EBV is spread through salivary contact. Most adults have been infected with the virus, and infectious transforming virus is readily detected in the saliva of most individuals throughout their lives. How the virus gains access to the circulatory system is not known. However, it is likely that the first infection is of B cells in mucosal epithelium.

2.1.1. Early Events in Infection in Vitro

What happens next *in vivo* is an area for speculation based heavily on what is observed *in vitro*. EBV can infect all B cells because of the ubiquitous expression of the viral receptor CD21. For reasons discussed below, however, it only establishes latency upon infecting resting B cells.[7] When EBV infects resting B lymphocytes *in vitro*, they become activated B cell blasts[8] and express at least nine virally encoded latent proteins, including six nuclear antigens (Epstein–Barr virus nuclear antigens - EBNA's) and three membrane proteins (latent membrane proteins - LMP's) (for a general review, see Kieff[9]). These cells proliferate indefinitely and are called immortalized or transformed lymphoblastoid cell lines (LCL) expressing the growth program[10] or latency III.[3]

CD21 is the receptor for the C3d component of complement. The major viral glycoprotein, gp350/220, recognizes and binds to the same motif on CD21 as C3d. Thus, the resting B cell is essentially tricked into believing that the virus is a component of the complement system binding to its cognate receptor.[11] CD21 forms part of a multimeric signal transduction complex with CD19, TAPA-1, and Leu-13. Cross-linking of this complex by gp350/220 provides the necessary signal to move the resting B cells out of G0 and, together with the earliest expressed latent proteins (EBNA2 and EBNA LP), to drive the cells through the first G1[12] on their way to entering the cell cycle.

Why does EBV immortalize only resting B cells? The reason for this is that a critical event in establishing latent infection can occur only as resting B cells transit into the cell cycle. This event is circularization of the linear viral genome.[13] The presence of a covalently closed circular viral episome, in the absence of linear viral DNA, is a hallmark of a latent herpesvirus infection. It is essential that the viral genome becomes circularized during a latent infection so that it is replicated along with the cellular DNA in the proliferating LCL.[14] Failure to achieve this results in loss of the viral DNA and death for the cell. This is the consequence when the virus infects already activated B cells. In the case of EBV, only one of the incoming linear genomes circularizes and is retained. This occurs when the newly infected B cell is in transition from G0 through the first G1 before entry into the cell cycle.[13] Subsequently, the single viral genome generally undergoes amplification to yield 10–50 copies per cell. Whether this is an active process or occurs through uneven segregation of viral genomes is unclear.

The activation and proliferation of B lymphocytes in response to an antigen during a normal immune response requires an interaction with antigen-specific helper T lymphocytes (Th cells). The Th cells provide signals to the B cell by engaging cell surface receptors called costimulatory molecules and by releasing soluble mediators called lymphokines, interleukins, or cytokines. The activation and growth of B cells upon infection with EBV *in vitro,* however, can occur without exogenous contact with Th cells or lymphokines. It is becoming apparent that the virus achieves this by encoding for a set of latent proteins that are expressed as the infected cell traverses and exits the first G1 before entry into the cell cycle. These proteins provide a series of signals that mimic/replace those delivered by antigen-specific Th cells. The result is that the B cell is tricked again, this time believing that it is responding to cognate antigen, and as a consequence the B cell expresses an activated phenotype and starts to proliferate. It is for this reason that the first markers of B cell activation were found on EBV-infected B cells.[8]

When a normal B cell becomes activated in response to antigen and Th cell signaling, it goes through several rounds of cell division, then exits the cell cycle, and terminally differentiates to become a plasma cell secreting large amounts of antigen-specific antibody. B cells activated through infection with EBV, however, proliferate indefinitely. Therefore, the virus must possess a molecular mechanism that prevents the cells from terminally differentiating because this would result in growth arrest. This differentiation block, together with the powerful selection of growth against cells that differentiate and growth arrest, probably explains why immortalized cells *in vitro* are remarkably resistant to terminal differentiation signals. However, as discussed later, these cells must respond to differentiation signals *in vivo.*

In summary, it is apparent that EBV achieves and maintains immortalization *in vitro* in a two-step process. In the first step, the virus infects and activates B cells analogously to normal B cell activation (the growth signal), and in the second step the B cell is blocked from terminal differentiation and so continues to proliferate (the differentiation block).[7] The molecular confirmation of this model is now beginning to take shape and is summarized below and in Fig. 1. One simple conclusion from this model is that B cells driven to proliferate by EBV are essentially normal B cells that have been tricked into growth as though they are responding to an antigen. They are not classically transformed cells, such as are obtained with other DNA tumor viruses, for example, SV40, papillomavirus, and adenovirus. This would explain why the cells are not tumorigenic in nude mice and why the functions of the tumor suppressor gene p53 appear normal.[15] The conclusion that EBV-immortalized cells are essentially normal has been challenged recently, based on the claim that EBV may encode a classical dominant oncogenic activity. Thus, it has been observed that, unlike normal proliferating B cells, LCL do not growth arrest in response to DNA damage.[16] Rather the LCL undergo p53-dependent apoptosis. This suggests that a constitutively active viral oncogene is driving the cells through arrest at the G1-S checkpoint imposed by p53, resulting in apoptosis. Furthermore, it has been shown that the EBV-encoded nuclear antigen EBNA3C is functionally analogous to the adenovirus E1A and papillomavirus E7 oncogenes,[17] in that it cooperates with the Ha-ras oncogene to transform primary rodent fibroblasts via a mechanism that involves neutralizing the tumor suppressor gene Rb. Whether these properties are essential components of EBV pathogenesis *in vivo* is unclear. It is also possible that they represent effects due to tissue culture selection and may be more pertinent to aberrant cellular growth caused by EBV in developing neoplastic disease.

2.1.a. The Growth Signal -LMP1. LMP1 is one of the virally encoded proteins expressed in latently infected B cells immortalized by EBV (for a general review and specific references, see Kieff[9]). It is a membranous protein whose structure resembles that of the classical seven transmembrane-spanning chemokine receptor family (reviewed in Horuk[18]) except that it lacks the most amino-terminal transmembranous sequence (Fig. 2). As a consequence, both amino- and carboxyl-terminal ends of the protein project into the cytoplasm. LMP1 is essential for B cell immortalization and has been shown to be an oncogene in classical transformation assays with rodent fibroblasts. It is probably not a coincidence that LMP1 is one of the most common of the latent proteins to be found in EBV-associated tumors, particularly those not of B cell origin, such as NPC.

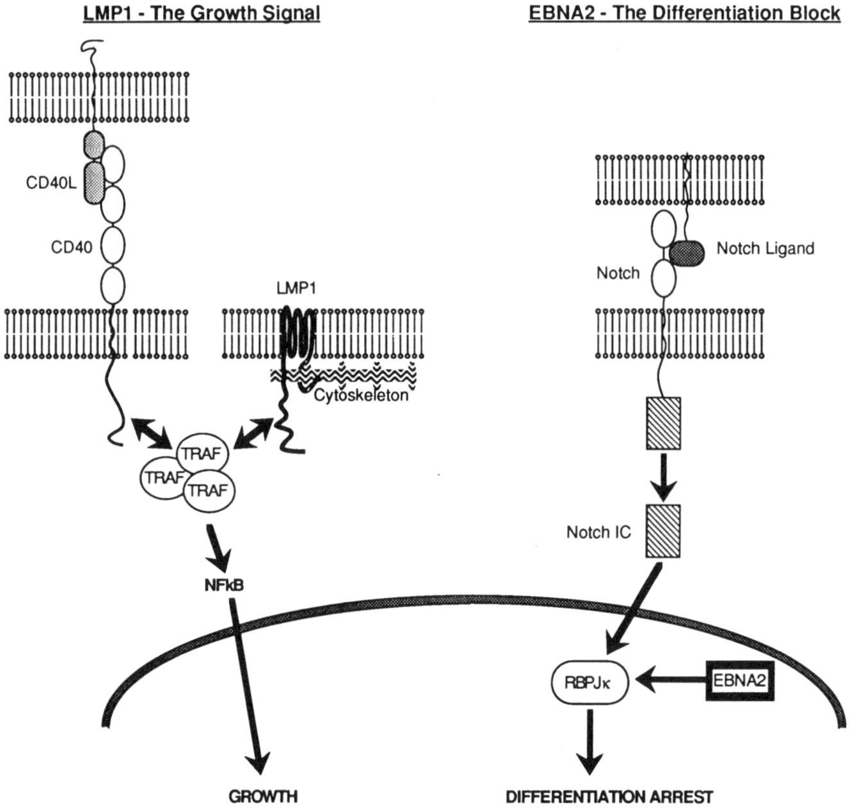

FIGURE 1. The two signaling events involved in the EBV-driven proliferation of LCLs. The growth signal is provided by LMP1. In a normal proliferating B cell, the CD40 molecule on the B cell surface interacts with its ligand CD40L on the surface of an antigen-specific helper T cell (Th cell). For the signal to be delivered into the B cell, it is believed that a trimer of CD40L molecules has to multimerize a trimer of CD40 molecules. This results in binding TRAF molecules to the cytoplasmic tail of CD40 which in turn leads to activation of NfκB and a growth signal. LMP1 usurps these functions by directly binding TRAF molecules, thereby removing the need for external signals. LMP1 is constitutively activated because it is multimerised by binding its amino terminus to the cytoskeleton. The differentiation block is provided by EBNA2. Based on work in other systems, particularly *Drosophila* muscle development, it is believed that proliferating cells arrest growth and begin to differentiate through a mechanism mediated by the Notch receptor and the DNA binding protein RBPJκ. Briefly, it is believed that interaction of Notch with its ligand causes a cleavage reaction that releases the cytoplasmic domain of Notch (Notch IC) which migrates to the nucleus to bind RBPJκ. RBPJκ mediates binding this complex to the regulatory region of various genes. The product of one of these genes blocks the expression of a master transcription factor, myo D in the case of muscle, which is required for the cells to arrest growth and differentiate. EBNA2 usurps this function by binding to the same region of RBPJκ as Notch. It is believed that the downstream consequence of this is that the B cell cannot arrest growth and continues to proliferate driven by LMP1.

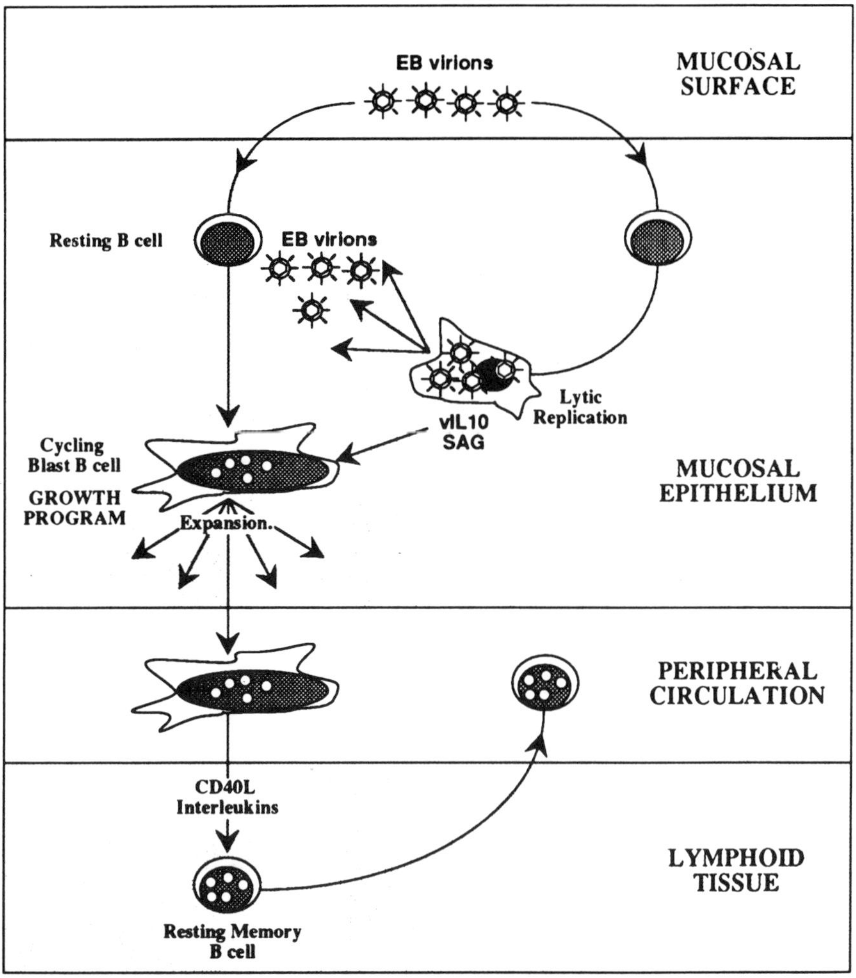

FIGURE 2. A schematic diagram of the processes believed to occur when EBV initially infects. The virus enters through salivary contact and infects B cells in the mucosal epithelium. A local nidus of infection is established involving lytic replication of the virus and the expansion of lymphoblasts expressing the growth program/latency III. The early stages of infection are aided by several events: (1) The virus encodes for an IL-10 analog that augments the efficiency of subsequent infection. (2) The virus expresses a superantigen. This is probably both immunosuppressive, delaying the onset of the immuneresponse and giving the virus time to establish itself, and stimulatory for the production of lymphokines that favor efficient infection and transformation. (3) The virus also stimulates high-level expression on the newly infected cells of a receptor (CD48) for heparin sulfate that likely serves to localize the infected cells to a limited area and also increases the efficacy of the viral IL-10. Eventually latently infected proliferating lymphoblasts enter the peripheral circulation. The role of these cells is to expand the number of viral genomes through cellular proliferation and to provide access for the virus to the B cell compartment. These cells are eventually eliminated by cytotoxic T cells. However, before this occurs some fraction of them become resting memory-like B cells, the form in which the virus persists for the lifetime of the host. It is hypothesized that this last transition occurs in the lymphoid tissue.

LMP1 is expressed in newly infected B cells in vitro just as they emerge from the first G1 into the S phase of the cell cycle.[19] It is almost certain that the role of LMP1 is to drive cells into S and through the cell cycle. The reason for believing this is that the signaling system employed by LMP1 is analogous to that of a very potent stimulator of B cell growth—CD40. CD40 is a member of a large family of receptors known as the tumor necrosis factor receptor (TNFR) family (reviewed in Armitage[20]). They are transmembranous receptors with large extracellular domains that bind ligands which themselves constitute a family of related molecules. The receptors and their ligands are widely expressed in both lymphoid and nervous tissues and are intimately involved in life and death decisions by cells (reviewed in Baker and Reddy[21]). The interaction of CD40, on an antigen activated B cell, with its ligand (CD40L), expressed on an antigen-specific Th cell, results in delivering a potent survival and growth signal. This is one of the most potent of the costimulatory signals mentioned above and is essential for expanding and differentiating antigen-specific B cells in lymphoid tissue *in vivo*. In the absence of this signal, the cells rapidly undergo apoptosis and die.[22,23] These events can be recapitulated *in vitro* where CD40L is a potent growth-promoting agent for freshly isolated peripheral B cells. When combined with the appropriate lymphokines, CD40L signaling causes the B cell to terminally differentiate and undergo switching of its immunoglobulin isotype.[22] B cells driven to proliferate by CD40L and the lymphokine interleukin 4 (IL-4) are phenotypically remarkably similar to B cells activated by EBV infection *in vitro*.

The parallels between signaling by LMP1 and CD40 are striking (Fig. 2). When CD40 is engaged by its ligand, it signals through association with members of a family of cytoplasmic molecules called TNFR-associated factors (TRAF's) because they were discovered through their association with members of the TNFR family (see Baker and Reddy[21] for a review). Association of TRAF's and CD40 leads to activation of the cytoplasmic transcription factor NFκB which then translocates to the nucleus resulting in survival and growth of the B cell. The amino-terminal sequence of LMP1 is tethered to the cytoskeleton and functions as a constitutively engaged receptor–ligand complex.The carboxyl terminus of LMP1 associates with TRAF molecules[24] and activates NFκB.[25] Unlike CD40, however, these two functions map to separate motifs within the LMP1 carboxy terminus.[26,27] It is likely though that the signaling mechanism used by LMP1 involves association with TRAF molecules and NFκB activation leading to B cell growth analogous to a constitutively activated CD40. Further support for this idea comes from recent observations that cell lines immortalized with EBV expressing a regulatable LMP1 continue to grow normally in the absence of LMP1 expression, if the cells are stimulated through CD40.[27a]

In summary, therefore, it appears that EBV has used LMP1 to trick the B cell into responding as though it were receiving a constitutive growth signal from antigen-specific Th cells through the interaction of CD40 with its ligand.

2.1.1b. The Differentiation Block—EBNA2. EBNA2 is a transcriptional transactivator that plays an essential and central role in EBV-driven immortalization[28] (for a general review and specific references, see Kieff[9]). It is the first latent protein detected in newly infected cells and appears as the cells begin to traverse the first G1. EBNA2-dependent elements are present in the promoters for all of the latent proteins. In addition to regulating the expression of the viral latent genes, EBNA2 also transactivates cellular genes and, in concert with LMP1, probably acts to induce the cellular genes, such as CD23, characteristically expressed by activated B cells.

EBNA2 does not bind directly to DNA but interacts with a cellular DNA-binding protein called RBPJκ or CBP1. This discovery gives deep insight into the potential role of EBNA2 because a great deal is known about RBPJκ/CBP1 from developmental biology (Fig. 2). In *Drosophila* muscle development a cell surface receptor termed Notch (reviewed in Artavanis-Tsakonas *et al.*[29]), upon interaction with its ligand, is cleaved on the cytoplasmic side to release a soluble produced called Notch IC. This moiety migrates to the nucleus where it interacts with RBPJκ/CBP1 through the same site as EBNA2.[30] Thus, EBNA2 and Notch IC can be thought of as functional analogs.

The interaction of Notch IC with RBPJκ/CBP1 leads to the transcription of a gene called HES (hairy enhancer of split) which in turn represses expression of the muscle differentiation master gene myoD. The role of myoD is to induce the activity of the cell-cycle inhibitor p21. Therefore, in the absence of myoD, p21 is not induced, the cells cannot arrest growth and differentiate, and therefore they keep growing. By analogy, we may speculate that, because Notch and EBNA2 are functional homologues, the association of EBNA2 with RBPJκ/CBP1 results in the differentiation arrest and continued proliferation observed when EBV immortalizes B cells. The prediction of this model is that a master gene(s), analogous to myoD in the muscle system, must exist in B cells. This master gene should be negatively regulated through RBPJκ/CBP1 and should be required for activated B cells to exit from the cell cycle and terminally differentiate into plasma cells.

In summary, EBNA2 may trick the B cell into continuing to cycle by constitutively binding to the regulatory factor RBPJκ/CBP1, thereby blocking the expression of genes essential for cell-cycle arrest and differentiation.

2.2. Establishment of the Infection

2.2.1. Getting a Foothold in the Mucosal Epithelium

Although we are learning a great deal about what occurs when EBV infects B cells *in vitro*, we know remarkably little about what occurs *in vivo*. It is likely that the situation is quite different. Virus enters through saliva exchange and presumably infects B cells in the epithelium. It is likely that the first events of infection involve establishing the infection in a local area. Although this could involve expansion of immortalized B cells into a nidus, the first round of infection probably involves binding of only a few virions to a small number of B cells. Thus it is unlikely that there is sufficient virus to cross-link enough of the CD21 receptor to provide the signal that pushes the resting cells into the cell cycle.

Therefore, it is quite likely that the first infections are lytic until sufficient concentrations of virus are produced to transform B cells. This probably explains the existence of a number of viral properties that are not essential for immortalization *in vitro* but play a key role in increasing the survival and expansion of infected cells *in vivo*. During the lytic cycle, the virus encodes for its own version of IL-10,[31] which is capable of suppressing the local immune response[32] and increasing the efficiency of viral transformation.[33] Eventually sufficient virus is released to cross-link enough of the viral receptors of newly infected cells to push them into the cycle and become lymphoblasts. To keep this small number of proliferating cells localized, the lymphoblasts express a number of adhesion molecules, most notably CD48.[8] CD48 binds to a specific form of heparin sulfate found in the basal epithelium and lamina propria of the mucosa,[33a] where EBV first establishes infection. The effect of this is that the infected proliferating lymphoblasts are preferentially retained at the site of the original infection in the mucosal epithelium. The viral IL-10 also interacts synergistically with heparin sulfate to increase the efficiency of transformation (Mackett, personal communication). Thus, the expression of CD48 on the surface of EBV-infected lymphoblastoid cells ensures that they expand in a localized area and that viral IL-10 is produced in an optimal environment to enhance further transformation. The synergy between heparin sulfate and IL-10 likely reflects the ability of glycosaminoglycan (GAG) structures like heparin sulfate to create a locally high concentration of a growth factor or interleukin through low-affinity binding to GAG.[34]

2.2.2. Breaking Out into the Peripheral Circulation

Another property of the virus that probably plays an important role early in establishing the infection is the expression of a superantigen during the lytic cycle.[35] In an antigen-independent fashion, superantigens stimulate T cells bearing receptors expressing specific Vβs (reviewed in Huber *et al.*[36]). The consequence of this is that a large fraction of the T cell repertoire becomes nonspecifically activated. The advantages of this to the virus are twofold. First, activated T cells produce lymphokines that could create a favorable environment for more efficient infection and transformation of B cells. Lymphokines could also push infected B cells more efficiently into the lytic cycle, thus driving the whole circuit of lytic replication and reinfection. Second, the large scale activation of antigen nonspecific T cells may lead to confusion and delay of the specific immune response, i.e., functional immunosuppression that provides time for the virus to become established.

Under the cover of this immunosuppression, proliferating infected lymphoblasts can expand and begin to migrate into the peripheral circulation. This is probably the major functional role of the growth program, to replicate and disseminate the viral DNA, not as free virus, but as viral DNA in proliferating cells. The best support for this scenario comes from observations on acute infectious mononucleosis (IM) if we assume that IM represents an exaggerated, but representative, version of primary EBV infection. IM is characterized by widespread activation of T cells and B cells. This is probably a downstream consequence of both specific activation by viral antigens on virions and infected cells and nonspecific activation by the viral superantigen and the appearance of EBV-infected lymphoblasts in the periphery. These lymphoblasts express the full panoply of latent genes and are proliferating.[37,38] The very existence of proliferating B cell blasts in the periphery speaks to the ability of EBV to disrupt the immune system because activated B cells would normally be restricted to the lymph nodes. Indeed the presence of these cells in the periphery probably reflects the tip of the iceberg, and the majority of infected cells reside in the lymph nodes. Immunohistological analysis of tonsils from acute IM cases reveals large numbers of virally infected B cells that express a wide range of phenotypes from blasts to fully differentiated cells indicating that viral infection *in vivo* does not simply give rise to the narrow range of phenotypes, i.e., only lymphoblastoids, as seen during infection *in vitro*.[6]

Although B cells that express the growth program can play a useful role in replicating and disseminating the virus and in giving the virus access to the B cell compartment, their usefulness is short-lived because specific T cell responses soon occur which aggressively destroy B cells expressing the growth program genes almost all of which, but particularly EBNA3a, 3b, and 3c, are potent targets for CTLs (see Khanna *et al.*[2] for a review and source of original references). Disruption of the T cell response due to immune suppression can lead to dangerous consequences in the form of aggressively growing lymphomas (see Rickinson[3] for a general review and specific references) expressing, and presumably driven by, the viral growth-promoting latent genes. The importance of the CTL response to the long-term survival of the infected host can be gauged from the observation that a large fraction of the memory CTL repertoire is given over to CTL precursors specific for EBV latent gene products.[39] Indeed, it appears that the virus itself has invested in helping to ensure the destruction of the life-threatening growth program B cells. Evidence for this comes from the observation that there is significant sequence conservation of the major immunogenic epitopes of the latent genes that are recognized by T cells[40] as though the virus has been selected for its ability to be killed! At first glance this seems counterintuitive and directly opposite to what is seen with HIV, for example, which can undergo rapid variation in its CTL epitopes in response to immune selection.[41] However, what has really been selected for in the case of EBV is the ability of the virus to be killed in a specific context—the growth program B cell. This no doubt reflects the fact that EBV depends on the long-term survival of the infected host to maximize the chances of infectious spread to new hosts. Thus, the growth program is required to establish the initial infection but thereafter is no longer useful because it poses a potential threat to the host which, in turn, is a threat to the long-term survival of the virus in the host. Therefore the virus has evolved to ensure that these cells are rapidly cleared by the immune response once the infection is established.

Now this discussion begs the question: if B cells expressing the growth program represent a potential and undesirable threat to the host, how does the virus persist once the acute infectious stage is over? This is the subject of the next section.

3. LONG-TERM LATENCY

3.1. The Site of Long-Term Persistence—Resting Memory B Cells

3.1.1. The Cell Type

Very recently a great deal of progress has been made toward an understanding of where and how EBV persists *in vivo*. It has been long known that virally infected cells are in the peripheral blood. Attempts to characterize the cells produced conflicting and confusing results because sensitive and reproducible assays were not available to detect them. This has changed, however, with the advent of highly sensitive PCR techniques and sophisticated fractionating methods involving the fluorescence-activated cell sorter or magnetic beads coupled to monoclonal antibodies against specific lymphocyte subsets. The first key observation was that the cell surface phenotype of the virally infected cells is not that of a lymphoblastoid B cell expressing the growth program because the activation markers CD23 and CD80(B7) are not expressed.[42] Subsequent analysis has shown that these cells are in G0, i.e., not in the cell cycle,[43] are tightly latent,[44] carry only a limited number of viral genomic copies (2–5),[42] and are surface-positive for either IgA, IgM, or IgG but do not express surface IgD.[44a] Thus, the cells have all of the characteristics of resting, long-term memory B cells. The lack of infected naive B cells in the periphery essentially precludes the idea that the virus is persisting in a stem cell population that continuously produces latently infected mature B cells since, if this were the case, the virus should continue to be detected in virgin B cells. Therefore, it appears that the virus has found a niche for lifetime persistence within B cell memory because immunologic memory is for life. The terms latency program[10] or latency 0 (Ernberg, personal communication) have been coined to describe this form of latent infection.

3.1.2. Viral Gene Expression

As mentioned before, B cells immortalized by EBV *in vitro* express at least nine virally encoded latent proteins including the six EBNAs and three LMPs (for a general review and original references, see Kieff[9]). Many of these, including EBNA2, 3a, 3c, and LMP1, are essential for immortalization *in vitro*. EBNA1 is also required, in the sense that it is essential for replicating the viral episome[14] (see below). The other latent genes (EBNA LP, 3b, and LMP2a and 2b) are not essential for immortalization but probably play essential roles *in vivo*.

Although viral gene expression has been studied in peripheral blood B cells,[38,45,46] definitive conclusions are difficult to draw because the RTPCR technology used in these studies is not quantitative enough to give unequivocal results. Some consensus has been reached, however, and the results are consistent with the conclusions derived from the quantitative DNA PCR approach used in the studies described in the section above. Specifically, the genes associated with the lytic cycle have not been detected,[38] consistent with the conclusion that the cells are tightly latent based on the absence of linear viral genomes in the peripheral blood.[44] Similarly, the growth-promoting viral latent genes, such as EBNA2 and LMP1, have not been detected[38,45,46] as expected, based on the observation that the cells are out of the cell cycle. There are conflicting reports on the expression of EBNA1. For reasons discussed below, however, it is unlikely that EBNA1 is expressed in the resting latently infected B cells. The unexpected result found in the first study[46] and subsequently confirmed is that the only viral latent gene to be consistently and readily detected is LMP2A. LMP2A possesses a functional ITAM motif (YXXL/I)2 on its cytoplasmic tail.[47] This motif is found in the Igα and Igβ chains of the B cell receptor (BCR) complex and mediates signaling by the BCR and other receptors by engaging tyrosine kinases of the src family. Because LMP2A has no external receptor domain to bind a ligand, the functional consequences of expressing LMP2A are for it to act as a dominant negative inhibitor, sequestering the tyrosine kinases and preventing the B cell from responding to exogenous signals.[48] This provides a mechanistic explanation for the observation that the cells are tightly latent in the peripheral circulation[44] and suggests that the virus has evolved a mechanism to specifically block unwanted signaling to the B cell that could lead to viral reactivation in the wrong location, i.e., the peripheral blood.

3.2. Maintenance of Long-Term Latency—Some Facts and Some Speculation

3.2.1. Stability of the Frequency of Virally Infected Cells and the Analogy to B Cell Memory

The number of virally infected cells in the peripheral blood of healthy persistently infected individuals varies between 5 and 500 per 10^7 B cells.[42,49]

This translates to 0.05–5 per ml of blood. Because we do not know the distribution of virally infected B cells throughout the lymphoid tissue, we cannot precisely calculate the total number of virally infected cells in an individual. However, assuming uniform distribution and approximately 2×10^{11} B cells per adult, we can estimate that there are approximately 10^6 virally infected cells per adult. This is more than enough to sustain infection.

One of the most striking observations on the numbers of infected cells is that the frequency in healthy donors is remarkably stable over several years.[49] Because the infected cells are phenotypically memory B cells[44a] and their numbers are stable over time, it is likely that the latently infected B cells in the periphery are being maintained as though they were normal memory B cells.

How then might EBV gain selective access to the peripheral memory B cell compartment? There is no direct information on this but the possible scenarios are constrained by what we already know.

It is conceivable that the virus could, under certain circumstances *in vivo,* selectively infect resting B cells and move directly into the latency program. This is unlikely since it would obviate the need for the growth program which in and of itself amplifies the latent viral genomes and provides access to the B cell compartment. Furthermore, the viral genome is in the form of the covalently closed circular episome[44] and it is known that the viral genome requires the cell to become activated in order to circularize.[13] This suggests that the resting, latently infected cells in the periphery proliferated when first infected. It is also clear that the virus does not preferentially infect memory cells *in vivo*. This is demonstrated by studies on tonsillar lymphoid tissue from healthy individuals.[44a] Unlike the peripheral blood, viral replication is frequently ongoing in the tonsils and both naive and memory B cells are infected. Therefore selectivity for the memory compartment in the periphery must arise through selective loss of the infected naive B cells found in the lymphoid tissue. This could occur if the naive cells are preferentially lysed by CTL. However, studies on allograft patients who are immunosuppressed indicate that this is not the mechanism. These individuals have a highly impaired CTL response that leads to a dramatic increase in the number of latently infected cells in the periphery but these cells are all resting memory cells, infected naive B cells do not appear (Babcock *et al.*, unpublished observations). An alternate possibility is that the infected naive B cells in the tonsils are a resident population of resting B cells. This explanation is also unlikely because the viral genome in these cells is also circularized and, as discussed before, this indicates that the cells, at some point, were activated. In order to become resident, resting cells they would need to dedifferentiate back to a naive resting cell—something that does not occur, at least with normally activated B cells. A third possibility is that infected naive B cells are short lived. Although the turnover of naive cells is rapid, those that survive can live for very long periods of time[50,51] therefore EBV infection would have to predispose the cells to be short lived. This seems counterintuitive since EBV infected naive and memory B cells survive equally well, at least *in vitro*. The most elegant explanation is that the virus infects and activates naive and memory B cells equally well and that these cells behave as normal activated B cells, that is, they exit the cell cycle by entering the memory compartment and recycling. This suggests that cells driven by the growth program *in vivo* respond to differentiation signals that allow them to become resting. They are not limited by the constraints on differentiation seen with immortalized cells *in vitro*. The analogies with normal B cell biology at this point are compelling [Fig. 3(a)]. Normal B cells circulate in the periphery and percolate through lymphoid tissue. If they encounter cognate antigen in the presence of T cells expressing the ligand for CD40, they respond by proliferating.[22,23] Then a number of different outcomes occur depending on the signals the B cell receives. Continued exposure to CD40 ligand in combination with various interleukins (see below) is believed to push the cells into becoming resting memory cells.[52] Immunohistochemical studies suggest that EBV-infected B cells are particularly prevalent in the interfollicular areas of the lymph nodes,[6] where the CD40 ligand is expressed,[23] raising the possibility that virally infected B cells are moved into the memory B cell pool through extended exposure to the CD40 ligand.

FIGURE 3. The parallels between B cell biology and EBV pathogenesis. (a) B cells circulate in the periphery as small resting cells. They occasionally percolate through lymphoid tissue and if they do not encounter cognate antigen, they stay out of the cell cycle and pass back into the periphery. If they encounter antigen, however, the cells are activated to proliferate and expand with the help of accessory signals supplied by antigen-specific Th cells. Recent work suggests that critical signals include CD40 ligand (CD40L) expressed on the surface of the T cells and the presence of lymphokines, such as interleukin-2, -4, -10 (IL-2, IL-4, IL-10) and transforming growth factor-β (TGFβ). Differential exposure to these signals has various outcomes. If insufficient signals are given, the cells swiftly die through apoptosis. If CD40L is no longer expressed, the B cells stop proliferating and terminally differentiate to become antibody-secreting plasma cells. The isotype of the antibody depends on the type of lymphokine. For example, IL-10 + TGFβ gives rise to cells that migrate to the mucosal epithelium and secrete IgA. Prolonged exposure to CD40L and lymphokines causes the B cell to cease growing and become a resting memory B cell which then can reenter the circulation. (b) EBV persists in resting, circulating, memory B cells, and its genome is present in limited copy number as covalently closed episomes. The virus expresses minimal genetic information probably limited to LMP2 (latency program). The resting cells are B7-negative so that they do not elicit a cytotoxic T cell (CTL) response. Upon entering lymphoid tissue, the cell may encounter no signals and pass back out into the peripheral circulation. Occasionally the cells pass through areas of active T cell signaling and fortuitously become activated to express the EBNA1 only program that allows the viral genome to replicate with the cell without being recognized by CTL. By analogy with normal B cells, the signals received determine whether the cells can return to a resting state and reenter the peripheral circulation to maintain the levels of latently infected cells, become terminally differentiated, or apoptose. The last two consequences could lead to viral reactivation (lytic program). EBV-infected B cells that express a terminally differentiated phenotype and replicate the virus have been observed in the mucosal epithelium of lymphoid tissue. This implies that activation of the cells leads to the expression of homing receptors that cause the B cells to enter the mucosal epithelium and release infectious virus on the mucosal surface. This is analogous to the homing of B cells that have preferentially switched to IgA through exposure to IL-10 + TGFβ.

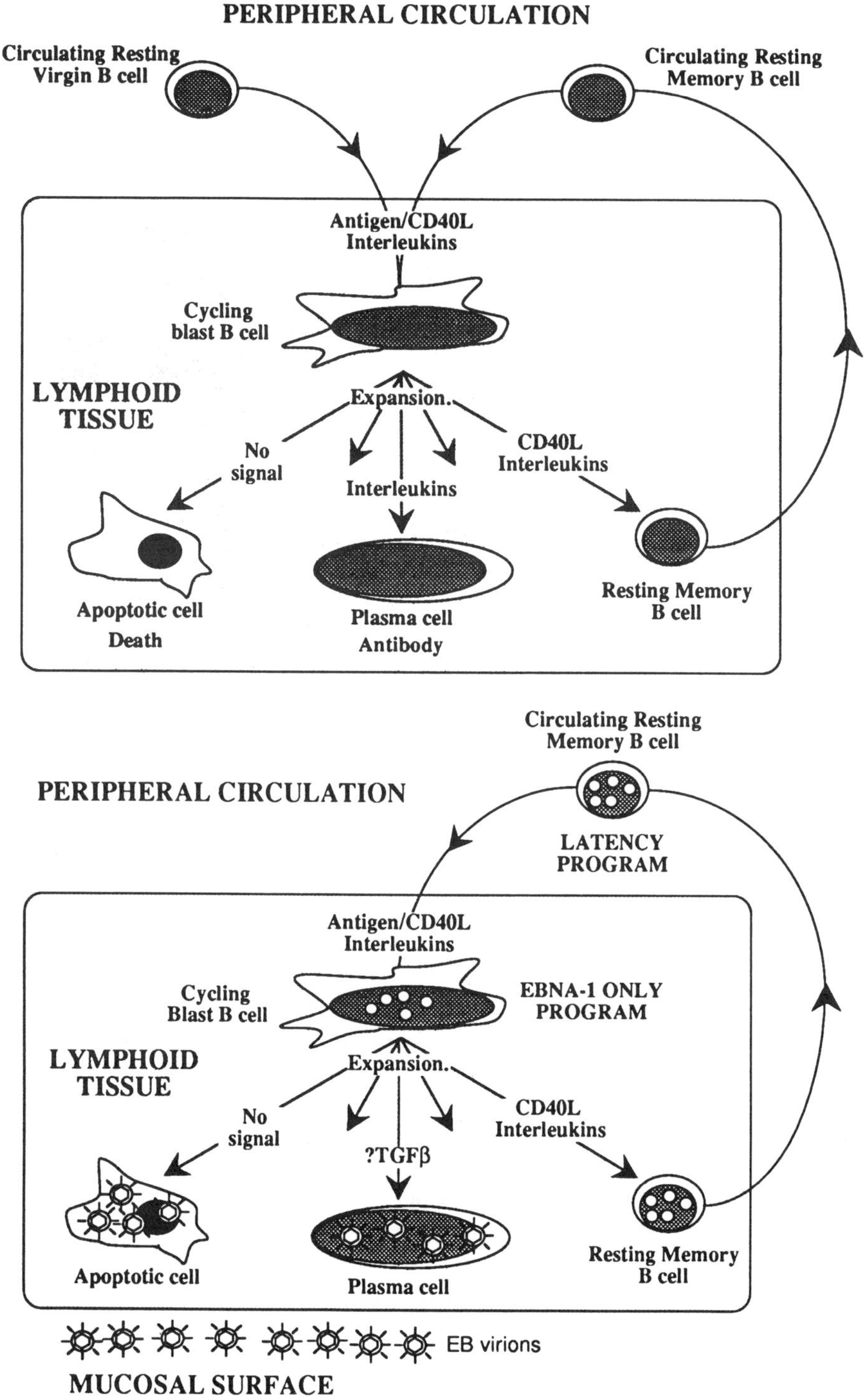
PERIPHERAL CIRCULATION
Circulating Resting Virgin B cell
Circulating Resting Memory B cell
Antigen/CD40L Interleukins
Cycling blast B cell
LYMPHOID TISSUE
Expansion.
No signal
Interleukins
CD40L Interleukins
Apoptotic cell
Death
Plasma cell
Antibody
Resting Memory B cell
Circulating Resting Memory B cell
PERIPHERAL CIRCULATION
LATENCY PROGRAM
Antigen/CD40L Interleukins
Cycling Blast B cell
EBNA-1 ONLY PROGRAM
LYMPHOID TISSUE
Expansion.
No signal
?TGFβ
CD40L Interleukins
Apoptotic cell
Plasma cell
Resting Memory B cell
EB virions
MUCOSAL SURFACE

3.2.2. Maintenance of the EBV "Memory" B Cell

Memory B cells are believed to require signaling through their BCR by continued exposure to antigen. Otherwise they perish after 10–12 weeks, at least in the mouse (for a review and specific references, see Ahmed[50] and Tough[51]). It is also believed that they occasionally need to traverse the cell cycle, but this occurs extremely infrequently, suggesting that exposure to antigen and BCR signaling does not necessarily result in proliferation but is required for survival. How might EBV replicate these conditions so as to persist in the memory B cell compartment? One possibility is that the virus infects and persists in bona fide antigen-specific memory B cells that carry the essentially dormant virus along for the ride. Another interesting possibility is that LMP2A expressed in the resting B cell can provide a positive survival signal. Signaling through the ITAM motif in LMP2A has not been demonstrated, but it is known to be functional. Under appropriate conditions the LMP2A motif may be capable of delivering a signal that mimics the BCR-mediated signaling required to keep resting memory B cells alive for long periods of time.

How do the latently infected resting cells maintain their numbers constantly over long periods of time? Suggestive evidence for a mechanism comes from studies on the EBNA1 protein. EBNA1 is essential for replicating the latent, episomal viral genome[14] and contains a domain that prevents it from being processed and presented as a CTL target.[52] Therefore, EBNA1 can be expressed in a proliferating B cell without being detected by the immune response. Thus, it is likely that the latently infected, resting B cells sporadically undergo periods of proliferation where only EBNA1 is expressed [Fig. 3(b)]. This type of latency is referred to as "EBNA1 only latency"[10] or latency I.[3] It was first described in Burkitt's lymphoma tumor cells (see Rickinson[3] for detailed references) but has yet to be detected in normal B cells. In EBNA1 only latency, the EBNA1 gene is transcribed from a promoter called Qp,[54,55] different from the promoters (Wp and Cp) used to transcribe the EBNAs that have the growth program. The fact that a specific promoter exists for EBNA1 only latency adds further credence to the idea that such a form of latency must exist in normal B cells. Even more satisfying is the recent observation that the Qp has enhancer elements that are responsive to the Rb and p107 proteins whose function is to regulate the transit of cells across the G1-S boundary of the cell cycle.[55a,b] This provides a simple on-off switch for the model of EBV persistence, namely, the resting cells do not express EBNA1 at all, as originally suggested by the first RTPCR studies,[46] but as soon as the cells become activated and begin to traverse the G1 boundary into S, Qp is activated in an Rb-dependent manner. This ensures that EBNA1 is synthesized, guaranteeing that the viral genomes are replicated, without expressing any of the other latent genes which would make the cell vulnerable to attack by CTL. When the cells reenter a quiescent state, EBNA1 transcription again ceases. One prediction of this model is that an opposite, cell-cycle-dependent switch may exist for LMP2A, guaranteeing that it is expressed when the cells are resting but not expressed when they are cycling in the EBNA1 only mode because then it would become a target for CTL.

3.2.3. Avoiding the Immune Response

If the virally infected cells are to persist for the lifetime of the infected host, they must do so in the face of a powerful CTL response that recognizes all of the latent proteins, except EBNA1, but including LMP2.[2] Then how do the resting, latently infected, memory cells avoid immunosurveillance? One possibility is that no latent genes are expressed at all, that only a small fraction of cells express LMP2A, and these cells are doomed to die. This type of latency, with no viral gene transcription, has been called latency 0 and it is claimed that it occurs in CD23-negative, high-density B cells (Ernberg, personal communication). A conflicting report, however, has shown that LMP2A is expressed in CD23-negative, resting B cells.[43] How these cells escape immunosurveillance is not completely clear. One part of the mechanism no doubt involves the observation that they do not express the costimulatory molecule B7.[42] It has been shown previously that B7 expression is required for a resting B cell to reactivate a memory T cell response.[56] Thus, the latently infected cells *in vivo*, which are B7-negative and resting, are shielded from the large numbers of memory CTL carried by healthy donors. Nevertheless, B7-negative cells activate CTL indirectly. This is believed to occur through a mechanism that involves the release of processed antigenic peptides which are picked up and presented by B7-positive professional antigen-presenting cells.[57] It is known that the latently infected resting B cells process and present antigen efficiently.[43] Then why are they not killed? A mundane explanation could be that there is insufficient processed peptide being produced to elicit a response. This could arise because the frequency of infected cells is very low, only small amounts of the LMP2A protein are made, and shed peptides have a short half-life. If correct, then the consistent failure to detect expression of the growth-promoting genes in the peripheral blood by RTPCR does not reflect a need for the cells to avoid immunosurveillance, as has long been supposed. Rather it may be a simple consequence of the fact that their expression is incompatible with the resting state of the cells.

3.3. Getting Back out Again: Viral Reactivation—More Facts and Speculation

3.3.1. Terminal Differentiation as a Signal for Viral Reactivation

Most healthy, persistently infected individuals shed virus in their saliva at a fairly constant rate for their whole lives.[3] How the virus makes the transition from the latently infected resting memory B cell to lytic replication is not completely clear. Along with persistent viral shedding, however, the infected host maintains a constant level of antibody to viral antigens expressed during lytic replication.[1] This suggests that some stochastic process is continuously sending a constant fraction of the latently infected cells into the lytic pathway. This results in the continuous production of infectious virus, viral antigens, and cell death. Immunohistochemical studies have shown that viral replication appears to be ongoing in B cells that have invaded the mucosal epithelium. Most striking was the observation that these cells are terminally differentiated, have lost their lineage markers, and express the plasmablast/plasma cell marker KiB3[6]. This raises the possibility that signals in the mucosal lymphoid tissue can drive the terminal differentiation of the virally infected cells, resulting in viral reactivation. The potential mechanisms to drive this differentiation are immediately suggested by continuing the analogy to normal B cell biology described above [Fig. 3(a) and (b)]. As already mentioned, prolonged exposure of an antigen-activated B cell to CD40 ligand is believed to cause it to enter the memory B cell compartment. Withdrawal of CD40 ligand in the sustained presence of lymphokines, however, also causes the activated B cells to stop proliferating, but in this case the cells terminally differentiate and begin to secrete immunoglobulin.[52] Which isotype of Ig the cells secrete depends on the lymphokine. Thus exposure to IL-4 results in switching to IgG1 or IgE, whereas TGFβ plus IL-10 induces switching to IgA.[22] The latter are particularly interesting because they would potentially cause switching of the B cell to the Ig isotype that would cause it to home to the mucosal epithelium at the same time that viral replication was being induced, thereby allowing release of the virus into saliva. Indeed, it has been observed that TGFβ can signal viral replication in Burkitt's lymphoma cells expressing the EBNA1 only phenotype.[58]

3.3.2. Apoptosis as a Signal for Viral Reactivation

By continuing the analogy with normal B cells, there is a third potential outcome for a germinal center B cell driven to proliferate by antigen, CD40 ligand, and lymphokines. We have already discussed that prolonged exposure to CD40 ligand drives the cells toward the memory cell compartment and prolonged exposure to lymphokines after withdrawal of CD40 ligand drives the cell toward terminal differentiation. The third outcome occurs when all signals are withdrawn [Fig. 3(a)] and under this circumstance the cells rapidly undergo apoptosis.[22,23] Although there is no direct evidence that apoptosis induces the viral lytic cycle, there are several reasons to believe that it does. The first is that apoptosis is a common mechanism for eliminating unwanted B cells and, if EBV has evolved to be highly sensitive to the status of the infected B cell, it is likely that it would respond to apoptosis by replicating and escaping. In support of this notion is the observation that EBV actually encodes for a homologue of the cellular antiapoptotic gene bcl-2 early during the lytic cycle,[59] suggesting that the virus has evolved to delay apoptosis while it is replicating. Lastly, it is well known that peripheral B cells placed in culture die through apoptosis, and it is also well documented that the latently infected cells in peripheral blood replicate the virus upon being placed into culture.[3,44] Although it has yet to be formally demonstrated that the virally infected cells behave like the bulk population, this circumstantial evidence also supports a role for apoptosis in inducing the viral lytic cycle.

4. FINAL REMARKS

4.1. Summing Up

During the course of this chapter, I have sought to build the case that EBV is not an aggressive pathogen, but a passive virus that has evolved so that its biology can neatly be encompassed within the biology of the normal B cell that is its natural target. It follows from this that the EBV-associated disease states occur when this highly developed interaction goes awry.

The virus establishes itself through the proliferation of activated latently infected B cells and in doing so has taken control of the normal signaling mechanisms that activate and drive that proliferation. However, the virus has evolved to make sure that these potentially life-threatening cells are rapidly destroyed once it has set up long-term latency in the memory B cell compartment. Once in this compartment, the virally infected cells are maintained as though they were normal memory B cells. This includes sporadic rounds of proliferation that replenish the pool of latently infected cells and provide a continuous supply of cells that replicate and release infectious virus.

This model still contains a deal of speculation but has the attraction that it explains most of the known properties of the virus and virally infected cells and provides a framework for experimentation. The discoveries of the last two years clearly show that the interaction of EBV with B cell biology *in vivo* is more complex than the simple derivation of immortalized LCL *in vitro* that we all know so well. This points to the necessity of applying new sensitive technologies to address questions of EBV pathogenesis *in vivo*. In the absence of an animal model for EBV, this is the only way that we will discover how EBV persists and what becomes altered in pathogenic states.

4.2. The Questions that Need to Be Addressed

The central questions that need to be addressed experimentally all involve the mechanism by which EBV affects the transitions from the different states of infection. The biggest gap in our knowledge is an understanding of the early events of infection *in vivo*. How does EBV gain access to B cells and what type of infection ensues when the first virion meets the first B cell? Of particular interest also is understanding the signaling mechanisms that cause the transition from the growth program to the latency program and from the latency program to the EBNA1 only program and back. Indeed it is critical to know if the EBNA1 only program exists *in vivo* and in what type of B cell. Lastly, we need to know the signals that regulate viral reactivation.

REFERENCES

1. Henle, W., and Henle, G., 1979, Seroepidemiology of the virus, in: *The Epstein–Barr Virus* (M. A. Epstein and B. G. Achong, eds.), Springer-Verlag, Berlin, pp. 61–78.
2. Khanna, R., Burrows, S. R., and Moss, D. J., 1995, Immune regulation in Epstein–Barr virus-associated diseases, *Microbiol. Rev.* **59:**387–405.
3. Rickinson, A. B., and Kieff, E., 1996, Epstein–Barr virus, in: *Virology* (B. N. Fields, D. M. Knipe, and P. M. Howley, eds.), Raven Press, New York, pp. 2397–2446.
4. Sixbey, J. W., Vesterinen, E. H., Nedrud, J. G., Raab, T. N., Walton, L. A., and Pagano, J. S., 1983, Replication of Epstein–Barr virus in human epithelial cells infected *in vitro, Nature* **306:** 480–483.
5. Rickinson, A. B., 1984, Epstein–Barr virus in epithelium, *Nature* **310:**99–100.

5a. Karajannis, M. A., Hummel, M., Anagnostopoulos, I., and Stein, H., 1997, Strict lymphotropism of Epstein–Barr virus during acute infectious mononucleosis in nonimmunocompromised individuals, *Blood* **89:**2856–2862.

6. Anagnostopoulos, I., Hummel, M., Kreschel, C., and Stein, H., 1995, Morphology, immunophenotype, and distribution of latently and/or productively Epstein–Barr virus-infected cells in acute infectious mononucleosis: Implications for the interindividual infection route of Epstein–Barr virus, *Blood* **85:**744–750.
7. Thorley-Lawson, D. A., and Mann, K. P., 1985, Early events in Epstein–Barr virus infection provide a model for B cell activation, *J. Exp. Med.* **162:**45–59.
8. Thorley-Lawson, D. A., Schooley, R. T., Bhan, A. K., and Nadler, L. M., 1982, Epstein–Barr virus superinduces a new human B cell differentiation antigen (B-LAST 1) expressed on transformed lymphoblasts, *Cell* **30:**415–425.
9. Kieff, E., 1996, Epstein–Barr virus and its replication, in: *Virology* (B. N. Fields, D. M. Knipe, and P. M. Howley, eds.), Raven Press, New York, pp. 2343–2396.
10. Thorley-Lawson, D. A., and Miyashita, E. M., 1996, Epstein–Barr virus and the B cell: That's all it takes, *Trends Microbiol.* **4:**204–208.
11. Nemerow, G. R., Houghten, R. A., Moore, M. D., and Cooper, N. R., 1989, Identification of an epitope in the major envelope protein of Epstein–Barr virus that mediates viral binding to the B lymphocyte EBV receptor (CR2), *Cell* **56:**369–377.
12. Sinclair, A. J., Palmero, I., Peters, G., and Farrell, P. J., 1994, EBNA-2 and EBNA-LP cooperate to cause G0 to G1 transition during immortalization of resting human B lymphocytes by Epstein–Barr virus, *EMBO J.* **13:**3321–3328.
13. Hurley, E. A., and Thorley-Lawson, D. A., 1988, B cell activation and the establishment of Epstein–Barr virus latency, *J. Exp. Med.* **168:**2059–2075.
14. Yates, J. L., Warren, N., and Sugden, B., 1985, Stable replication of plasmids derived from Epstein–Barr virus in various mammalian cells, *Nature* **313:**812–815.
15. Allday, M. J., Sinclair, A., Parker, G., Crawford, D. H., and Farrell, P. J., 1995, Epstein–Barr virus efficiently immortalizes human B cells without neutralizing the function of p53, *EMBO J.* **14:**1382–1391.

16. Allday, M. J., Inman, G. J., Crawford, D. H., and Farrell, P. J., 1995, DNA damage in human B cells can induce apoptosis, proceeding from G1/S when p53 is transactivation competent and G2/M when it is transactivation defective, *EMBO J.* **14:**4994–5005.
17. Parker, G. A., Crook, T., Bain, M., Sara, E. A., Farrell, P. J., and Allday, M. J., 1996, Epstein–Barr virus nuclear antigen (EBNA)3C is an immortalizing oncoprotein with similar properties to adenovirus E1A and papillomavirus E7, *Oncogene* **13:**2541–2549.
18. Horuk, R., 1994, Molecular properties of the chemokine receptor family, *Trends Pharm. Sci.* **15:**159–165.
19. Mann, K. P., Staunton, D., and Thorley-Lawson, D. A., 1985, Epstein–Barr virus-encoded protein found in plasma membranes of transformed cells, *J. Virol.* **55:**710–720.
20. Armitage, R. J., 1994, Tumor necrosis factor receptor superfamily members and their ligands, *Curr. Opinion Immunol.* **6:**407–413.
21. Baker, S. J., and Reddy, E. P., 1996, Transducers of life and death: TNF receptor superfamily and associated proteins, *Oncogene* **12:**1–9.
22. Banchereau, J., Bazan, F., Blanchard, D., Briere, F., Galizzi, J. P., van, K. C., Liu, Y. J., Rousset, F., and Saeland, S., 1994, The CD40 antigen and its ligand, *Ann. Rev. Immunol.* **12:**881–922.
23. MacLennan, I. C., 1994, Germinal centers, *Ann. Rev. Immunol.* **12:**117–139.
24. Mosialos, G., Birkenbach, M., Yalamanchili, R., Van Arsdale, T., Ware, C., and Kieff, E., 1995, The Epstein–Barr virus transforming protein LMP1 engages signaling proteins for the tumor necrosis factor receptor family, *Cell* **80:**389–399.
25. Hammarskjold, M. L., and Simurda, M. C., 1992, Epstein–Barr virus latent membrane protein transactivates the human immunodeficiency virus type 1 long terminal repeat through induction of NF-kappa B activity, *J. Virol.* **66:**6496–6501.
26. Sandberg, M., Hammerschmidt, W., and Sugden, B., 1997, Characterization of LMP-1's association with TRAF1, TRAF2, and TRAF3. *J. Virol.* **71:**4649–4656.
27. Brodeur, S. R., Cheng, G., Baltimore, D., and Thorley-Lawson, D. A., 1997, Localization of the major NF-kappaB-activating site and the sole TRAF3 binding site of LMP-1 defines two distinct signaling motifs [in process citation], *J. Biol. Chem.* **272:**19777–19784.
27a. Kilger, E., Kieser, A., Baumann, M., and Hammerschmidt, W., 1998, Epstein–Barr virus-mediated B-cell proliferation is dependent upon latent membrane protein 1, which simulates an activated CD40 receptor, *Embo J.* **17:**1700–1709.
28. Hammerschmidt, W., and Sugden, B., 1989, Genetic analysis of immortalizing functions of Epstein–Barr virus in human B lymphocytes, *Nature* **340:**393–397.
29. Artavanis-Tsakonas, S., Matsuno, K., and Fortini, M. E., 1995, Notch signaling, *Science* **268:**225–232.
30. Hsieh, J. J., Henkel, T., Salmon, P., Robey, E., Peterson, M. G., and Hayward, S. D., 1996, Truncated mammalian Notch1 activates CBF1/RBPJk-repressed genes by a mechanism resembling that of Epstein–Barr virus EBNA2, *Mol. Cell. Biol.* **16:**952–959.
31. MacNeil, I. A., Suda, T., Moore, K. W., Mosmann, T. R., and Zlotnik, A., 1990, IL-10, a novel growth cofactor for mature and immature T cells, *J. Immunol.* **145:**4167–4173.
32. de Waal Malefyt, R., Haanen, J., Spits, H., Roncarolo, M. G., te Velde, A., Figdor, C., Johnson, K., Kastelein, R., Yssel, H., and de Vries, J. E., 1991, Interleukin 10 (IL-10) and viral IL-10 strongly reduce antigen-specific human T cell proliferation by diminishing the antigen-presenting capacity of monocytes via downregulation of class II major histocompatibility complex expression, *J. Exp. Med.* **174:**915–924.
33. Stuart, A. D., Stewart, J. P., Arrand, J. R., and Mackett, M., 1995, The Epstein–Barr virus encoded cytokine viral interleukin-10 enhances transformation of human B lymphocytes, *Oncogene* **11:**1711–1719.
33a. Ianelli, C. J., DeLellis, R., and Thorley-Lawson, D. A., 1998, CD 48 binds to heparan sulfate on the surface of epithelial cells, *J. Biol. Chem.* **273**.

34. Gallagher, J. T., 1994, Heparan sulphates as membrane receptors for the fibroblast growth factors, *Eur. J. Clin. Chem. Clin. Biochem.* **32:**239–247.
35. Sutkowski, N., Palkama, T., Gong, Y., Ciurli, C., Sekaly, R., Thorley-Lawson, D. A., and Huber, B. T., 1996, An Epstein–Barr virus associated superantigen, *J. Exp. Med.* **184:**971–980.
36. Huber, B. T., Hsu, P. N., and Sutkowski, N., 1996, Virus encoded superantigens, *Microbiol. Rev.* **60:**473–482.
37. Robinson, J., Smith, D., and Niederman, J., 1980, Mitotic EBNA-positive lymphocytes in peripheral blood during infectious mononucleosis, *Nature* **287:**334–335.
38. Tierney, R. J., Steven, N., Young, L. S., and Rickinson, A. B., 1994, Epstein–Barr virus latency in blood mononuclear cells: Analysis of viral gene transcription during primary infection and in the carrier state, *J. Virol.* **68:**7374–7385.
39. Bourgault, I., Gomez, A., Gomard, E., and Levy, J. P., 1991, Limiting-dilution analysis of the HLA restriction of anti-Epstein–Barr virus-specific cytolytic T lymphocytes, *Clin. Exp. Immunol.* **84:**501–507.
40. Khanna, R., Slade, R. W., Poulsen, L., Moss, D. J., Burrows, S. R., Nicholls, J., and Burrows, J. M., 1997, Evolutionary dynamics of genetic variation in Epstein–Barr virus isolates of diverse geographical origins: Evidence for immune pressure-independent genetic drift, *J. Virol.* **71:**8340–8346.
41. Phillips, R. E., Rowland-Jones, S., Nixon, D. F., Gotch, F. M., Edwards, J. P., Ogunlesi, A. O., Elvin, J. G., Rothbard, J. A., Bangham, C. R., Rizza, C. R. *et al.*, 1991, Human immunodeficiency virus genetic variation that can escape cytotoxic T cell recognition, *Nature* **354:**453–459.
42. Miyashita, E. M., Yang, B., Lam, K. M., Crawford, D. H., and Thorley-Lawson, D. A., 1995, A novel form of Epstein–Barr virus latency in normal B cells *in vivo, Cell* **80:**593–601.
43. Miyashita, E. M., Yang, B., Babcock, G. J., and Thorley-Lawson, D. A., 1997, Identification of the site of Epstein–Barr virus persistence *in vivo* as a resting B cell, *J. Virol.* **71:**4882–4891.
44. Decker, L. L., Klaman, L. D., and Thorley-Lawson, D. A., 1996, Detection of the latent form of Epstein–Barr virus DNA in the peripheral blood of healthy individuals, *J. Virol.* **70:**3286–3289.
44a. Babcock, G. J., Decker, L. L., Volk, M., and Thorley-Lawson, D. A., 1998, EBV Persistence in Memory B cells *in vivo, Immunity* **8**.
45. Chen, F., Zou, J. Z., di Renzo, L., Winberg, G., Hu, L. F., Klein, E., Klein, G., and Ernberg, I., 1995, A subpopulation of normal B cells latently infected with Epstein–Barr virus resembles Burkitt lymphoma cells in expressing EBNA-1 but not EBNA-2 or LMP1, *J. Virol.* **69:**3752–3758.
46. Qu, L., and Rowe, D. T., 1992, Epstein–Barr virus latent gene expression in uncultured peripheral blood lymphocytes, *J. Virol.* **66:**3715–3724.
47. Beaufils, P., Choquet, D., Mamoun, R. Z., and Malissen, B., 1993, The (YXXL/I)2 signaling motif found in the cytoplasmic segments of the bovine leukemia virus envelope protein and Epstein–Barr latent membrane protein 2A can elicit early and late lymphocyte activation events, *EMBO J.* **12:**5105–5112.
48. Miller, C. L., Burkhardt, A. L., Lee, J. H., Stealey, B., Longnecker, R., Bolen, J. B., and Kieff, E., 1995, Integral membrane protein 2 of Epstein–Barr virus regulates reactivation from latency through dominant negative effects on protein-tyrosine kinases, *Immunity* **2:**155–166.
49. Khan, G., Miyashita, E. M., Yang, B., Babcock, G. J., and Thorley-Lawson, D. A., 1996, Is EBV persistence *in vivo* a model for B cell homeostasis?, *Immunity* **5:**173–179.
50. Ahmed, R., and Gray, D., 1996, Immunological memory and protective immunity: Understanding their relation, *Science* **272:**54–60.
51. Tough, D. F., and Sprent, J., 1995, Lifespan of lymphocytes, *Immunol. Res.* **14:**1–12.
52. Arpin, C., Dechanet, J., Van, K. C., Merville, P., Grouard, G., Briere, F., Banchereau, J., and Liu, Y. J., 1995, Generation of memory B cells and plasma cells *in vitro, Science* **268:**720–722.

53. Levitskaya, J., Coram, M., Levitsky, V., Imreh, S., Steigerwald, M. P., Klein, G., Kurilla, M. G., and Masucci, M. G., 1995, Inhibition of antigen processing by the internal repeat region of the Epstein–Barr virus nuclear antigen-1, *Nature* **375:**685–688.
54. Tsai, C. N., Liu, S. T., and Chang, Y. S., 1995, Identification of a novel promoter located within the Bam HI Q region of the Epstein–Barr virus genome for the EBNA 1 gene, *DNA Cell Biol.* **14:**767–776.
55. Schaefer, B. C., Strominger, J. L., and Speck, S. H., 1995, Redefining the Epstein–Barr virus-encoded nuclear antigen EBNA-1 gene promoter and transcription initiation site in group I Burkitt lymphoma cell lines. *Proc. Natl. Acad. Sci. USA* **92:**10565–10569.
55a. Sung, N. S., Wilson, J., Davenport, M. Sista, N. D., and Pagano, J. S., 1994, Reciprocal regulation of the Epstein–Barr virus BamHI-F promoter by EBNA-1 and an E2F transcription factor, *Mol. Cell. Biol.* **14:**7144–7152.
55b. Nonkwelo, C., Ruf, I. K., and Sample, J., 1997, The Epstein–Barr virus EBNA-1 promoter Qp requires an initiator-like element, *J. Virol.* **71:**354–361.
56. Azuma, M., Cayabyab, M., Buck, D., Phillips, J. H., and Lanier, L. L., 1992, CD28 interaction with B7 costimulates primary allogeneic proliferative responses and cytotoxicity mediated by small, resting T lymphocytes, *J. Exp. Med.* **175:**353–360.
57. Wu, T. C., Huang, A. Y., Jaffee, E. M., Levitsky, H. I., and Pardoll, D. M., 1995, A reassessment of the role of B7-1 expression in tumor rejection, *J. Exp. Med.* **182:**1415–1421.
58. di Renzo, L., Altiok, A., Klein, G., and Klein, E., 1994, Endogenous TGF-beta contributes to the induction of the EBV lytic cycle in two Burkitt lymphoma cell lines. *Int. J. Canc.* **57:**914–919.
59. Henderson, S., Huen, D., Rowe, M., Dawson, C., Johnson, G., and Rickinson, A., 1993, Epstein–Barr virus-coded BHRF1 protein, a viral homologue of Bcl-2, protects human B cells from programmed cell death, *Proc. Natl. Acad. Sci. USA* **90:**8479–8483.

10

Epstein–Barr Virus Vaccines

HANS J. WOLF and ANDREW J. MORGAN

1. INTRODUCTION

Epstein–Barr virus (EBV) is one of the eight known human herpesviruses. Infectious mononucleosis (IM) is typically associated with primary infection of EBV.[6] This is the second most frequent disease in adolescents in the U.S.[27] and likely in most industrialized countries. When infection occurs in early childhood the clinical symptoms are weak to undetectable in persons in most developing countries and persons living in low socioeconomic milieus. Spontaneous, fatal and prolonged phases of IM are rare but documented. Specific acute genetic disposition leads invariably to fatal consequences of EBV infection (XLP). Unlike most other human herpesviruses, except for the more recently described relationship of HHV-8 to Kaposi's sarcoma, EBV clearly has an oncogenic potential. There is a strong link between EBV and endemic Burkitt's lymphoma (BL)[1] and undifferentiated nasopharyngeal carcinoma (NPC)[22]. Furthermore, it is clear that a major proportion of lymphomas in immunocompromised patients are caused by EBV[3,4], and more recently an association has been demonstrated between certain forms of Hodgkin's lymphomas and EBV.[5,88] The biology of EBV and its association with these diseases are discussed at length elsewhere.[7,89] NPC is a major world health problem with at least 80,000 new cases reported per year.[8] In some regions of Southern China up to 1% of the population will die of NPC. One

HANS J. WOLF • Institut für Mikrobiologie und Hygiene der Medizinischen, Fakultät der Universität Regensberg, D-93053 Regensberg, Germany. ANDREW J. MORGAN • Department of Pathology and Microbiology, School of Medical Science, University of Bristol, Bristol BS8 1TD, England.

Herpesviruses and Immunity, edited by Medveczky *et al.* Plenum Press, New York, 1998.

observation of particular interest which underlies the apparent association between EBV and NPC is the rapid rise of serum IgA antibodies against EBV viral antigens (MA, VCA, EA, and EBNA-1) at the onset of NPC.[9,73] Despite the variety of diseases linked to EBV, it is important to keep a perspective on the significance of EBV as a human pathogen because more than 95% of the world's population is infected with EBV.

The types of cells that are infected by EBV are very restricted and include B lymphocytes and certain epithelial cells. The infection is primarily mediated by the CD21 complement receptor which binds the virus to the major envelope glycoprotein gp340.[10,11] The route of viral transmission is oral. Viral replication is restricted to organs not accessible to the immune system, such as the parotid gland,[74] the oropharynx, or other body cavities. One indication for this is the frequent absence of antibodies to essential replication-associated antigens, such as early antigens, in persons who shed EBV in their saliva.

B cell infection *in vivo* is almost completely restricted to latent infection. It is currently believed that the latently infected B lymphocyte, wherever present in the circulation or in the bone marrow, is the reservoir of EBV infection during lifelong persistent infection.[12,13] The population of latently infected B lymphocytes is regulated by cell-mediated immune responses[81] through the recognition of certain EBV immediate early[75] and possibly also some latent[14] antigens, as far as the latter are expressed.

2. IS AN EBV VACCINE DESIRABLE?

Initial suggestions for an EBV vaccine[15] focused on BL and NPC and were aimed to prevent these neoplasias and also to substantiate the pivotal role of EBV in their development. The cost of developing an EBV vaccine, however, together with the lack of a strong commercial incentive to develop a vaccine against NPC and BL have delayed EBV vaccine development until recent times. The realization that IM is a very significant disease in the West and thus offers far greater commercial incentives for EBV vaccine development has given rise to a renewed interest in this area.[20] A cost–benefit analysis for the development and production of an EBV vaccine[21] clearly shows that such a vaccine would pay for itself in a relatively short time from savings in health care costs in the West. Any vaccine developed to prevent IM will also have widespread application in the prevention of NPC and possibly BL, assuming that prevention or modification of primary infection by vaccination to prevent IM is effective. This is so despite the fact that it has been clear for many years that EBV cannot be the single causal factor of NPC or Burkitt's lymphoma (BL) and also that holoendemic malaria has been identified as a probable cofactor for BL.[16] The cofactors involved, at least in NPC, are more complex[17] with consumption of salted fish during childhood,[18] the exposure to certain medicinal herbs and possibly other environmental cofactors, and a genetic component[80,81] to list those which are best characterized. EBV seems to be an indispensable, yet by itself an insufficient risk factor, and therefore the easiest to control.

NPC was the first malignant tumor of humans whose close association with a virus was substantiated by the repeated demonstration of the presence of EBV DNA sequences in tumor cells.[22,79] Although the factors determining the remarkable geographical clustering of NPC in North Africa, Greenland, and particular regions of southern China are not understood, the suggested pivotal role of EBV can be proven only by vaccination. The rationale for a prophylactic EBV vaccine to control NPC must allow for the fact that most primary infections occur in the very first few years of life, and almost the entire population is infected sooner or later. The prevention of EBV infection would perhaps be worth attempting in these regions. In other areas where NPC is far less frequent and also for those who are already infected, the possibility of postinfection and therapeutic vaccination should be given serious consideration. The distinction between the aims and possible methods of postinfection and therapeutic vaccination should be made clear. Postinfection vaccination will aim to modify the immune status with respect to available EBV immunogens (presently only gp340) of an already infected individual, such that NPC will not develop in later life. This approach is quite distinct from the therapeutic one which aims to enhance the immune response against the tumor itself once it has begun to develop. Here it would seem that strategies involving EBV latent genes which may be expressed in the tumors themselves, such as Epstein–Barr nuclear antigens (EBNAs), latent membrane protein LMP1 and LMP 2a/2b, but possibly also additional genes, such as BZLF-1, would also be candidates.[82]

2.1. Rationale for Vaccination of Infants

It has been established in several animal models that the development of tumors associated with viruses, to an extent, depends on the age at which the host is first infected. Examples are (murine) Moloney leukemia virus[90,91] and avian myeloblastosis viruses.[92,93] Such a situation might not even require lifelong protection from viral infection, but rather a delay, possibly followed by a subclinical infection, as is observed in humans when secondary EBV strains infect persons already EBV-positive. Apart from sporadic cases, there are no reports of second episodes of clinical IM. The strongest argument for early childhood vaccination against EBV infection, however, is the prevention of NPC and also of IM and diseases now seen as possible late complications of EBV infection (Table 1). Because NPC in most Chinese is detected in adults aged 40 and older, the strategy of vaccination to prevent infection by EBV does not apply to the major

TABLE I
Diseases Associated with EBV Infection

Conditions associated with EBV replication
Nonapparent primary infection
Primary infection with clinical symptoms
Infectious mononucleosis (IM)
Spontaneous fatal IM
Fatal IM with genetic predisposition (XLP)
Chronic course oral hairy leukoplakia (OHL)
Conditions associated with cell proliferation
B cell lymphoma (LPD)
Burkitt's lymphoma (BL)
T cell lymphoma (TCL)
Angioimmunoblastic lymphadenopathy (AILD)
Hodgkin's lymphoma (HD) (nodular sclerosing)
Lethal midline granuloma
Nasopharyngeal carcinoma (NPC)
Gastric carcinoma
Smooth muscle carcinomas

part of the present population. The modification of infection by postinfection vaccination may be feasible and possibly of immediate value.

2.2. Rationale for Vaccination of Persons Already Infected with EBV

It has been suggested that persistent and excessive replication of EBV, as reflected in serum IgA levels to MA, VCA, EA, or EBNA-1 before clinically apparent NPC, could itself be the basis for development of the disease. Mechanistic explanations may involve EBV-induced cell fusion[23] or IgA-mediated infection[24] of atypical target cells for EBV along with insufficient clearance of the infected target cells because of inhibition of immune mechanisms, such as ADCC.[25] Such a scenario is consistent with the antibody titers to EBV replicative antigens, which are persistently high in populations at risk, and high levels of IgA antibodies against the same antigens marking the onset and actually preceding the clinical detection of NPC.

If this or similar mechanisms are involved, then a postexposure vaccine that controls the lytic cycle of EBV better could be expected to have positive effects. Stimulation of cytotoxic T-cell (CTL) responses and modulation of the humoral responses could be the levels at which a postexposure vaccine might act, i.e., by changing the quality and quantity of specific IgG and IgA antibodies, respectively.

2.3. Testing an EBV Vaccine

Animal systems are of variable use in evaluating vaccines when the infectious agent has a different pathogenesis in the experimental host. It is necessary, however, that the immunogenicity and induction of unwanted side effects by new vaccines are evaluated in laboratory animals, such as mice and rabbits. The choice of adjuvants and, to a greater extent, the protective potential of a given vaccine are difficult to evaluate. Two primate model systems have become established in EBV vaccine research, the cottontop tamarin (*Saguinus oedipus*) and the common marmoset (*Calithrix jaccus*).

Inoculation of EBV into the common marmoset gives rise to a poorly defined mononucleosislike syndrome.[26,27] Secondary diseases, such as Hodgkin's lymphoma, Burkitt's lymphoma, and nasopharyngeal carcinomalike malignancies, cannot be studied in this model. Recent work on the common marmoset indicates that this animal model of EBV infection might be much more useful than previously thought. EBV infection with the M81 strain of EBV in the common marmoset gives rise to long-term maintenance of antibodies to viral antigens without clinical disease. The presence of EBV DNA has now been reliably demonstrated in tissues and oral fluids by PCR analysis. When infected common marmosets were paired with uninfected animals, the uninfected animals seroconverted within four to six weeks.[28]

Malignant lymphoma is induced in the cottontop tamarin by injecting large doses of EBV, and these tumors have been studied in some detail. The lymphoid lesions induced in this animal by the virus are clearly genuine tumors because they are mono- or oligoclonal and arise independently at different sites in the injected animal.[29] A number of EBV vaccines based on the envelope glycoprotein gp340 have been developed in recent years and evaluated in this model.[30] As is the case with most animal models, the cottontop tamarin is less than ideal. First, the tamarin has never been infected by the oral route and, unlike humans, does not sustain a persistent infection, at least not at the same level as in humans. Recent work, however, shows that the tamarin can sustain latent infection because small numbers of EBV+B cells have been detected in animals that were immunized and challenged with a lymphomogenic dose of EBV.[31] Furthermore, the tamarin has a very restricted major histocompatibility complex class I polymorphism.[32]

Other animal studies have included infection of rabbits with an EBV-like virus derived from the baboon.[33] A third, yet unexploited possibility is the use of chimpanzees[34] which can be infected by human EBV. This phenomenon happens in play units for young animals where exposure to humans as a source of EBV infection is high. Their limited availability makes it almost impossible to use them for initial development work but does not preclude their use in advanced trials. The rabbit/baboon EBV system still awaits exploitation. Much depends on the cross-reactivity of human- and baboon-derived antigens and challenge viral stocks, respectively.

In the absence of a fully satisfactory animal system, it is reasonable to progress directly to human trials after initial immunogenicity and toxicity have been evaluated in animals. This strategy is even more attractive if human target populations are identified where, within a short period, a high proportion become naturally infected with EBV. In this case protection from natural infection can be monitored by the absence of immune responses to EBV replicative antigens other than the antigen used as a vaccine. It would be an advantage if such a susceptible group of volunteers were also located in an area where EBV-related diseases are a major public health problem. According to WHO guidelines, a vaccine should be evaluated only in populations which ultimately benefit from its successful development.[86] Further conditions for such a trial are that the vaccine would be made available to the population in question only after evaluation in informed, consenting volunteers and that the product meets the quality and safety standards set by that country's authorities.

3. CHOICE OF IMMUNOGEN

The validity of the concept that herpesvirus tumors can be prevented by vaccination was first demonstrated with Marek's disease of chickens[35] and with Herpes saimiri in nonhuman primates.[36] The major EBV envelope glycoprotein gp340 was originally selected as a candidate subunit vaccine to prevent EBV infection on the grounds that antibodies raised against this molecule are virus-neutralizing *in vitro*. It was shown that gp340 is a protective immunogen in the tamarin lymphoma model.[37] That antibodies against gp340 are virus neutralizing is consistent with the fact that gp340 is the virus ligand which binds to the host cell complement component receptor CD21. Of particular relevance to gp340 vaccine development is the recent observation that there is very little sequence variation in gp340 genes taken from various EBV isolates around the world.[38] Gp340 contains up to 50% carbohydrate, much of which is 0-linked.[39,40] Very little is yet known about the contribution that carbohydrate makes to the immunologic profile of gp340. Although at least 20 open reading frames in the EBV genome potentially code for glycoproteins, only a few of these have been identified and characterized, and their possible role in making an effective EBV vaccine has not received serious consideration to date.

The correlates of protective immunity in humans that either prevent EBV infection or provide resistance against EBV associated diseases are not known. Viral infection persists for life in the face of considerable humoral and cell-mediated immune responses. Although the tamarin lymphoma model is not an ideal model of EBV infection and consecutive diseases, it became clear in protection experiments with this species that humoral immune responses were not necessary in preventing EBV-induced lymphoma and that protection is provided by cell-mediated responses. Whether this observation also holds up for humans remains to be seen. Until human trials have taken place, it is unlikely that any correlates of protective immunity will become known.

3.1. Production Systems

The first experiments on gp340 as an effective subunit vaccine were carried out using material isolated from infected cells grown in bulk culture.[41] The fact that only very small quantities of protein could be isolated and that phorbol esters and sodium butyrate were used to induce the lytic cycle proteins in culture meant that this particular product was totally unsuitable for human use. The difficulties encountered in purifying authentic natural product gp340 for bulk EBV-infected cell cultures has led to the development of effective purification procedures now being used to isolate recombinant products.[51]

The complete nucleotide sequence of the B958 strain of EBV was determined some years ago[43], and the gene encoding for gp340 and its spliced companion gp220 has long been mapped.[44a,b] Fragments of the gene have been expressed in bacteria,[45–47] and the complete gene has been expressed in yeast.[48] Unfortunately bacterial gp340 products did not induce a virus-neutralizing antibody, and at that time the induction of virus-neutralizing antibodies was considered of paramount importance in inducing protective immunity against the virus. It is not at all clear whether the induction of virus-neutralizing antibodies in humans will be an essential requirement of any EBV vaccine. The gp340 gene has been expressed by our group (HJW) and others in a baculovirus system[49] but has not been characterized for its immunogenic or protective potential.

A number of mammalian cell expression systems have been used to express the gp340 gene where glycosylation and posttranslational modifications found in the natural product are assumed to have taken place.[50,52,53,76] With one or two exceptions it has not been possible to distinguish between the natural gp340 product and recombinant gp340 products expressed in mammalian cells by a range of monoclonal antibodies.[50,53] Large-scale production of recombinant products expressed in mammalian cells markedly advanced with the discovery that removal of the 21 amino acids encompassing the membrane anchor dramatically enhances the yield.[76] Two mammalian expression systems, in particular, have been used to produce relatively large quantities of gp340. These include a bovine papilloma virus (BPV) system[50] and a Chinese hamster ovary (CHO) cell system.[51] The use of recombinant systems to produce gp340 as a subunit vaccine bypasses a major obstacle in generating EBV vaccines, the presence of oncogenic

EBV, even in vanishingly small quantities, in any vaccine. From a regulatory point of view, the gp340 produced in a bovine papilloma virus system using C127 mouse fibroblasts likewise seems less than ideal because papilloma viruses themselves have an oncogenic potential and also the C127 cell line is tumorogenic in nude mice.

3.2. Choice of Adjuvant

Subunit vaccines invariably require an adjuvant to induce a sufficiently powerful immune response, and gp340 is not different in this respect. The only adjuvant currently licensed for widespread human use is alum. Compared with a range of recently developed adjuvants, alum is relatively weak and unsatisfactory. Adjuvants are also crucial in determining the type of immune response induced against the subunit antigen. It may be desirable in some cases to favor the Th1 end of the spectrum of immune responses instead of the Th2, and the choice of adjuvant can certainly affect the distribution of immune responses between the two extremes of Th1 and Th2. A range of adjuvants has been employed in the tamarin lymphoma model, including iscoms,[54] artificial liposomes,[55] threonyl muramyl dipeptide,[56] and more recently alum.[57] Remarkably, an SDS-treated preparation of gp220/350 gave protection in the tamarin model whereas a carefully produced native preparation did not always do so. A recent observation[77] that SDS treatment of protein antigens converts them into good antigens for induction of CTL might give a lead in understanding this phenomenon.

4. LIVE RECOMBINANT VIRUS VECTOR VACCINES

The benefits of using live recombinant virus vector vaccines to present a vaccine antigen to the immune system are considerable. Adjuvants are not required and a more broad ranging immune response can be induced.[58,59a] This, of course, depends on the degree to which the recombinant virus vector is attenuated in humans. Both adenovirus[60] and vaccinia virus gp340 recombinants[59b] have been used successfully in the tamarin lymphoma model and both induce complete protective immunity against a lymphomogenic challenge dose of EBV. A vaccinia gp340 recombinant has also been used to immunize common marmosets before challenge with the M81 human EBV strain,[61] and viral replication in the challenged and vaccinated animals was reduced. Gp340 has also been expressed in Tien Tan Chinese strain vaccinia. This Chinese vaccinia strain has been used to vaccinate small groups of EBV-seropositive adults and seropositive and seronegative children in China[62,63,64] (see below). The adenovirus recombinants have not been further developed for evaluation in humans.

5. CELL-MEDIATED IMMUNE RESPONSES TO gp340

Knowledge of cell-mediated immune responses to gp340 is extremely limited. The induction of protective immunity in the cottontop tamarin against EBV-induced lymphomas by vaccinia gp340 recombinants, which did not induce antibody responses, suggested that protective immunity specific for gp340 on a cellular basis could be induced. Studies of cellular immunity to EBV have been limited in humans and have initially centered on the latent viral antigens, including the EBNAs and latent membrane proteins.[72] Only recently study was expanded to the group of immediate early antigens which are believed to play a pivotal role in the controlling lytic replication of EBV in the peripheral blood of immune individuals.[75] Limited studies on cell-mediated immune responses to gp340 in humans have demonstrated that seropositive individuals possess T helper cells specific for gp340[65,66] and that gp340 iscoms stimulate the production of T cells that prevent EBV transformation *in vitro*.[67] Gp340 can also be a target antigen in HLA Class II restricted $CD4^+$ T cell recognition and cytotoxicity.[68] It has been shown that cells expressing gp340 are good targets for antibody-dependent cellular cytotoxicity (ADCC).[69,70]

6. HUMAN TRIALS

Chinese strain recombinant vaccinia expressing gp340[64] and protein harvested from CHO cells[76] were suggested to Chinese authorities as candidate antigens for human trials. The vaccinia-based approach was permitted in a stepwise process for evaluation. After appropriate testing in mice and rabbits, adult volunteers (EBV-positive, vaccinia-positive) were immunized with 10^8 pfu/ml. No unexpected side effects were observed. There was no measurable effect, however on the titers of neutralizing antibodies in the vaccinees. Because previous exposure to vaccinia virus may have been the reason for weak or absent responses to gp340, a further evaluation followed in ten children aged about ten years. In these children a clear gp340-specific stimulation of the humoral immune response could be observed, again without any unexpected side effects. This encouraged a third step of the trial. Altogether 19 infants (9–12 months of age), negative for both EBV and vaccinia virus antibodies, were recruited and vaccinated once with 10^7 pfu/ml.[62,64]

The results show that all ten infants not vaccinated were infected through natural routes with EBV during the 12 months following vaccination. In contrast only two children of the group vaccinated just one time were infected during the first year and one more in the following year. After three years, four out of nine vaccinated children were EBV-positive whereas five remained negative. Lower antibody titers to non-vaccine-related EBV antigens (VCA and EA) in the vaccinated group could be interpreted as a partial protection which might ultimately reduce clinical symptoms in infected vaccinees. The small number of subjects in the trial, however, should caution against any firm conclusion. In summary and although only small numbers are available, these data suggest that gp340/220 is a candidate immunogen which, following appropriate presentation, might induce a protective immune response. Clearly the next logical step is evaluating purified gp340/220 with appropriate powerful adjuvants that are now available. If successful, such a formulation would be clearly favorable over recombinant live vectors for reasons of safety, the ability to perform successive immunizations in individuals, because of the product's stability, and the potential advantage of being combinable with other vaccine antigens to induce protection simultaneously against other disease agents. The latter aspect is increasingly important for financial and logistic reasons. Furthermore, these experiments provide evidence that prior exposure to a carrier virus, such as the vaccinia strain used, negatively influences the immune response to the additionally expressed antigen even several decades after the prior contact.

7. EBV LATENT ANTIGEN VACCINES

EBV infection of B lymphocytes gives rise mainly to latent infection and switches to productive infection only under certain circumstances. Eleven EBV genes are expressed in latent infection, including EBNAs 1–6. In addition, the three latent membrane proteins (LMP) 1, 2a, and 2b are also expressed.[7] It is believed that the numbers of EBV infected cells expressing these latent genes in the circulation of human seropositives is very low and these are regulated by CTLs specific for EBNA 3 and LMP. A recent approach to EBV vaccination has been based on these factors. Epitopes recognized by CD8+CTLs against particular MHC backgrounds have been documented. Peptide vaccines based on these epitopes could be used to vaccinate a significant proportion of a given population providing the distribution of particular MHC restrictions is known. Phase One trials of an EBNA 3 peptide, restricted through the HLA B8 allele, are in progress. These vaccines could elicit T cell memory which would be activated by natural EBV challenge to produce specific CTLs.[71,72] However, EBV persists in cells not expressing any or only EBNA-1 in small lymphocytes of the periphery[84] and can enter into a productive cycle of viral replication directly from these cells without the need for prior differentiation into EBNA 2 or 3 producing cells. Hence an effective vaccine would have to control the incoming virus or freshly infected cells via membrane proteins, such as the gp220/350, the glycoprotein gp110 encoded by BALF4,[25] the more recently discovered fusion-inducing gp85,[87] or replication-associated viral genes, such as immediate early antigens.

8. CONCLUSIONS

The major EBV envelope glycoprotein gp340, expressed as a genetically engineered product in CHO or other mammalian cell expression systems, is the prime candidate to be evaluated in larger human trials. Because the correlates of protective immunity in humans against EBV and EBV-related diseases are largely unknown, there is a limited message to be learned from the results of protection experiments in tamarins, although a number of recombinant subunit and recombinant viral vector constructs induced protective immunity against lymphomogenic doses of EBV. For a variety of reasons a subunit preparation of recombinant gp340 combined with an appropriate adjuvant is the vaccine formulation of choice. Although initial human trials with a live vaccinia recombinant vector expressing gp340 in a small group of children in Southern China were certainly promising, few application-oriented conclusions can yet be drawn. It does seem unlikely that vaccinations with a gp340 based vaccine will induce lifelong sterilizing immunity protective against EBV infection. Whether the immune status of a vaccinated individual will be sufficiently altered to offer protection against EBV associated diseases will take some time to determine, probably about 20 years. However, a postinfection strategy could give results in only a few years.

The readouts of a thoroughly conducted human trial will have to be the following: (a) generating anti-gp340 EBV-neutralizing antibodies; (b) inducing types of cell-mediated immune response specific for gp340; and (c) preventing or delaying the onset of IM in adolescents.

The results of the human trials will undoubtedly lead to a revision of current strategies for handling people at risk of serious complications from EBV infections, including heavily immunocompromised children, persons with Duncan's syndrome and EBV-negative transplant patients. Such trials will change the perspective for still EBV-negative adolescents, and in the long term may also alter the incidence of Hodgkin's lymphoma. Furthermore it does not seem unreasonable to expect that a postinfection vaccination could alter the immune status of a seropositive individual so that individuals at high risk for developing NPC may gain greater resistance to it.

REFERENCES

1. Magrath, I., 1990, The pathogenesis of Burkitt's lymphoma, *Adv. Cancer Res.* **55:**133–270.
2. Henle, W., Henle, G., Ho, H-C., Burtin, P., Cachin, Y., Clifford, P., de Schryver, A., de-Thé, G., Diehl, V., and Klein, G., 1970, Antibodies to Epstein–Barr virus in nasopharyngeal carcinoma, other head and neck neoplasms, and control groups, *J. Natl. Cancer Inst.* **44:**225–231.
3. Purtilo, D. T., Tatsumi, E., Manolov, G., Manolova, Y., Harada, S., Lipscomb, H., and Krueger, G., 1985, Epstein–Barr virus as an etiological agent in the pathogenesis of lymphoproliferative and aproliferative diseases in immune deficient patients, *Int. Rev. Exp. Pathol.* **27:**113–183.

4. Thomas, J. A., and Crawford, D. H., 1989, Epstein–Barr virus associated B-cell lymphomas in AIDS and after organ transplantation, *Lancet:* **i:**1075–1076.
5. Niedobitek, G., 1996, The role of Epstein–Barr virus in the pathogenesis of Hodgkin's disease, *Ann. Oncol.* **7:**S11–S17.
6. Henle, G., Henle, W., and Diehl, V., 1968, Relation of Burkitt's tumor-associated herpes-type virus to infectious mononucleosis, *Proc. Natl. Acad. Sci. USA* **59:**94–101.
7. Rickinson, A. B., and Kieff, E., 1996, Epstein–Barr virus, in *Fields Virology,* Volume 2 (B. N. Fields, D. M. Knipe, P. M. Howley *et al.,* eds.), Lippincot-Raven, New York, pp. 2397–2446.
8. Parkin, D. M., Stjernsward, J., and Muir, C. S., 1984, Estimates for the worldwide frequency of twelve major cancers, *Bull. World Health Org.* **62:**163–182.
9. Zeng, Y., Pi, G. H., Deng, H., Zhang, J. M., Wang, P. C., Wolf, H., DeThé, G., 1986, Epstein–Barr virus seroepidemiology in China, *AIDS Res.* **2**(supplement 1)**:**7–15.
10. Tanner, J., Whang, Y., Sample, J., Sears, A., and Kieff, E., 1988, Soluble gp350/220 and deletion mutant glycoproteins block Epstein–Barr virus adsorption to lymphocytes, *J. Virol.* **62:**4452–4464.
11. Roberts, M. L., Luxemberg, A. T., and Cooper, N. R., 1996, Epstein–Barr virus binding to CD21, the virus receptor, activates resting B-cells via an intracellular pathway that is linked to B-cell infection, *J. Gen. Virol.* **77:**3077–3085.
12. Decker, L. L., Klaman, L. D., and Thorley-Lawson, D. A., 1996, Detection of the latent form of Epstein–Barr virus DNA in the peripheral blood of healthy individuals, *J. Virol.* **70:**3286–3289.
13. Khan, G., Miyashita, E. M., Yang, B., Babcock, G. J., and Thorley-Lawson, D. A., 1996, Is EBV persistence *in vivo* a model for B cell homeostasis?, *Immunity* **5:**173–179.
14. Khanna, R., Burrows, S. R., and Moss, D. J., 1995. Immune regulation in Epstein–Barr virus-associated diseases, *Microbiol. Rev.* **59:**387–405.
15. Epstein, M. A., 1976, Epstein–Barr virus—Is it time to develop a vaccine program?, *J. Natl. Cancer Inst.* **56:**697–700.
16. Facer, C. A., and Playfair, J. H., 1989, Malaria, Epstein–Barr virus, and the genesis of lymphomas, *Adv. Cancer Res.* **53:**33–72.
17. Chen, C. J., Liang, K. Y., Chang, Y. S., Wang, Y. F., Hsieh, T., Hsu, M. M., Chen, J. Y., and Liu, M. Y., 1990, Multiple risk factors of nasopharyngeal carcinoma: Epstein–Barr virus, malarial infection, cigarette smoking and familial tendency, *Anticancer Res.* **10:** 547–553.
18. Poirier, S., Bouvier, G., Malaveille, C., Ohshima, H., Shao, Y. M., Hubert, A., Zeng, Y., de Thé, G., and Bartsch, H., 1989, Volatile nitrosamine levels and genotoxicity of food samples from high-risk areas for nasopharyngeal carcinoma before and after nitrosation, *Int. J. Cancer* **44:**1088–1094.
19. Lu, S., Day, N. E., Degos, L., Lepage, V., Wang, P-C., Chan, S-H., Simons, M., McKnight, B., Easton, D., Zeng, Y., and de Thé, G., 1990, Linkage of a nasopharyngeal carcinoma susceptibility locus to the HLA region, *Nature* (London) **346:**470–471.
20. Spring, S., Hascall, G., and Gruber, J., 1996, Issues related to development of Epstein–Barr virus vaccines, *J. Natl. Cancer Inst.* **88:**1436–1441.
21. Evans, A. S., 1993, Epstein–Barr vaccine: Use in infectious mononucleosis, in *The Epstein–Barr Virus and Associated Diseases* (T. Tursz, J. S. Pagano, D. V. Ablashi, G., de Thé, G. Lenoir, and G. R. Pearson, eds.), J. Libbey, London/INSERM, Paris, pp. 593–598.
22. Wolf, H., Zur Hausen, H., and Becker, V., 1973, EP viral genome in epithelial nasopharyngeal carcinoma cells, *Nature New Biol.* **244:**245–247.
23. Bayliss, G. J., and Wolf, H., 1980, Epstein–Barr virus-induced cell fusion, *Nature* **287:**164–165.
24. Sixbey, J. W., and Yao, Q. Y., 1992, Immunoglobulin A-induced shift of Epstein–Barr virus tissue tropism, *Science* **255:**1578–1580.
25. Jilg, W., Bogedain, C., Mairhofer, H., Gu, S. Y., and Wolf, H., 1994, The Epstein–Barr virus-

encoded glycoprotein gp 110 (BALF 4) can serve as a target for antibody-dependent cell-mediated cytotoxicity (ADCC), *Virology* **202:**974–977.

26. Emini, E. A., Luka, J., Armstrong, M. E., Banker, F. S., Provost, P. J., and Pearson, G. R., 1986, Establishment and characterization of a chronic infectious mononucleosislike syndrome in common marmosets, *J. Med. Virol.* **18:**369–379.
27. Wedderburn, N., Edwards, J. M. B., Desgranges, C., Fontaine, C., Cohen, B., de Thé, G., 1984, Infectious mononucleosis-like response in common marmosets infected with Epstein–Barr virus, *J. Infect. Dis.* **150:**878–882.
28. Cox, C., Chang, S., Karran, L., Griffin, B., and Wedderburn, N., 1996, Persistent Epstein–Barr virus infection in the common marmoset (Callithrix jacchus), *J. Gen. Virol.* **77:**1173–1180.
29. Cleary, M. L., Epstein, M. A., Finerty, S., Dorfman, R. F., Bornkamm, G. W., Kirkwood, J. K., Morgan, A. J., and Sklar, J., 1985, Individual tumors of multifocal EB virus-induced malignant lymphomas in tamarins arise from different B-cell clones, *Science* **228:**722–724.
30. Morgan, A. J., 1992, Epstein–Barr virus vaccines, *Vaccine* **10:**563–571.
31. Niedobitek, G., Agathanggelou, A., Finerty, S., Tierney, R., Watkins, P., Jones, E. L., Morgan, A., Young, L. S., and Rooney, N., 1994, Latent Epstein–Barr virus infection in cottontop tamarins. A possible model for Epstein–Barr virus infection in humans, *Am. J. Pathol.* **145:**969–978.
32. Watkins, D. I., Hodi, F. S., and Letvin, N. L., 1988, A primate species with a limited major histocompatability complex class I polymorphism, *Proc. Natl. Acad. Sci. USA* **85:**7714–7718.
33. Wutzler, P., Meerbach, A., Farber, I., Wolf, H., and Scheibner, K., 1995, Malignant lymphomas induced by an Epstein–Barr virus-related herpesvirus from Macaca arctoides—a rabbit model, *Arch. Virol.* **140**(11)**:**1979–1995.
34. Levine, P. H., Leiseca, S. A., Hewetson, J. F. *et al.,* 1980, Infection of rhesus monkeys and chimpanzees with Epstein–Barr virus, *Arch Virol.* **66:**341–351.
35. Kaaden, O. R., and Dietzschold, B., 1974, Alterations of the immunological specificity of plasma membranes of cells infected with Marek's disease and turkey herpes viruses, *J. Gen. Virol.* **25:**1–10.
36. Laufs, R., and Steinke, H., 1975, Vaccination of non-human primates against malignant lymphoma, *Nature* (London) **253:**71–72.
37. Epstein, M. A., Morgan, A. J., Finerty, S., Randle, B. J., and Kirkwood, J. K., 1985, Protection of cottontop tamarins against Epstein–Barr virus-induced malignant lymphoma by a prototype subunit vaccine, *Nature* (London) **318:**287–289.
38. Lees, J. F., Arrand, J. E., Pepper, S. D. *et al.,* 1993, The Epstein–Barr virus candidate vaccine antigen gp340/220 is highly conserved between virus types A and B, *Virology* **195:**578–586.
39. Morgan, A. J., Smith, A. R., Barker, R. N., and Epstein, M. A., 1984, A structural investigation of the Epstein–Barr (EB) virus membrane antigen glycoprotein, gp340, *J. Gen. Virol.* **65:**397–404.
40. Serafini-Cessi, F., Malagolini, N., Nanni, M., Dall'Olio, F. Campadelli-Fiume, G., Tanner, J., and Kieff, E., 1989, Characterization of N- and O-linked oligosaccharides of glycoprotein 350 from Epstein–Barr virus, *Virology* **170:**1–10.
41. Morgan, A. J., North, J. R., and Epstein, M. A., 1983, Purification and properties of the gp340 component of Epstein–Barr virus membrane antigen in an immunogenic form, *J. Gen. Virol.* **64:**455–460.
42. David, E. M., and Morgan, A. J., 1988, Efficient purification of Epstein–Barr virus membrane antigen gp340 by fast protein liquid chromatography, *J. Immunol. Methods* **108:**231–236.
43. Baer, R., Bankier, A. T., Biggin, M. D., Deininger, P. L., Farrell, P. J., Gibson, T. J., Hatfull, G., Hudson, G. S., Satchwell, S. C., Seguin, C., Tuffrrell, P. S., and Barrell, B. G., 1984, DNA sequence and expression of the B95-8 Epstein–Barr virus genome, *Nature* (London) **310:**207–211.

44a. Beisel, C., Tanner, J., Matsuo, T., Thorley-Lawson, D., Kezdy, F., and Kieff, E., 1985, Two major outer envelope glycoproteins of Epstein–Barr virus are encoded by the same gene, *J. Virol.* **54:**665–674.

44b. Seibl, R., Wolf, H., 1985, Mapping of Epstein–Barr virus proteins on the genome by translation of hybrid-selected RNA from induced P3HR1 cells and induced Raji cells, *Virology* **141:**1–13.

45. Pither, R. J., Zhang, C. X., Wallace, L. E., Rickinson, A. B., and Morgan, A. J., 1991, Mapping of B and T cell epitopes on the Epstein–Barr major envelope glycoprotein gp340, in *Vaccines 91* (R. A. Lerner, H. Ginsberg, R. M. Chanock, and F. Brown, eds.), Cold Spring Harbor Laboratory Press, Cold Spring Harbor, New York, pp. 197–201.

46. Pither, R. J., Zhang, C. X., Shiels, C., Tarlton, J., Finerty, S., and Morgan, A. J., 1992, Mapping of B-cell epitopes on the polypeptide chain of the Epstein–Barr virus major envelope glycoprotein and candidate vaccine molecule gp340, *J. Virol.* **66:**1246–1251.

47. Zhang, P. F., Klutch, M., Armstrong, G., Qualtiere, L., Pearson, G., and Marcus-Sekura, C. J., 1991, Mapping of the epitopes of Epstein–Barr virus gp350 using monoclonal antibodies and recombinant proteins expressed in *Escherichia coli* defines three antigenic determinants, *J. Gen. Virol.* **72:**2747–2755.

48. Schultz, L. D., Tanner, J., Hofmann, K. J., Emini, E. A., Condra, J. H., Jones, R. E., Kieff, E., and Ellis, R. W., 1987, Expression and secretion in yeast of a 400-kDa envelope glycoprotein derived from Epstein–Barr virus, *Gene* **54:**113–123.

49. Nuebling, C. M., Buck, M., Boos, H., von Deimling, A., and Mueller-Lantzsch, N., 1992, Expression of Epstein–Barr virus membrane antigen gp350/220 in *E. coli* and in insect cells, *Virology* **191:**443–447.

50. Madej, M., Conway, M. J., Morgan, A. J., Sweet, J., Wallace, L., Arrand, J., and Mackett, M., 1992, Purification and characterisation of Epstein–Barr virus gp340/220 produced by a bovine papilloma virus vector system, *Vaccine* **10:**777–782.

51. Hessing, M., van Schijndel, H. B., van Grunsven, W. M., Wolf, H., and Middeldorp, J., 1992, Purification and quantification of recombinant Epstein–Barr viral glycoproteins gp350/220 from Chinese hamster ovary cells, *J. Chromatogr.* **599:**267–272.

52. Motz, M., Deby, G., Jilg, W., and Wolf, H., 1986, Expression of the Epstein–Barr virus major membrane proteins in Chinese hamster ovary cells, *Gene* **44:**353–359.

53. Whang, Y., Silberklang, M., Morgan, A., Munshi, S., Lenny, A. B., Ellis, R. W., and Kieff, E., 1987, Expression of the Epstein–Barr virus gp350/220 gene in rodent and primate cells, *J. Virol.* **61:**1796–1807.

54. Morgan, A. J., Finerty, S., Lovgren, K., Scullion, F. T., and Morein, B., 1988, Prevention of Epstein–Barr (EB) virus-induced lymphoma in cottontop tamarins by vaccination with the EB virus envelope glycoprotein gp340 incorporated into immune-stimulating complexes, *J. Gen. Virol.* **69:**2093–2096.

55. North, J. R., Morgan, A. J., Thompson, J. L., and Epstein, M. A., 1982, Purified Epstein–Barr virus M_r 340,000 glycoprotein induces potent virus-neutralizing antibodies when incorporated in liposomes, *Proc. Natl. Acad. Sci. USA* **79:**7504–7508.

56. Morgan, A. J., Allison, A. C., Finerty, S., Scullion, F. T., Byars, N. E., and Epstein, M. A., 1989, Validation of a first-generation Epstein–Barr virus vaccine preparation suitable for human use, *J. Med. Virol.* **29:**74–78.

57. Finerty, S., Mackett, M., Arrand, J. R., Watkins, P. E., Tarlton, J., and Morgan, A. J., 1994, Immunisation of cottontop tamarins and rabbits with a candidate vaccine against the Epstein–Barr virus based on the major viral envelope glycoprotein gp340 and alum, *Vaccine* **12:**1180–1184.

58. Moss, B., 1996, Genetically engineered poxviruses for recombinant gene expression, vaccination, and safety, *Proc. Natl. Acad. Sci. USA* **93:**11341–11348.

59a. Mackett, M., and Arrand, J. A., 1985, Recombinant vaccinia virus induces neutralising antibodies in rabbits against Epstein–Barr virus membrane antigen gp340, *EMBO J.* **4:**3229–3234.

59b. Morgan, A. J., Mackett, M., Finerty, S., Arrand, J. R., Scullion, F., and Epstein, M. A., 1988,

Recombinant vaccinia virus expressing Epstein–Barr virus glycoprotein gp340 protects cotton-top tamarins against EB virus-induced malignant lymphomas, *J. Med. Virol.* **25:**189–195.
60. Ragot, T., Finerty, S., Watkins, P. E., Perricaudet, M., and Morgan, A. J., 1993, Replication-defective recombinant adenovirus expressing the Epstein–Barr virus (EBV) envelope glycoprotein gp340/220 induces protective immunity against EBV-induced lymphomas in the cottontop tamarin, *J. Gen. Virol.* **74:**501–507.
61. Mackett, M., Cox, C., de V, Pepper, S., Lees, J. F., Naylor, B. A., Wedderburn, N., and Arrand, J. R., 1996, Immunisation of common marmosets with vaccinia virus expressing Epstein–Barr virus (EBV) gp340 and challenge with EBV, *J. Med. Virol.* **50:**263–271.
62. Gu, S., Huang, T., Ruan, L., Miao, Y., Lu, Y., Chu, H., Motz, C. M., and Wolf, H., 1993, On the first EBV vaccine trial in humans using recombinant vaccinia virus expressing the major membrane antigen, in *Epstein–Barr Virus and Associated Diseases* (T. Tursz, J. S. Pagano, D. V. Ablashi, G. de Thé, G. Lenoir, and G. R. Pearson, eds.), Libbey, London/INSERM, Paris, pp. 579–584.
63. Gu, S., Huang, T., Miao, Y., Ruan, L., Zhao, Y., Han, C., Xiao, Y., Zhu, J., and Wolf, H., 1991, A preliminary study on the immunogenicity in rabbits and in human volunteers of a recombinant vaccinia virus expressing Epstein–Barr virus membrane antigen, *Chin. Med. Sci. J.* **6:**241–243.
64. Gu, S. Y., Huang, T. M., Ruan, L., Miao, Y. H., Lu, H., Chu, C. M., Motz, M., and Wolf, H., 1995, First EBV vaccine trial in humans using recombinant vaccinia virus expressing the major membrane antigen, *Dev. Biol Stand.* **84:**171–177.
65. Ulaeto, D., Wallace, L., Morgan, A. J., Morein, B., and Rickinson, A. B., 1988, *In vitro* T cell responses to a candidate Epstein–Barr virus vaccine: Human $CD4^+$ T-cell clones specific for the major envelope glycoprotein gp340, *Eur. J. Immunol.* **18:**1689–1697.
66. Wallace, L. E., Wright, J., Ulaeto, D. O., Morgan, A. J., and Rickinson, A. B., 1991, Identification of two T-cell epitopes of the candidate Epstein–Barr virus vaccine glycoprotein gp340 recognized by $CF4^+$ T-cell clones, *J. Virol.* **65:**3821–3828.
67. Bejarano, M. T., Masucci, M. G., Morgan, A. J., Morein, B., Klein, G., and Klein, E., 1990, Epstein–Barr virus (EBV) antigens processed and presented by B cells, B blasts and macrophages trigger T-cell-mediated inhibition of EBV-induced B-cell transformation, *J. Virol.* **64:**1398–1401.
68. Lee, S. P., Wallace, L. E., Mackett, M., Arrand, J. R., Searle, P. F., Rowe, M., and Rickinson, A. B., 1993, MHC class II-restricted presentation of endogenously synthesized antigen: Epstein–Barr virus transformed B cell lines can present the viral glycoprotein gp340 by two distinct pathways, *Int. Immunol.* **5:**451–460.
69. Khyatti, M., Patel, P. C., Stefanescu, I., and Menezes, J., 1991, Epstein–Barr virus (EBV) glycoprotein gp350 expressed on transfected cells resistant to natural killer cell activity serves as a target antigen for EBV-specific antibody-dependent cellular cytotoxicity, *J. Virol.* **65:**996–1001.
70. Khyatti, M., Stefanescu, I., Blagdon, M., and Menezes, J., 1994, Epstein–Barr virus gp350-specific antibody titers and antibody-dependent cellular cytotoxic effector function in different groups of patients: A study using cloned gp350-expressing transfected human T cell targets, *J. Infect. Dis.* **170:**1439–1447.
71. Khanna, R., Burrows, S. R., Kurilla, M. G., Jacob, C. A., Misko, I. S., Sculley, T. B., Kieff, E., and Moss, D. J., 1992, Localization of Epstein–Barr virus cytotoxic T cell epitopes using recombinant vaccinia: Implications for vaccine development, *J. Exp. Med.* **176:**169–176.
72. Moss, D. J., Schmidt, C., Elliot, S., Suhrbier, A., Burrows, S., Khanna, R., 1996, Strategies involved in developing an effective vaccine for EBV-associated diseases, *Adv. Cancer Res.* **69:**213–246.
73. Zeng, Y., Wolf, H., Takada, K., Arrand, J. R., De Thé, G., 1988, Detection of EBV-specific IgA antibodies to EA, MA and EBNA-1 recombinant proteins in NPC patients and controls, in: *Epstein–Barr Virus Hum. Dis.* (O. V. Ablashi, ed.), Clifton, NJ, Humana Press, pp. 309–313.
74. Wolf, H., Haus, M., and Wilmes, E., 1984, Persistence of Epstein–Barr Virus in the parotid gland, *J. Virol.* **51:**795–798.

75. Bogedain, C., Wolf, H., Modrow, S., Stuber, G., and Jilg, W., 1995, Specific cytotoxic T-lymphocytes recognize the immediate-early transactivator Zta of Epstein–Barr virus, *J. Virol.* **69:**4872–4879.
76. Motz, M., Deby, G., and Wolf, H., 1987, Truncated versions of the two major Epstein–Barr viral glycoproteins (gp250/350) are secreted by recombinant Chinese hamster ovary cells, *Gene* **58:**149–154.
77. Schirmbeck, R., Deml, L., Melber, K., Wolf, H., Wagner, R., and Reimann, J., 1995, Priming of class I-restricted cytotoxic T-lymphocytes by vaccination with recombinant protein antigens, *Vaccines* **13:**857–865.
78. Jilg, W., Bogedain, C., Mairhofer, H., Gu, S. Y., and Wolf, H., 1994, The Epstein–Barr virus-encoded glycoprotein gp110 (BALF4) can serve as a target for antibody-dependent cell-mediated cytotoxicity (ADCC), *Virology* **202:**974–977.
79. Desranges, C., Wolf, H., De Thé, G., Shanmugaratnam, K., Cammoun, N. Ellouz, R., Klein, G., Lennert, K., Munoz, N., and zur Hausen, H., 1975, Nasopharyngeal carcinoma. X. Presence of Epstein–Barr genomes in separated epithelial cells of tumors in patients from Singapore, Tunisia and Kenya, *Int. J. Cancer* **16:**7–15.
80. Simons, M. J., Wee, G. B., and Goh, E. H., 1976, Immunogenetic aspects of nasopharyngeal carcinoma. IV. Increased risk in Chinese of nasopharyngeal carcinoma associated with a Chinese-related HLA profile (A2, Singapore 2), *J. Natl. Cancer Inst.* **57:**977–980.
81. Lu, Sh-J., Day, N. E., Degos, L., Lepage, V., Wang, P-C., Chan, S-H., Simons, M., McKnight, B., Easton, D., Zeng, Y., and De Thé, G., 1990, Linkage of a nasopharyngeal carcinoma susceptibility locus to the HLA region, *Nature* **346:**470–471.
82. Cochet, C., Martel-Renoir, D., Grunewald, V., Bosq, J., Cochet, G., Schwaab, G., Bernaudin, J-F., and Joab, I., 1993, Expression of the Epstein–Barr virus immediate early gene, BZLF1, in nasopharyngeal carcinoma tumor cells, *Virology* **197:**358–365.
83. Chen, F., Zou, J-Z., diRenzo, L., Hu, L. F., Martony, A., Ehlin-Henriksson, B., Klein, G., and Ernberg, I., 1994, EBNA1 positive cells detected in resting B-cells from normal healthy donors, *Epstein–Barr Virus & Associated Diseases, Cold Spring Harbor Meeting on Cancer Cells,* Cold Spring Harbor, New York, November 7–11, 1994, p. 88 (Abstract).
84. Decker, L. L., Klamen, L. D., and Thorley-Lawson, D. A., 1996, Detection of the latent form of Epstein–Barr virus DNA in the peripheral blood of healthy individuals, *J. Virol.* **70:**3286–3289.
85. Miyashita, E. M., Yang, B., Lam, K. M., Crawford, D. H., Thorley-Lawson, D. A., 1995, A novel form of Epstein–Barr virus latency in normal B cells *in vivo, Cell* **80:**593–601.
86. Equitable distribution of burdens and benefits, Guideline 10, in *International Ethical Guidelines for Biomedical Research Involving Human Subjects,* Organizations of Medical Sciences (CIOMS), Geneva, 1993.
87. Miller, N., and Hutt-Fletcher, L. M., 1988, A monoclonal antibody to glycoprotein gp85 inhibits fusion but not attachment of Epstein–Barr virus, *J. Virol.* **62:**2366–2372.
88. Mueller, N. E., 1997, Epstein–Barr virus and Hodgkin's disease: An epidemiological paradox, *Epstein–Barr Virus Report* **4**(1)**:**1–2.
89. Wolf, H., Bogedain, C., Schwarzmann, F., 1993, Epstein–Barr virus and its interaction with the host, *Intervirology* **35:**26–39.
90. Nobis, P., and Jaenisch, R., 1980, Passive immunotherapy prevents expression of endogènous Moloney virus and amplification of proviral DNA in BALB/Mo mice, *Proc. Natl. Acad. Sci. USA* **77**(6)**:**3677–3681.
91. Fiedler, W., Nobis, P., Jahner, D., and Jaenisch, R., 1982, Differentiation and virus expression in BALB/Mo mice: Endogenous Moloney leukemia virus is not activated in hematopoietic cells, *Proc. Natl. Acad. Sci. USA* **79**(6)**:**1874–1878.
92. Purchase, H. G., and Burmester, B. R., 1973, *Diseases of Poultry,* 6th ed., Iowa State University Press, Ames, Iowa, pp. 502–567.
93. Graf, T., and Beug, H., 1978, Avian leukemia viruses: Interaction with their target cells *in vivo* and *in vitro, Biochim. Biophys. Acta* **516**(3)**:**269–299.

11

Inhibition of MHC Class I Function by Cytomegalovirus

HARTMUT HENGEL and ULRICH H. KOSZINOWSKI

1. INTRODUCTION

Cytomegaloviruses (CMVs) represent a subgroup of beta herpesviruses that are widely distributed in nature. As members of the herpesvirus family CMVs are enveloped viruses which share large double-stranded DNA genomes of about 235 kb in size. Phenotypically, cytomegaloviruses are characterized by their strict species specificity, slow replication in a limited number of cell types, and their typical cytopathology.[1] CMVs genomes contain more than 200 separate genes likely to encode protein. Their genomic structure is characterized by a number of conserved gene blocks arranged in the central 100 kb of the genome ("core") that are closely related among CMVs and more distantly related to genes from other herpesviruses. This core is flanked at either end of the genome by regions of diverging gene families without homologues in other herpesviruses.[2,3] As in other herpesviruses, CMV replication is tightly regulated in a multistep process and can be subdivided into the immediate early (IE), early (E), and late (L) phases of gene expression.[1] CMVs establish acute, persistent, and latent infections. After clearance of productive infection, the viral genome persists in a latent state for the lifetime of the infected host. Reactivation from latency to productive infection, however, occurs frequently in a multitude of organs.[4,5]

HARTMUT HENGEL and ULRICH H. KOSZINOWSKI • Max von Pettenkofer-Institut für Hygiene und Medizinische Mikrobiologie, Lehrstuhl Virologie, Ludwig-Maximilians-Universität München, D-80336 München, Germany.

Herpesviruses and Immunity, edited by Medveczky *et al.* Plenum Press, New York, 1998.

In humans, persistent infection with human CMV (HCMV) is usually not associated with clinical symptoms unless the host is immunosuppressed. In transplant patients, in AIDS patients, or following intrauterine infection, CMV frequently progresses to organ- or even life-threatening disease.[6] The notion that the status of the host's immune system largely defines the outcome of CMV infection is based on solid experimental evidence and broadly fits clinical experience.

2. PRINCIPLES OF THE IMMUNE CONTROL OF CMV INFECTION

Understanding the host response to cytomegaloviral infections has been furthered by experiments in murine models. Mouse CMV (MCMV) is a species-specific member of the CMV family but shares important properties with HCMV. The antiviral defense against CMVs includes several immune effector functions. Natural killer (NK) cells forming a part of the nonadaptive immunity significantly contribute to virus control, although to a variable extent depending on the genetic background of the specific mouse strain analyzed,[7,8] On its own, this lymphocyte subset failed to protect against lethal CMV infection.[9] Antibodies contribute to the limitation of viral spread in recurrent infection but have no role in the clearance of primary infected animals.[5] *In vivo* depletion experiments and adoptive transfer studies in MCMV-infected mice identified the $CD8^+$ T cell subset as essential and sufficient to clear acute MCMV infection in spleen, lungs, and liver and to mediate protection from otherwise lethal infection.[10–12] Priming of $CD8^+$ T cells with a single epitope from a MCMV IE antigen renders mice resistant against a lethal challenge with virus.[13] In addition, an important role for $CD4^+$ T cells was demonstrated for controlling MCMV infection in the salivary gland.[14] A profound antiviral effect *in vivo* was demonstrated for certain cytokines. In particular, depletion of endogenous IFNγ and TNFα resulted in an increase in virus replication and severe CMV disease.[15,16] The influence of IFNγ on MCMV control and survival of disease was linked to the effects of this cytokine on antigen processing and presentation. IFNγ combats viral escape from $CD8^+$ immunity by restoring major histocompatibility complex (MHC) I-dependent antigen presentation[17] and, in addition, increases the efficiency of processing viral proteins into peptides.[18]

The efficacy of adoptively transferred CMV-specific $CD8^+$ T cells for controlling CMV infection in mice can be confirmed in humans. Transfer of $CD8^+$ CMV-specific CTL lines expanded *in vitro* from MHC identical donors into patients undergoing bone marrow transplantation prevents CMV disease despite their compromised immune systems and concurrent treatment with immunosuppressive drugs.[19] In conclusion, the MHC class I-restricted immunity has a pivotal role in controlling CMV infection in both mice and humans.

3. THE MHC CLASS I PATHWAY OF ANTIGEN PROCESSING AND PRESENTATION

T lymphocytes recognize peptides derived from CMV proteins presented by either MHC class I or class II molecules. Although MHC class I molecules are expressed on virtually all cells, MHC class II molecules are constitutively expressed only on few cell types like B cells and "professional antigen-presenting cells." Each type of MHC molecule is recognized by a distinct T lymphocyte subpopulation. Activated $CD8^+$ T cells have cytolytic activity against virally infected cells and release cytokines. They are called cytotoxic T lymphocytes (CTL). $CD4^+$ T cells, although capable of cytolytic effector functions, are mainly producers of cytokines. $CD4^+$ T cells recognize peptides presented by MHC class II molecules, whereas $CD8^+$ T cells recognize peptides associated with MHC class I molecules. Therefore, two principal pathways exist by which proteins are processed and presented. Usually, exogenous proteins taken up by the cell are processed mainly in the endosomal pathway and presented by MHC class II molecules, whereas proteins processed through the cytosolic pathway are presented by MHC class I molecules. The biosynthesis of viral proteins as endogenous proteins favors processing them within the cytosolic pathway which perhaps explains the dominance of $CD8^+$ T cells in antiviral control. In this pathway, MHC class I molecules, composed of a 44–49-kDa polymorphic transmembranous heavy chain, a 12-kDa soluble light chain, β_2-microglobulin (β_2m), and an 8–10 residue peptide, are assembled in ternary complexes in the endoplasmic reticulum (ER) for transport to the cell surface[20] (see Fig. 1). Association of the N-linked glycosylated MHC class I heavy chain with β_2m and peptide is sequentially assisted by transient interactions with molecular chaperones. Calnexin interacts with free class I heavy chains,[21,22] and calreticulin binds human class I/β_2m dimers.[23] MHC class I heterodimers assemble with the transporter associated with antigen processing (TAP) via the TAP1 subunit[24–26] and are mediated by an 48-kDa ER glycoprotein, tapasin.[23] Binding of high-affinity peptides to class I molecules leads to dissociation of TAP–class I complexes and the exit of the heterotrimeric class I complex from the ER.[24,26] MHC class I molecules lacking peptide display a different conformation and are deficient in surface transport and stability.[27,28] Antigenic peptides are processed from viral proteins by cytosolic proteases. For assembly into ternary MHC class I complexes, peptides have to be translocated across the endoplasmic reticulum (ER) membrane.[20,29] The TAP transporter is a heterodimer composed of two homologous proteins, TAP1 and TAP2, both encoded in the MHC. It is predicted that both

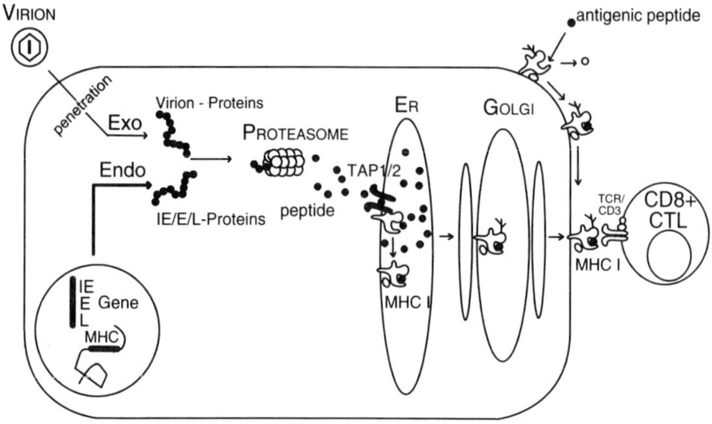

FIGURE 1. The MHC class I pathway of antigen presentation. Viral proteins *de novo* synthesized in the infected cell (endogenous proteins) or structural proteins of the virion (exogenous proteins) are processed into peptides (black circles) by the proteasome. Peptides are translocated by TAP1/2 (transporters associated with antigen processing; shown in grey) into the lumen of the endoplasmic reticulum (ER) for binding to MHC class I/β_2m heterodimers (white). Ternary MHC class I complexes are transported via the Golgi to the cell surface for presentation of the peptide to $CD8^+$ T cells. Experimentally, peptides can also be loaded onto MHC class I molecules on the cell surface. Abbreviations: IE: immediate-early; E: early; L: late; TCR/CD3: T cell receptor complex.

subunits span the ER membrane six to ten times with small loops penetrating the cytosol and the ER lumen and possess a large cytosolic domain containing an ATP-binding cassette. The transport of peptides by TAP requires two events. In the first step, the peptide is bound to the cytosolic face of TAP before it is subsequently translocated in an ATP-dependent manner and released into the lumen of the ER.[30–33] The processing of a viral protein to antigenic peptides for presentation to CTL can be experimentally bypassed *in vitro* by externally loading peptides onto MHC class I molecules that are already expressed on the cell surface (Fig. 1). Altogether, cytolytic T cells play a decisive role in eliminating virally infected cells, and MHC class I molecules are required to guide the attack.

4. CMV STRATEGIES FOR IMMUNE ESCAPE

The lifelong persistence of CMV in the face of a vigorous and repeatedly boosted immune system must require an intricate balance between virus and host. Different mechanisms allow cytomegaloviruses to escape eradication by the host's immunity. First, CMVs can abortively infect cells and establish latency. During the latent state of infection, minimal gene expression prevents immune recognition. Secondly, cytomegaloviruses replicate at privileged sites in the body, i.e., productive infection is maintained for longer periods of time in specific tissues that are poorly controlled by CD8$^+$ T cells, e.g., salivary gland tissue. Thirdly, highly adapted viral gene functions allow the virus to actively subvert detection by the immune system.

Because CMV infection is most tightly controlled by the MHC class I-restricted immunity, it is readily clear that CMVs have been under evolutionary pressure to develop selective functions which counteract MHC class I expression and antigen presentation in the MHC class I pathway. On the other hand, it has become increasingly clear that MHC class I expression provides a protective signal for cells against lysis by natural killer (NK) cells.[34] Thus, it is not excluded that, by escaping the immune attack of CD8$^+$ T cells, CMV may render infected cells sensitive to attack by NK cells. In this article we review work devoted to the identification and molecular analysis of cytomegalovirus gene functions which interfere with the MHC class I pathway of antigen presentation. Common to both cytomegaloviruses, this pathway is targeted by more than one viral protein which prevent surface expression of MHC class I–peptide complexes. The molecular principles of the mechanisms evolved by MCMV and HCMV, however, differ markedly and provide valuable tools for studying novel aspects of the biogenesis and function of MHC class I molecules. The functions employed by CMVs are compared with analogous strategies followed by adenoviruses[35] and herpes simplex virus (HSV).[36–40]

5. MCMV GENE FUNCTIONS AFFECTING THE MHC CLASS I PATHWAY OF ANTIGEN PRESENTATION

5.1. The *m152* Encoded Glycoprotein Retains MHC Class I Complexes in the ERGIC

MCMV was the first herpesvirus suggesting direct interference with the MHC class I pathway of antigen presentation.[41,42,43] When the presentation of the immunodominant MCMV IE antigen pp89 was tested, the absence of target cell recognition by CD8$^+$ T cells was already noted 30 min after the onset of early gene expression although pp89 synthesis and peptide processing was not affected.[42,44] When the fate of MHC class I molecules was studied, the formation and stability of class I complexes was found intact, but the maturation of the MHC class I glycoprotein and its transport into the *medial*-Golgi compartment was inhibited.[44] The limited knowledge of the MCMV genome slowed down identification of the responsible gene, but a combined approach of constructing deletion mutant viruses and injecting cloned MCMV DNA fragments into cells and screening them for MHC class I molecule retention by immunofluorescence microscopy succeeded in identifying the *m152* gene that mediates this effect.[45,46] The gene encodes a type I transmembrane glycoprotein of 37 and 40 kDa, respectively, that arrests the export of mouse class I complexes from the ER-Golgi intermediate compartment (ERGIC)/*cis*-Golgi-compartment. Notably, the plasma membrane transport of human MHC class I molecules is not affected. Deletion of the cytoplasmic tail of gp40 did not lift its effect on MHC retention, indicating that gp40 differs in its function from the E3/19K protein of Adenoviruses.[47]

Several lines of evidence suggested the presence of additional MCMV genes that affect the MHC class I pathway of antigen presentation: (1) *m152* deletion mutant viruses still down-regulated MHC class I surface expression;[45] (2) the gene expression of *m152* during the MCMV replication cycle is limited to a maximum after 4–6 hours p.i. and cannot explain the continuous effect on MHC surface expression late after infection.[46]

5.2. gp34 of MCMV Binds to MHC Class I Complexes

Within a second set of MCMV early genes expressed later than *m152,* a type I glycoprotein of 34 kDa is expressed by the gene *m04* which tightly associates with folded and β_2m-associated MHC class I molecules in the ER.[48] gp34/MHC class I complexes are also found on the cell surface of MCMV-infected cells, suggesting that this complex may escape the *m152*/gp40-mediated transport block in the ERGIC. Combined with the observations that MHC class I molecule retention in MCMV-infected cells is incomplete and that expression of *m152* precedes that of *m04,* the data were interpreted as showing that gp34 counteracts the class I MHC retention mechanism and allows class I molecules to escape to the cell surface.[48] The display of the gp34/class I complex on the cell surface may prevent recognition of the CMV-infected cell by cytotoxic effector cells, e.g., MHC class I-restricted CTL and NK cells.[48] This attractive hypothesis awaits further explanation.

5.3. gp48 of MCMV Targets MHC Class I Complexes to the Lysosome for Destruction

Besides the *m152* gene product, the extended analysis of MCMV deletion mutant viruses identified a further protein implicated in the loss of MHC class I surface expression by infected fibroblasts. This glycoprotein of 48 kDa binds directly to the MHC class I complex.[49] Both components of the complex, gp48 and MHC class I, leave the ER and are degraded by a leupeptin and ammonium chloride-sensitive mechanism. In the presence of leupeptin, gp48 accumulates and can be co-localized with LAMP-1, a marker of the late endosome and lysosome. During the MCMV replication cycle, the expression of gp48 starts later than the transcription of gene *m152* but is maintained throughout in the late phase of infection.[49] Although the interplay between both viral proteins has not yet been investigated, it is tempting to speculate that the sequential mode of gene expression indicates a cooperative function between both glycoproteins. The *m152* encoded polypeptide mediates retention of stable MHC class I complexes in the ERGIC from the very beginning of the early phase.[46] gp48 disposes of MHC class I complexes to the endolysosome. Therefore gp48 may also transport MHC class I complexes accumulated by *m152* to the endolysosome for degradation.

6. MCMV GENE FUNCTIONS AFFECTING THE MHC CLASS I PATHWAY OF ANTIGEN PRESENTATION

6.1. The HCMV UL18 Glycoprotein Binds β_2m and Peptides

The *UL18* gene of HCMV attracted attention because of its homology with human polymorphic MHC class I genes, about 21% identity, and the formation of three putative domains representing a typical feature of class I molecules.[50] Expression of this highly glycosylated 69-kDa glycoprotein from a recombinant vaccinia virus resulted in binding β_2m.[51] Deletion of *UL18* from the HCMV genome did not affect MHC class I down-regulation in HCMV-infected cells, indicating that *UL18* is not responsible for this effect by catching β_2m.[52] The MHC class-I-like properties of the UL18 glycoprotein were further confirmed by eluting endogenous peptides from purified gpUL18 whose characteristics are similar to those bound to conventional class I molecules.[53] Provided that UL18 is expressed at the cell membrane of HCMV-infected cells, this MHC class I molecule may serve as a surrogate and may protect HCMV-infected cells lacking surface MHC class I molecules from lysis by NK cells by engaging NK cell inhibitory receptors.

6.2. The HCMV US11 Glycoprotein Dislocates MHC Class I Heavy Chains to the Cytosol

Infection of fibroblasts with HCMV results in a drastic decrease of MHC class I complex formation and a reduction in the stability of free, unassembled MHC class I heavy chains.[54–57] By screening a bank of HCMV mutants with deletions in the S component of the genome, Jones and co-workers identified two independent loci associated with down-regulating MHC class I heavy chains, one of which is *US11*.[58] In permissive cells, *US11* is transcribed with early kinetics, and the mRNA is most abundant 8–16 hours p.i.[59] Biochemical studies by Ploegh and coworkers[60] revealed that the ER resident type I transmembrane glycoprotein encoded by HCMV *US11* can dislocate the MHC class I heavy chain back to the cytosol. In the presence of gpUS11, MHC class I molecules become core-glycosylated and therefore must have been inserted into the ER membrane before export from this compartment. After entering the cytosol, the heavy chain is immediately deglycosylated and degraded by a multisubunit cytosolic proteolytic complex, the proteasome. This mechanism shortens the half-life of MHC class I heavy chains to less than one minute.[60] This immune evasion mechanism of HCMV points to a general principle by which misfolded ER proteins can be degraded.

6.3. The HCMV US2 Glycoprotein Transfers MHC Class I Heavy Chains via Sec61 to the Cytosol

Similarly to gpUS11, the 24-kDa glycoprotein encoded by *US2* transports the MHC class I heavy chain from the ER lumen back to the cytosol where the molecules are destroyed by the proteasome.[61] Most remarkably, dislocation of the glycosylated MHC class I heavy chain is mediated by the translocon, a structure in the ER membrane by which nascent proteins normally begin their journey in the secretory pathway. The translocon forms a cylindrical structure with a central pore and is composed of 3–4 copies of the Sec61 heterotrimeric complex. The reverse translocation of the MHC heavy chain includes gpUS2 itself which becomes deglycosylated in the cytosol to a 21-kDa protein and thereafter is degraded.[61] Details of the reverse translocation specified by gpUS2 and gpUS11 may differ because both proteins exhibit different abilities to attack allelic forms of murine MHC class I heavy chains.[62] Thus the dual recognition by either gpUS2 or gpUS11 may allow HCMV to control a broader range of class I molecules than might be possible with either gpUS11 or gpUS2 alone.

6.4. The HCMV US3-Encoded Glycoprotein Inhibits MHC Class I Transport

US3 represents the only IE gene within the US region and encodes an ER-resident glycoprotein of 32 and 33 kDa, respectively.[63] *US3* is transcribed abundantly under immediate early conditions but is eventually abolished at early times after infection. The US3 glycoprotein binds to β_2m-associated MHC class I molecules that are stable at 37 °C in 1% NP40 which indicates peptide loading.[63] In contrast to *US11*-expressing cells, MHC class I heterodimers in US3-transfectants are not degraded but stably retained.[63,64] The maturation of the N-linked glycan is impaired, indicating inhibition of its intracellular transport.[63,64] Because the gpUS11-mediated effect includes MHC class I/β_2m-heterodimers,[60] it may well be that HCMV inhibits antigen presentation by a sequential multistep process which is initiated by gpUS3-mediated MHC I retention followed by subsequent degradation due to *US11* and *US2*.

6.5. HCMV US6 Blocks Peptide Translocation by the MHC-Encoded Peptide Transporter TAP1/2

HCMV interferes with MHC class I complex formation by minimizing the supply of MHC class I heavy chains and also by inhibiting the import of peptides into the ER by TAP1/2.[65] The inactivation of peptide transport in HCMV-infected cells functions despite augmented TAP expression during HCMV infection. The inhibition of peptide translocation is mediated by the HCMV *US6* gene encoding a 21-kDa glycoprotein.[66,67] The expression kinetics of the early US6 protein correlates with the inhibition of peptide transport and peaks 72 hours p.i. in the late phase of the replication cycle.[66] Although *US2* and *US11* are abundantly expressed up to 24 hours p.i.,[59] the appearance of gpUS6 is maximal 48 to 96 hours p.i. when other genes interfering with the MHC class I pathway of antigen presentation become almost silent. The subcellular localization of gpUS6 is ER-restricted and identical with that of TAP1. gpUS6 is found in association with the the TAP1/2–MHC class I heavy chain–calreticulin–tapasin assembly complex and also with calnexin, an ER chaperone.[66] Inhibition of the peptide transporter by gpUS6 is independent of the presence of class I heavy chain and tapasin.[66]

6.6. HCMV Prevents Antigen Presentation of the 72-kDa IE Protein

The relatively poor recognition of the abundantly expressed HCMV immediate early 72-kDa transcription factor by $CD8^+$ T cells was not attributed to the viral proteins that globally block MHC class I antigen presentation but to a structural protein of the virus, the 65-kDa matrix phosphoprotein.[68] This matrix protein exhibits associated kinase activity. It has been suggested that the matrix protein selectively abrogates the presentation of 72-kDa IE-derived peptides to CTL by phosphorylation of the 72-kDa protein substrate which could result in inhibition of peptide processing.[68] The mechanism requires further analysis. Because the pp65 matrix protein is abundantly incorporated in HCMV virions, this structural protein might modify endogenously synthesized viral antigens immediately after being released into the cytosol.

7. DISCUSSION

Among the herpesviruses and viruses in general, cytomegaloviruses have apparently evolved the most extensive genetic repertoire to evade the MHC class I-restricted T lymphocyte response of the host by direct protein–protein interaction. These strategies probably reduce its risk of immune destruction during its life cycle, e.g., by the production of viral progeny upon reactivation from the latent state when an antiviral $CD8^+$ T cell memory already exists in the host.

According to present knowledge, MCMV expresses three early gene functions that interfere with the MHC class I pathway of antigenic presentation (see Table I and Fig. 2),[45,46,48] and HCMV expresses a cascade of even four consecutive US gene functions interrupting the class I pathway of antigen presentation

TABLE 1
Cytomegaloviral Gene Functions Interfering with the MHC Class I Pathway of Antigen Presentation

Gene	Replication phase	Encoded protein	Mechanism employed	References
MCMV m152	E	gp37/40	Retention of MHC class I complexes in the ERGIC/ *cis*-Golgi	46
MCMV m04	E	gp34	Binding and transporting MHC class I complexes	48
MCMV m06	E	gp48	Binding and targeting MHC class I complexes to the lysosome	49
HCMV UL18	n.d.	gp69	Binding β_2 and peptide	51, 53
HCMV US11	E	gp32	Directing MHC class I heavy chains into the cytosol	60
HCMV US2	E	gp24, p21	Directing MHC class I heavy chains into the cytosol	61
HCMV US3	IE	gp32/33	Retention of MHC class I complexes	63, 64
HCMV US6	E/L	gp21	Binding and inactivation to TAP1/2	66, 67
HCMV UL83	L	pp65	Blocking of antigen presentation of HCMV IE antigen	68

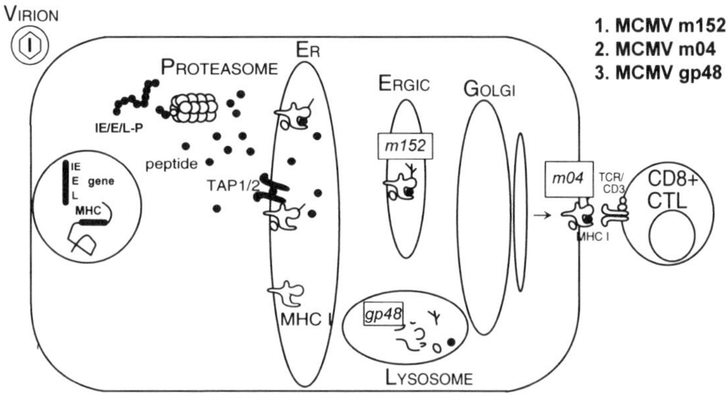

FIGURE 2. Mechanisms of MCMV to evade MHC class I restricted immunity. The points of viral interference are indicated as explained in Table 1 and in the text. Abbreviations: ERGIC: endoplasmic reticulum Golgi intermediate compartment.

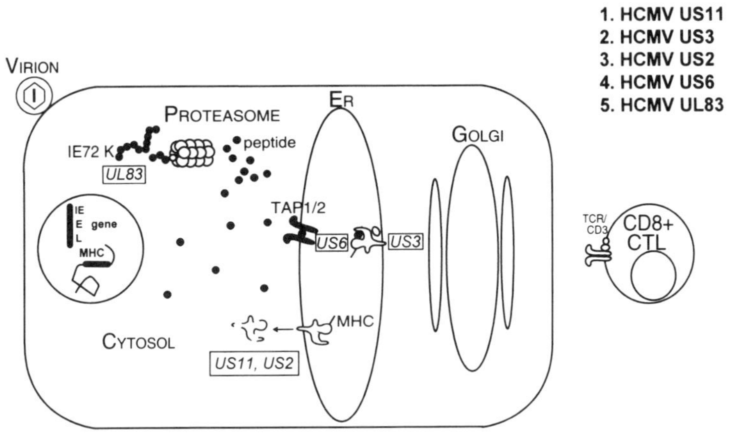

FIGURE 3. Mechanisms of HCMV to evade MHC class I restricted immunity. The points of viral interference are indicated as explained in Table 1 and in the text.

in a general manner (see Table 1 and Fig. 3).[60,61,63,64,66,67] All of these viral factors are MHC class I-specific, i.e., they affect the fate of MHC class I molecules but not the function of other glycoproteins. In both viruses, the genes are expressed temporarily and reduce antigen presentation during the viral replication cycle. All CMV genes dedicated to interfere with the MHC class I function are dispensable for viral replication *in vitro*.[45,58] They code for glycoproteins operating within the secretory pathway at multiple checkpoints because every single step of the biogenesis of MHC class I molecules is fair game for the virus. So far, there are no cellular genes homologous to these CMV genes sequences available in the gene bank. The genes also differ between MCMV and HCMV by their sequence and mechanism of function. Furthermore, although the studies so far are limited, their function appears to be species-specific reflecting the intimate coevolution of cytomegaloviruses with their hosts for millions of years. Although the factors of HCMV and MCMV used for immune escape have a different molecular basis for the majority of proteins, the general strategy results in an identical phenotype, i.e., retention of MHC class I complexes within the secretory pathway or proteolytic degradation of MHC class I molecules. In addition, HCMV has evolved a third mechanism represented by the *US6* gene function that prevents the import of antigenic peptides into the ER and thereby withholds peptide from MHC class I molecule.[65–67] Although this phenotype of HCMV resembles the HSV-1 product ICP47,[36–41] again both polypeptides utilize entirely different mechanisms to inhibit TAP. In contrast to the soluble 9-kDa protein ICP47 that binds to the cytosolic face of TAP1/2 dimers and inhibits peptide association with the transport in a competitive fashion,[39,40] gpUS6 does not interfere with peptide binding to TAP but binds to the luminal face of the transporter.[66,67] Likewise, the inhibition of MHC class I transport by HCMV *US3* and MCMV *ml52* is not identical with the first reported example, the E3/19K protein of adenoviruses. E3/19K binds to MHC class I molecules and mediates their retention by a carboxy-terminal, double-lysin motif (KKXX) recognized by coatomer proteins that retrieve ER-escaped proteins from the Golgi compartment.[47] *US3* and *ml52* lack a KKXX retrieval motif, and *m152* does not require the carboxy terminus for its function.[46] Notably, both CMVs also evolved different means to target MHC class I molecules for proteolytic destruction by cellular proteases. In HCMV, this is achieved by the immediate dislocation of MHC class I molecules to the cytosol before dissociating from the translocon.[60,61] In MCMV-infected cells class I complexes assemble but are subsequently tightly bound to gp48 and are guided to another principal proteolytic compartment of the cell, the lysosome.[49]

Why do CMVs harbor such a bewildering array of gene functions all of which interfere with one target, MHC class I? The multitude of stealth genes in CMV's genomes may be required for several reasons: first, to cover the protracted replication cycle which takes at least 72 hours in the case of HCMV during which over 200 antigenic proteins are synthesized within the infected cell; secondly, to compensate for the opposite effects of cytokines on MHC class I[17,69] that are produced in infected tissues and which reconstitute antigen presentation despite the expression of viral immune evasion proteins. Finally, the great demand to regulate the presentation function of a high number of MHC class I alleles and their peptides in different cell types may have favored the diversification of MHC-reactive genes and their functions. On the other hand, despite the CMV genes, a vigorous CMV-specific $CD8^{4+}$ T cell response is, nevertheless, generated that is effective in controlling the spread of CMVs *in vivo*.[11,13,19,70] How can we solve this paradox? The protective properties of $CD8^+$ T cells imply that host mechanisms must exist that can overcome the viral effects on MHC class I and regulate the antigen presentation function *in vivo*. Indeed, the immunoevasive functions of both MCMV and HCMV on MHC class I can be compensated *in vitro* by the activity of cytokines like IFNγ, type I IFNs, and TNFα, which restore the lysis of CMV-infected fibroblasts by CTL.[17,69] In addition, quantitative analysis of CMV-derived antigenic peptides extracted from MCMV-infected organs revealed that IFNγ is a potent regulator of antigenic processing *in vivo* which determines the yield of processed peptides.[18] Furthermore, the antigen-presentation function by MHC class I molecules in certain professional antigen-presenting cells, like macrophages, is maintained during permissive MCMV-infection,[71] indicating that some cell types may be constitutively resistant to viral effects on MHC class I. In an immunocompetent host, such cells may initiate a vigorous $CD8^+$ T lymphocyte response. Thus, is seems most likely that the degree of MHC class I inhibition operative *in vivo* is subject to regulation by host factors and is part of the delicate balance between CMVs and their hosts.

Besides the presentation of peptides to $CD8^+$ T cells, MHC class I molecules also play a role in activating natural killer cells.[34] For NK cells, MHC class I molecules deliver an inhibitory signal which is lost if MHC class I molecules disappear from the cell surface. Clearly, NK cells contribute to MCMV control,[7–9] but whether viral proteins like HCMV UL18 or MCMV gp34 function as decoy MHC class I molecules and modify susceptibility to NK cells is an area of further investigation. So far it is not known whether the viral genes reproduce their *in vitro* effect *in vivo* and what their impact on persistent CMV infection is. Although CMV genes accomplishing immune escape from MHC class I restricted immunity have been valuable tools that extended our knowledge of antigen presentation *in vitro,* further experiments using virus deletion mutants are required to teach us the next lesson—how these genes operate *in vivo.*

8. SUMMARY

Cytomegaloviruses (CMV) are subject to stringent $CD8^+$ T cell control in an immunocompetent host. Disease conditions caused by CMV are associated with an immature or an immunodeficient cellular immune response. CMV persists for life in an infected host despite an active immune system. This selective pressure has led to the selection of a set of immune evasion functions that are focused on avoiding cytolytic attack by $CD8^+$ T cells. The target of mouse and human CMV genes is the class I antigenic presentation pathway. In this pathway the peptide products of proteins after cytosolic degradation are translocated by the transporter associated with antigen processing, TAP1/2, into the ER lumen where they bind to the MHC class I heavy chain and, together with the β_2m light chain form a trimeric complex which is exported to the plasma membrane for T cell recognition. So far, several CMV proteins have been identified that interfere with this pathway. The common phenotype induced in cells by most of these functions is the loss of MHC class I molecules at the plasma membrane. The underlying molecular mechanisms differ. The earliest function with respect to this pathway is an ER resident glycoprotein encoded by HCMV gene *US6* which blocks peptide transport by TAP1/2. The glycoproteins encoded by genes *US2* and *US11* act at a very early step in MHC molecule formation and dislocate the nascent heavy chain back to the cytosol where it is degraded by the proteasome. The glycoprotein encoded by HCMV *US3* blocks the export of MHC complexes from the ER. At about this stage the MCMV gene products target the class I presentation pathway. The glycoprotein specified by *m152* arrests MHC complexes in the ERGIC/*cis*-Golgi. A 48-kDa glycoprotein binds to MHC complexes and diverts them to lysosomal degradation. Another glycoprotein encoded by the MCMV gene *m04* binds to the complex and comigrates to the cell surface. The redundancy of the evasion functions and their potential cooperation, the precise mechanism of action of each of the proteins, and their usage as investigatory tools in cell biology are attractive aspects of further research.

ACKNOWLEDGMENTS. Our work is supported by the Deutsche Forschungsgemeinschaft. We thank Dr. K. Früh, La Jolla, California, for sharing unpublished results.

REFERENCES

1. Mocarski, E. S., 1996, Cytomegaloviruses and their replication, in *Virology*, Volume 2 (B. N. Fields, D. M. Knipe, and P. M. Howley, eds.), Lippincott-Raven, Philadelphia, pp. 2447–2492.
2. Chee, M. S., Bankier, A. T., Beck, S., Bohni, R., Brown, C. M., Cerny, R., Horsnell, T., Hutchison, C. A. III, Kouzarides, T., Martignetti, J. A., Preddie, E., Satchwell, S. C., Tomlinson, P., Weston, K. M., and Barrell, B. G., 1990, Analysis of the protein-coding content of the sequence of human cytomegalovirus strain AD169, *Curr. Top. Microbiol. Immunol.* **54:**125–169.
3. Rawlinson, W. D., Farrell, H. E., and Barrell, B. G., 1996, Analysis of the complete DNA sequence of murine cytomegalovirus, *J. Virol.* **70:**8833–8849.
4. Reddehase, M. J., Balthesen, M., Rapp, M., Jonjic, S., Pavic, I., and Koszinowski, U. H., 1994, The condition of primary infection defines the load of latent viral genome in organs and the risk of recurrent cytomegalovirus disease, *J. Exp. Med.* **179:**185–193.
5. Jonjic, S., Pavic, I., Polic, B., Crnkovic, I., Lucin, P., and Koszinowski, U. H., 1994, Antibodies are not essential for the resolution of primary cytomegalovirus infection but limit dissemination of recurrent virus, *J. Exp. Med.* **179:**1713–1717.
6. Ho, M., 1995, Cytomegalovirus, in *Principles and Practice of Infectious Diseases*, Volume 2 (G. L. Mandell, J. H., Bennett, and R. Dolin, eds.), Churchill Livingstone, New York, pp. 1351–1364.
7. Shellam, G. R., Allan, J. E., Papidimitriou, J. M., and Bancroft, G. J., 1981, Increased susceptibility to cytomegalovirus infection in beige mutant mice, *Proc. Acad. Natl. Sci. USA* **78:**5104–5108.
8. Scalzo, A. A., Fitzgerald, N. A., La Vista, A. B., and Shellam, G. A., 1990, Cmv-1, a genetic locus that controls murine cytomegalovirus replication in the spleen, *J. Exp. Med.* **171:**1469–1483.
9. Welsh, R. M., Brubaker, J. O., Vargas-Cortes, M., and O'Donnell, C. L., 1991, Natural killer (NK) cell response to virus infections in mice with severe combined immunodeficiency. The stimulation of NK cells and the NK cell-dependent control of virus infections occur independently of T and B cell function, *J. Exp. Med.* **173:**1053–1063.
10. Reddehase, M. J., Weiland, F., Münch, K., Jonjic, S., Lüske, A., and Koszinowki, U. H., 1985, Interstitial murine cytomegalovirus pneumonia after irradiation: Characterization of cells that limit viral replication during established infection of the lungs, *J. Virol.* **55:**264–273.
11. Reddehase, M. J., Mutter, W., Münch, K., Bühring, H. J., and Koszinowski, U. H., 1987, CD8 positive T lymphocytes specific for murine cytomegalovirus immediate-early antigens mediate protective immunity, *J. Virol.* **61:**3102–3108.
12. Reddehase, M. J., Jonjic, S., Weiland, F., Mutter, W., and Koszinowski, U. H., 1988, Adoptive immunotherapy of murine cytomegalovirus adrenalitis in the immunocompromised host: CD4-helper-independent antiviral function of CD8-positive memory T lymphocytes derived from latently infected donors, *J. Virol.* **62:**1061–1065.

13. Jonjic, S., Del Val, M., Keil, G. M., Reddehase M. J., and Koszinowski, U. H., 1988, A nonstructural viral protein expressed by a recombinant vaccinia virus protects against lethal cytomegalovirus infection, *J. Virol.* **62:**1653–1658.
14. Jonjic, S., Mutter, W., Weiland, F., Reddehase, M. J., and Koszinowski, U. H., 1989, Site-directed persistent cytomegalovirus infection after selective long-term depletion of CD4-positive T lymphocytes, *J. Exp. Med.* **169:**1199–1212.
15. Lucin, P., Pavic, I., Polic, B., Jonjic, S., and Koszinowski, U. H., 1992, Gamma interferon-dependent clearance of cytomegalovirus infection in salivary glands, *J. Virol.* **66:**1977–1984.
16. Pavic, I., Polic, B., Crnkovic, I., Lucin, P., Jonjic, S., and Koszinowski, U. H., 1993, Participation of endogenous tumour necrosis factor α in host resistance to cytomegalovirus infection, *J. Gen. Virol.* **74:**2215–2223.
17. Hengel, H., Lucin, P., Jonjic, S., Ruppert, T., and Koszinowski, U. H., 1994, Restoration of cytomegalovirus antigen presentation of gamma interferon combats viral escape, *J. Virol.* **68:**289–297.
18. Geginat, G., Ruppert, T., Hengel, H., Holtappels, R., and Koszinowski, U. H., 1997, Interferon-γ is a prerequisite for efficient antigen processing of viral peptides *in vivo*, *J. Immunol.*, in press.
19. Riddell, S. R., Watanabe, K. S., Goodrich, J. M., Li, C. R., Agha, M. E., and Greenberg, P. D., 1992, Restoration of viral immunity in immunodeficient humans by the adoptive transfer of T cell clones, *Science* **257:**238–241.
20. Heemels, M. T., and Ploegh, H. L., 1995, Generation, translocation, and presentation of MHC class I-restricted peptides, *Annu. Rev. Biochem.* **64:**463–491.
21. Degen, E., and Williams, D. B., 1991, Participation of a novel 88-kDa protein in the biogenesis of murine class I histocompatibility molecules, *J. Cell. Biol.* **112:**1099–1115.
22. Rajagopalan, S., and Brenner, M. B., 1994, Calnexin retains unassembled major histocompatibility complex class I free heavy chains in the endoplasmic reticulum, *J. Exp. Med.* **180:**407–412.
23. Sadasivan, B., Lehner, P. J., Ortmann, B., Spies, T., and Cresswell, P., 1996, Roles for calreticulin and a novel glycoprotein, tapasin, in the interaction of MHC class I molecules with TAP, *Immunity* **5:**103–114.
24. Ortmann, B., Androlewicz, M., and Cresswell, P., 1994, MHC class I/β_2-microglobulin complexes associate with TAP transporters before peptide binding, *Nature* **368:**864–867.
25. Androlewicz, M. J., Ortmann, B., van Endert, P. M., Spies, T., and Cresswell, P., 1994, Characteristics of peptide and major histocompatibility complex class I/β2-microglobulin binding to the transporters associated with antigen processing (TAP1 and TAP2), *Proc. Natl. Acad. Sci. USA* **91:**12176–12720.
26. Suh, W.-K., Cohen-Doyle, M. F., Fruh, K., Wang, K., Peterson, P. A., and Williams, D. B., 1994, Interaction of MHC class I molecules with the transporter associated with antigen processing, *Science* **264:**1322–1326.
27. Baas, E. J., van Santen, H.-M., Kleijmeer, M. J., Geuze, H. J., Peters, P. J., and Ploegh, H. L., 1992, Peptide-induced stabilization and intracellular localization of empty HLA class I complexes, *J. Exp. Med.* **176:**147–156.
28. Bluestone, J. A., Jameson, S., Miller, S., and Dick, R., 1992, Peptide-induced conformational changes in class I heavy chains alter major histocompatibility complex recognition, *J. Exp. Med.* **176:**1757–1761.
29. Koopmann, J.-O., Hämmerling, G. J., and Momburg, F., 1997, Generation, intracellular transport and loading of peptides associated with MHC class I molecules. *Curr. Opinion Immunol.* **9,** in press.
30. Neefjes, J. J., Momburg, F., and Hämmerling, G. H., 1993, Selective and ATP-dependent translocation of peptides by the MHC-encoded transporter, *Science* **261:**769–771.

31. Shepherd, J. C., Schumacher, T. N. M., Ashton-Richardt, P. G., Imaeda, S., Ploegh, H. L., Janeway, C. A., and Tonegawa, S., 1993, TAP1-dependent peptide translocation *in vitro* is ATP dependent and peptide selective, *Cell* **74:**577–584.
32. Androlewicz, M. J., Anderson, K. S., and Cresswell, P., 1993, Evidence that transporters associated with antigen processing translocate a major histocompatibility complex class I-binding peptide into the endoplasmic reticulum in an ATP-dependent manner, *Proc. Natl. Acad. Sci. USA* **90:**9130–9134.
33. van Endert, R., Tampe, R., Meyer, T. H., Tisch, R., Bach, J.-F., and McDevitt, H. O., 1994, A sequential model for peptide binding and transport by the transporters associated with antigen processing, *Immunity* **1:**491–500.
34. Kärre, K., 1995, Express yourself or die: Peptides, MHC molecules, and NK cells, *Science* **267:**978–979.
35. Burgert, H.-G., and Kvist, S., 1985, An adenovirus type 2 glycoprotein blocks cell surface expression of human histocompatibility class I antigens, *Cell* **41:**987–997.
36. York, I. A., Roop, C., Andrews, D. W., Riddell, S. R., Graham, F. L., and Johnson, D. C., 1994, A cytosolic herpes simplex virus protein inhibits antigen presentation to $CD8^+$ T lymphocytes, Cell **77:**525–535.
37. Früh, K., Ahn, K., Djaballah, H., Sampe, P., van Endert, P. M., Tampe, R., Peterson, P. A., and Yang, Y., 1995, A viral inhibitor of peptide transporters for antigen presentation, *Nature* **375:**415–418.
38. Hill, A., Jugovic, P., York, I., Russ, G., Bennink, J., Yewdell, J., Ploegh, H., and Johnson, D., 1995, Herpes simplex virus turns off the TAP to evade host immunity, *Nature* **375:**411–415.
39. Ahn, K., Meyer, T. H., Uebel, S., Sempé P., Djaballah, H., Yang, Y., Peterson, P. A., Früh, K., and Tampé, R., 1996, Molecular mechanism and species specificity of TAP inhibition by herpes simplex virus protein ICP47, *EMBO J.* **15:**3247–3255.
40. Tomazin, R., Hill, A. B., Jugovic, P., York, I., van Endert, P. Ploegh, H. L., Andrews, D. W., and Johnson, D. C., 1996, Stable binding of the herpes simplex virus ICP47 protein to the peptide binding site of TAP, *EMBO J.* **15:**3256–3266.
41. Reddehase, M. J., Fibi, M. R., Keil, G. M., and Koszinowski, U. H., 1986, Late-phase expression of a murine cytomegalovirus immediate-early antigen recognized by cytolytic T lymphocytes, *J. Virol.* **60:**1125–1129.
42. Del Val, M., Münch, K., Reddehase, M. J., and Koszinowski, U. H., 1989, Presentation of CMV immediate-early antigen to cytolytic T lymphocytes is selectively prevented by viral genes expressed in the early phase, *Cell* **58:**305–315.
43. Campbell, A. E., Slater, J. S., Cavanaugh, V.J., and Stenberg, R. M., 1992, An early event in murine cytomegalovirus replication inhibits presentation of cellular antigens to cytotoxic T cells, *J. Virol.* **66:**3011–3017.
44. Del Val, M., Hengel, H., Häcker, H., Hartlaub, U., Ruppert, T., Lucin, P., and Koszinowski, U. H., 1992, Cytomegalovirus prevents antigen presentation by blocking the transport of peptide-loaded major histocompatibility complex class I molecules into the medial-Golgi compartment, *J. Exp. Med.* **172:**729–738.
45. Thäle, R., Szepan, U., Hengel, H., Geginat, G., Lucin, P., and Koszinowski, U. H., 1995, Identification of the mouse cytomegalovirus genomic region affecting major histocompatibility complex class I molecule transport, *J. Virol.* **69:**6098–6105.
46. Ziegler, H., Thäle, R., Lucin, P., Muranyi, W. H., Flohr, T., Hengel, H., Farrell, H., Rawlinson, W., and Koszinowski, U. H., 1997, A mouse cytomegalovirus glycoprotein retains MHC class I complexes in the ERGIC/*cis*-Golgi compartments, *Immunity* **6:**57–66.
47. Jackson, M. R., Nilsson, T., and Peterson, P. A., 1990, Identification of a consensus motif for retention of transmembrane proteins in the endoplasmic reticulum, *EMBO J.* **9:**3153–3162.
48. Kleijnen, M., Huppa, J. B., Lucin, P., Mukherjee, S., Farrell, H., Campbell, A., Koszinowski,

U. H., Hill, A. B., and Ploegh, H. L., 1997, A mouse cytomegalovirus glycoprotein, gp34, forms a complex with folded class I MHC molecules in the ER which is not retained but transported to the cell surface, *EMBO J.* **16:**685–694.
49. Reusch, U., Lucin, P., Muranyi, W., Thäle, R., Yamashita, Y., Hengel, H., and Koszinowski, U. H., 1997, A cytomegalovirus glycoprotein targets MHC class I complexes to the lysosome for degradation, manuscript in preparation.
50. Beck, S., and Barrell, B. G., 1988, Human cytomegalovirus encodes a glycoprotein homolgoous to MHC class-I antigen, *Nature* **331:**269–272.
51. Browne, H., Smith, G., Beck, S., and Minson, T., 1990, A complex between the MHC class I homologue encoded by human cytomegalovirus and β_2 microglobulin, *Nature* **347:**770–772.
52. Browne, H., Churcher, M., and Mison, T., 1992, Construction and characterization of a human cytomegalovirus mutant with the UL18 (class I homolog) gene deleted. *J. Virol.* **66:**6784–6787.
53. Fahnestock, M. L., Johnson, J. L., Feldman, R. M. R., Neveu, J. M., Lane, W. S., and Bjorkman, P. J., 1995, The MHC class I homolog encoded by human cytomegalovirus binds endogenous peptides, *Immunity* **3:**583–590.
54. Barnes, P. D., and Grundy, J. E., 1992, Down-regulation of the class HLA heterodimer and β_2-microglobulin on the surface of cells infected with cytomegalovirus, *J. Gen. Virol.* **73:**2395–2403.
55. Beersma, M. F. C., Bijlmakers, M. J. E., and Ploegh, H. L., 1993, Human cytomegalovirus down-regulates HLA class I expression by reducing the stability of class I H chains, *J. Immunol.* **151:**4455–4464.
56. Yamashita, Y., Shimokata, K., Mizuno, S., Yamaguchi, H., and Nishiyama, Y., 1993, Down-regulation of the surface expression of class I MHC antigens by human cytomegalovirus, *Virology* **193:**727–736.
57. Warren, A. P., Ducroq, D. H., Lehner, P. J., and Borysiewicz, L. K., 1994, Human-cytomegalovirus-infected cells have unstable assembly of major histocompatibility complex class I complexes and are resistant to lysis by cytotoxic T lymphocytes, *J. Virol.* **68:**2822–2829.
58. Jones, T. R., Hanson, L. K., Sun, L., Slater, J. S., Stenberg, R. S., and Campbell, A. E., 1995, Multiple independent loci within the human cytomegalovirus unique short region down-regulate expression of major histcompatibility complex class I heavy chains, *J. Virol.* **69:**4830–4841.
59. Jones, T. R., and Muzithras, V. P., 1991, Fine mapping of transcripts expressed from the US6 gene family of human cytomegalovirus strain AD169, *J. Virol.* **65:**2024–2036.
60. Wiertz, E. J. H. J., Jones, T. R., Sun, L., Bogyo, M., Geuze, H. J., and Ploegh, H. L., 1996. The human cytomegalovirus US11 gene product dislocates MHC class I heavy chains from the endoplasmic reticulum to the cytosol, *Cell* **84:**769–779.
61. Wiertz, E. J. H. J., Tortorella, D., Bogyo, M., Yu, J., Mothes, W., Jones, T. R., Rapoport, T. A., and Ploegh, H. L., 1996, Sec61-mediated transfer of a membrane protein from the endoplasmic reticulum to the proteasome for destruction, *Nature* **384:**432–438.
62. Machold, R. P., Wiertz, E. J. H. J., Jones, T. R., and Ploegh, H. L., 1997, The HCMV gene products US11 and US2 differ in their ability to attack allelic forms of murine major histocompatibility complex (MHC) class I heavy chains, *J. Exp. Med.* **185:**363–366.
63. Ahn, K., Angulo, A., Ghazal, P., Peterson, P. A., Yang, Y., and Früh, K., 1996, Human cytomegalovirus inhibits antigen presentation by a sequential multistep process, *Proc. Natl. Acad. Sci. USA* **93:**10990–10995.
64. Jones, T. R., Wiertz, E. J. H. J., Sun, L., Fish, K. N., Nelson, J. A., and Ploegh, H. L., 1996, Human cytomegalovirus US3 impairs transport and maturation of major histocompatibility complex class I heavy chains, *Proc. Natl. Acad. Sci. USA* **93:**11327–11333.
65. Hengel, H., Flohr, T., Hämmerling, G. J., Koszinowski, U. H., and Momburg, F., 1996, Human cytomegalovirus inhibits peptide translocation into the endoplasmic reticulum for MHC class I assembly, *J. Gen. Virol.* **77:**2287–2296.

66. Hengel, H., Koopmann, J.-O., Flohr, T., Muranyi, W., Goulmy, E., Hämmerling, G. J., Koszinowski, U. H., and Momburg, F., 1997, A viral ER resident glycoprotein inactivates the MHC encoded peptide transporter, *Immunity* **6:**623–632.
67. Ahn, K., Gruhler, A., Galocha, B., Jones, T. R., Wiertz, E. J. H. J., Ploegh, H. L., Peterson, P. A., Yang, Y., and Früh, K., 1997, The ER luminal domain of the HCMV glycoprotein US6 inhibits peptide translocation by TAP, *Immunity* **6:**613–621.
68. Gilbert, M. J., Riddell, S. R., Plachter, B., and Greenberg, P. D., 1996, Cytomegalovirus selectivity blocks antigen processing and presentation of its immediate-early gene product, *Nature* **383:**720–722.
69. Hengel, H., Eßlinger, C., Pool, J., Goulmy, E., and Koszinowski, U. H., 1995, Cytokines restore MHC class I complex formation and control antigen presentation in human cytomegalovirus-infected cells. *J. Gen. Virol.* **76:**2987–2997.
70. Borysiewicz, L. K., Hickling, J. K., Graham, S., Sinclair, J., Cranage, M. P., Smith, G. L., and Sissons, J. G. P., 1988, Human cytomegalovirus-specific cytotoxic T cells. Relative frequency of stage-specific CTL recognising the 72-kDa immediate early protein and glycoprotein B expressed by recombinant vaccinia viruses, *J. Exp. Med.* **168:**919–931.
71. Hengel, H., Geginat, G., and Koszinowki, U. H., Unpublished observation.

12

Cell-Mediated Immunity against Varicella-Zoster Virus

ANN M. ARVIN

1. INTRODUCTION

Varicella-zoster virus (VZV) is a herpesvirus which is pathogenic only in the human host.[1] VZV has a unique pattern of infectivity because it is lymphotropic for $CD4^+$ and $CD8^+$ T cells, as established by our recent experiments in the SCID-hu mouse model, and also exhibits the neurotropism that is a defining characteristic of the alpha herpesviruses.[2] An effective host response to VZV requires limiting its capacity to cause lymphocyte-associated viremia and maintaining viral latency in cells of the dorsal root ganglia. Better understanding of VZV immunity has been achieved in parallel with advances in knowledge about the molecular virology of the virus and the basic immunobiology of virus–host interactions.[3,4] The live attenuated varicella vaccine is the first effective vaccine against a human herpesvirus.[5] Its development has provided new opportunities for evaluating the components of protective immunity against VZV in healthy and immunocompromised individuals.[6,7]

ANN M. ARVIN • Department of Pediatrics and Microbiology/Immunology, Stanford University School of Medicine, Stanford, California 94305.

Herpesviruses and Immunity, edited by Medveczky *et al.* Plenum Press, New York, 1998.

2. THE VIRUS

Among the human herpesviruses, VZV is most similar to herpes simplex virus (HSV) types 1 and 2. VZV is also related to simian varicella virus, pseudorabies virus, and other mammalian alpha herpesviruses.[4,8–11] VZV replication is presumed to follow the pattern of sequential expression of α, β, and γ genes, which is characteristic of herpesviruses.[4] The VZV virion consists of DNA contained within an icosahedral nucleocapsid. The capsid is surrounded by tegument proteins and enclosed in a lipid membrane envelope. The linear, double-stranded DNA genome consists of approximately 125,000 base pairs, arranged in long and short unique sequences with terminal repeat regions and codes for at least 68 viral gene products. Although there are no known subgroups, DNA profiles of unrelated isolates differ some as shown by restriction enzyme analysis.[12,13] Nevertheless, conserved DNA sequences allow the detection of VZV strains by *in situ* hybridization or PCR.[14–17] Genetic stability, demonstrated by consistency in restriction enzyme patterns, is maintained after multiple passages of VZV isolates *in vitro*. Transfection of intact VZV genomic DNA into permissive cells yields infectious virus, and VZV is recovered after transfection of overlapping fragments of viral DNA from cosmids that constitute the complete genome sequence.[18,19] VZV infectivity is highly temperature sensitive. The virus requires storage at −70 °C or lyophilization to preserve it, and the virus is inactivated by incubation at 56–60 °C.

Many VZV proteins are detected within 4–10 hours after inoculation of permissive cells *in vitro*. VZV causes syncytial changes. Many multinucleated giant cells and other cytopathic effects are observable within 2–7 days. Unrelated VZV isolates produce proteins that have many common antigenic regions, as demonstrated by their reactivity with polyclonal antisera and monoclonal antibodies to glycoproteins, but more subtle antigenic differences may exist.[20] The products of six of the eight or more possible glycoprotein genes of VZV, including gB (gpII), gC (gpV), gE (gp I), gH (gp III), gI (gp IV) and gL, have been characterized. The VZV envelope glycoproteins mediate viral attachment and cell entry. During the infection of permissive cells, glycoprotein insertion into cellular membranes promotes fusion, permitting cell-to-cell spread of the virus.[20–22] Disruption of the envelope by detergents or lipid solvents eliminates infectivity. gE, the most abundant glycoprotein in virions and in VZV-infected cells, is found as a complex with gI in infected cells, and has Fc receptor function on infected cell membranes.[23] The binding of gE to mannose 6-phosphate receptors is involved in VZV pathogenesis because blocking these receptors prevents VZV infection of tissue culture cells.[23,24] gE also affects intracellular trafficking of newly synthesized virions, directing some progeny virus into degradative endosomes.[25] New data indicate that gE may also have signal sequences that localize viral proteins for assembly into virions in the *trans*-Golgi network.[26] Our recent experiments using recombinant VZV strains from which gI was deleted demonstrate significant changes in the phenotype of VZV replication in tissue culture and altered patterns of gE expression within the infected cell.[27] VZV gB is the target of neutralizing antibodies and by analogy with other herpesviruses is presumed to be essential for infectivity. Antibodies against VZV gB and HSV gB are cross-reactive. As is true of HSV gC, VZV gC protein is not essential for VZV replication *in vitro*.[28,29] VZV gH is a target of neutralizing antibodies and induces membrane fusion, allowing cell-to-cell spread of the virus. gH complexes with gL in infected cells.[30]

The VZV tegument proteins, which are located within the envelope and surround the capsid, have regulatory and structural functions. The product of ORF 62, the IE62 protein (originally designated p170), is the most abundant of these proteins and acts as the initial transactivating protein of VZV. Other regulatory proteins are encoded by ORFs 4, 10, 61, and 63.[31–4] The regulatory proteins in the tegument initiate VZV replication without requiring new viral DNA or protein synthesis. Among the viral enzymes produced by VZV, protein kinases are encoded by ORFs 47 and 66, and most isolates produce thymidine kinase.[4] Thymidine kinase expression makes VZV isolates susceptible to inhibition by acyclovir (ACV) and other nucleoside analogs, but it is not required for VZV replication *in vitro* or *in vivo*. Variants of VZV that do not make thymidine kinase or gC (gp V) have been recovered by plaque purification from wild-type stains but whether other genetic mutations occur is not known.[28,29] The Oka strain, which is used to manufacture varicella vaccines, was derived by passage of a clinical isolate in a nonhuman cell line consisting of guinea pig embryo fibroblasts, and further passage in human fibroblasts.[34]

3. METHODS FOR ASSESSING CELLULAR IMMUNITY TO VZV

Antiviral immunity against VZV is mediated by $CD4^+$ and $CD8^+$ T cells. The presence of T cells specific for VZV antigens can be demonstrated by stimulating peripheral blood mononuclear cells (PBMC) with VZV-infected cell extract and measuring proliferation by the uptake of ^{3}[H]-thymidine.[3] Incubation of PBMC with unfractionated extracts of tissue culture cells infected with the virus, usually called whole VZV antigen, allows detecting T cells that recognize any of the proteins in these extracts, including glycoproteins, regulatory/structural proteins of the virion, and nonstructural proteins required for VZV replication. Specific protein reagents are made by immunoaffinity purification from VZV-infected cell extracts or by using expression vectors to produce a single VZV gene product, as has been accomplished with vaccinia recombinants or GST fusion. Assessing cellular immunity to specific VZV proteins is complicated by the fact that VZV has many ORFs: several that are representative of various classes of viral proteins or that have a defined role in viral pathogenesis have been selected for evaluation.[35] Purified proteins are made by using VZV monoclonal antibodies coupled to cyanogen bromide-activated Sepharose 4B and incubated with a solubilized extract of VZV-infected cells or vaccinia recombinants that express VZV proteins. Antigenic preparations enriched for viral glycoproteins have been made with lectin columns.[36] Synthetic peptides corresponding to sequences of VZV proteins can also be used to stimulate PBMC *in vitro* because T cells recognize linear residues of viral proteins composed of only a few amino acids.[37,38]

VZV-specific T cell proliferation is detected by calculating the ratio of cpm in antigen-stimulated and control wells following five to seven days of incubation of PBMC with whole VZV antigen or a viral protein. A stimulation index (SI) $\geq$ 2.0–3.0 is expected in healthy immune individuals.[39] VZV-specific T cells that proliferate under these conditions are predominantly from the $CD4^+$ subset and release Th-1 cytokines, such as interleukin 2 (IL-2) and gamma interferon (IFN-γ). The frequency of circulating T cells that are specific for VZV antigen or a single VZV protein can be calculated when the proliferation assay is carried out using limiting dilution conditions.[40] T cell recognition of VZV antigens in the standard proliferation assay indicates prior sensitization of the immune system, but a positive response is detected in individuals who have a broad range of numbers of circulating VZV-specific T cells. Quantitative differences in T cell responses to VZV are established by responder cell frequency assays, and this method is also used to identify the numbers of T cells that release particular cytokines after stimulation with VZV antigen.

T cells that mediate antiviral cytotoxicity (CTL) against VZV-infected cells are also present in PBMC cultures after secondary *in vitro* stimulation with VZV antigen or viral proteins. Cytotoxic function is detected by incubating effector T cells in chromium release assays with virally infected autologous lymphoblastoid cells, which express both class I and class II major histocompatibility complex (MHC) antigens.[41–43] $CD8^+$-mediated CTL activity against VZV is detectable by using VZV-infected fibroblasts that express only MHC class I antigen as targets or by testing highly purified subpopulations of $CD8^+$ T cells as effectors against lymphoblastoid cells that express VZV proteins.[41,44] $CD4^+$ CTL derived by secondary *in vitro* stimulation with VZV antigen are cytotoxic against autologous lymphoblastoid cells infected with VZV or against vaccinia recombinants that express VZV proteins.[41] The frequencies of cytotoxic T cells that recognize VZV antigens and the protein specificity of the CTL response to VZV are assessed by expanding effector T cells under limiting dilution conditions. The efficiency and reproducibility of preparing targets infected with VZV–vaccinia recombinants makes it feasible to determine the usual frequencies of memory CTLs specific for particular VZV proteins. Delayed hypersensitive responses to VZV, which indicate the memory T cell immunity to the virus, can be demonstrated *in vivo* by using whole VZV antigen or viral glycoproteins as skin test reagents.[45,46] Most recently, it has been possible to induce a primary T cell response to VZV *in vitro* by stimulating T cells from naive donors with dendritic cells that were pulsed with synthetic peptides corresponding to amino acid residues of the IE62 protein.[47]

4. CELL-MEDIATED IMMUNITY IN THE CONTROL OF PRIMARY VZV INFECTION

Although other human herpesviruses are often acquired subclinically, primary infection with VZV almost always produces the clinical symptoms of varicella, commonly called chicken pox.[1] Primary VZV infection begins when susceptible individuals are exposed to an infected person and mucous membranes are inoculated with infectious virus in respiratory droplets or vesicular fluid from skin lesions. Inoculation is followed by a prolonged incubation period, during which the virus remains undetected by the immune system of the host. During this interval of 10 to 21 days, the virus spreads to regional lymph nodes and probably causes a primary viremia resulting in transport of the virus to reticuloendothelial cells of the liver and other sites.[48] Another phase of cell-associated viremia occurs in the last few days of the incubation period. This capacity to cause viremia is an essential element in VZV pathogenesis, differentiates it from primary HSV infection, and accounts for the transport of the virus to scattered skin sites. Infectious VZV is recovered from PBMC, and VZV is present in 67–74% of PBMC samples as shown by *in situ* hybridization or PCR.[15,16,49] During acute varicella, from 1:30,000–1:100,000 PBMC contain VZV as demonstrated by *in situ* hybridization.[15] *In vitro,* activated T cells can be infected with VZV, and the IE62 protein transactivates all classes of VZV genes in a human T cell line.[50] Our recent experiments in the SCID-hu mouse model demonstrate that VZV is lymphotropic for human $CD4^+$ and $CD8^+$ T cells and that human T cells release infectious virus.[2] Unless the host generates an effective immune response, cell-associated viremia persists, allowing continued formation of new skin lesions and dissemination of the virus to the lungs and other organs.

The host response can modify the extent of the exanthem, but some VZV replication at skin sites is observed during all primary infections.[51] Histologically, the initial skin lesions of varicella are characterized by vasculitis, dilated lymphatic vessles, and multinucleated epithelial cells with intranuclear eosinophilic inclusions. The vesicular phase of the skin lesion is associated with epithelial cell degeneration, coalescence of fluid-filled vacuoles, and increased numbers of infected cells within the lesion. A marked inflammatory infiltrate, consisting predominantly of lymphocytes, is evident at this stage. In contrast to HSV, VZV causes infection and necrosis of the deeper, germinal layer of the dermis and the epidermis. Infection of superficial keratinocytes is associated with the release of cell-free VZV into vesicular fluid, perhaps because fully differentiated keratinocytes lack lysosomal pathways for degradation. Characteristic pathological changes are also apparent when human skin implants in SCID-hu mice are infected with VZV.

Nonspecific cell-mediated immune responses against VZV can be demonstrated using PBMC from susceptible individuals and may limit VZV replication in the initial phase of infection. VZV-infected fibroblasts are lysed by natural killer (NK) cells, and cytotoxicity is enhanced by incubation of NK cells with IL-2.[52,53] PBMC from nonimmune subjects also release IFN-α after *in vitro* stimulation with VZV antigen. The functional significance of this response was demonstrated by the reduced severity of varicella in immunocompromised children who received exogenous IFN-α within 72 hours after the cutaneous rash appeared.[54]

The self-limited course of varicella in a healthy host is associated with rapid induction of T cells that recognize VZV antigens[55] (Table I). Although the virus escapes immune surveillance during the incubation period, individuals who have detectable VZV-specific T cells within 72 hours after the onset of varicella experience mild primary VZV infection. The mean SI to VZV antigen was 7.5 ± 10.43 SD among immunocompetent individuals who had very few cutaneous lesions ($<$100 lesions/m^2) compared with 1.4 ± 1.85 SD for those with $>$400 lesions/m^2 ($p < 0.05$). The early T cell proliferative response to VZV antigen is accompanied by the release of cytokines of Th-1 type, including IL-2 and IFN-γ, which enhance the clonal expansion of virus-specific T cells. Like IFN-α, IFN-γ also restricts VZV replication by its direct antiviral effects and is made in sufficient quantities to be detected in the serum of healthy subjects with acute varicella. Adults, who are more susceptible to prolonged, extensive varicella, are less likely to have detectable serum IFN-γ than children.[56] The current evidence is

TABLE I
Cell-Mediated Immunity in the Control of Varicella-Zoster Virus Infection

Pathogenic event	Clinical correlate	Role of cell-mediated immunity
Primary infection	Varicella	Terminate VZV infection of CD4$^+$ and CD8$^+$ T cells
		Terminate VZV replication at cutaneous sites
Latency	No symptoms	Block VZV reactivation in dorsal root ganglia cells
		Limit repliction to ganglion cells if reactivation occurs
Reactivation	Herpes zoster	Terminate VZV replication in ganglion cells and at cutaneous sites
		Prevent or limit VZV infection of CD4$^+$ and CD8$^+$ T cells
Reexposure	No symptoms	Block viral replication at muscosal sites of inoculation

TABLE II
Viral Protein Targets of the Cell-Mediated Immune Response to Varicella-Zoster Virus

Viral protein targets	Class	Location	Functions
Immediate early 62 protein	α	Tegument	Gene regulation: major transactivating protein
Immediate early 63 protein	α	Tegument	Gene regulation
Glycoproteins gB, gC, gE, gH, gI	γ	Envelope	Virus entry Cell-to-cell spread

Cell-mediated immune responses
$CD4^+$ T-cell proliferation
Th-1 cytokine production
$CD4^+$ T cell-mediated, class II restricted cytoxicity
$CD8^+$ T cell-mediated, class I restricted cytoxicity
$CD4^+$ T cell-mediated B cell helper function

that little or no TH-2 cytokines, such as IL-4, are produced in response to VZV antigen.

An initial cell-mediated response during primary VZV infection is directed against two major VZV glycoproteins gE and gH and the IE62 protein[35] (Table II). The glycoproteins of herpesviruses are likely to be important targets of the initial host response because these proteins are components of the virion envelope and are expressed on the membranes of virally infected cells. Immunity elicited by the IE62 protein is of interest because it is a major component of the virion tegument, which is released early after viral entry into target cells and is required to initiate viral replication.[57] Other VZV proteins are also likely to be targets of early cell-mediated immunity but have not been evaluated. In our experiments, the kinetics of induction of T cells that recognized gE, gH and the IE62 protein varied but 67% of subjects had early proliferative responses to gE, 71% to gH, and 57% to IE62 protein.[35] The variability of the protein specificity of early cellular immunity in individuals with uncomplicated varicella indicates that T cell recognition of any of several VZV proteins is equally effective for terminating primary infection.

The CTL response to primary VZV infection is of interest because of the evidence that viral clearance correlates with the induction of T cells that disrupt virally infected target cells. T cells from individuals with acute varicella mediate specific killing of VZV-infected targets.[42] CTL recognition of IE62 protein and gE was demonstrated by limiting dilution culture of PBMC from two individuals recovering from varicella. The frequencies of CTL specific for IE62 protein and gE were 1:69,000 and 1:57,000 in the first case and 1:173,000 and 1:166,000 in the second case.[58]

In the absence of an effective host response, VZV can pathogenically infect lungs, liver, brain and other organs. Among immunodeficient children, an absolute lymphopenia of $< 500/mm^3$ at the onset of varicella is associated with a significant risk of life-threatening varicella. Lymphopenia is accompanied by a failure to develop VZV-specific T cell immunity. In our experiments, only one (7.7%) of 13 immunocompromised patients had early VZV-specific T cell proliferation compared with 19 (42%) of 45 healthy subjects ($p < 0.05$).[55] Diminished cell-mediated immunity is associated with a prolonged phase of secondary viremia, increased cutaneous virus replication, and high risk of viral spread to lungs and other sites.[59,60] Varicella pneumonia is characterized by infection of pulmonary alveolar epithelium, and hepatitis reflects hepatocellular destruction by the virus. Other complications of varicella, such as encephalitis, cerebellar ataxia, and thrombocytopenia, may be mediated in part by the immune response rather than by a direct cytopathic effect. Nevertheless, these complications are rare, and the acquisition of cell-mediated immunity is associated with the complete resolution of disease in most cases.

5. COMPONENTS OF THE MEMORY T CELL RESPONSE TO VZV AND ITS PROTEIN SPECIFICITY

In the course of primary VZV infection, the virus reaches cells in the dorsal root ganglia and establishes latency.[61,62] Memory immune responses protect the host from symptomatic reinfection with VZV as a result of a new exogenous exposure to the virus. In addition, the capacity of the virus to persist after primary infection means that memory immunity is also necessary to preserve a lifelong equilibrium between the virus and the host. Limited clinical evidence suggests that VZV reaches its site of latency at a time before virus-specific immunity is elicited and that access to ganglia is not blocked by the developing host response. VZV may be carried to dorsal root ganglia as a result of the cell-associated viremia that precedes the development of a cutaneous rash, or it may reach these sites by ascending by retrograde transport along neuronal cell axons that innervate infected skin. The neural cell tropism of VZV has not been fully defined, but current observations indicate that it persists in both neuronal and satellite cells of the sensory ganglia.[63–65] VZV infects both neuronal and nonneuronal cells, including Schwann cells and astrocytes, in cultures of fetal ganglia tissue. By analogy with pseudorabies virus, the VZV glycoproteins genes gE and gI, which are located in the short unique sequence, may be important for neurovirulence. Whereas VZV lacks the antisense, latency-associated RNA transcripts and the neurovirulence-related ICP34.5 gene of HSV-1 and HSV-2, transcription of the

ORFs 4, 10, 29, 61, 62, and 63 persists in dorsal root ganglia cells during latency.[4] Recent experiments indicate that the transcription of ORF63 results in protein synthesis in cells of the dorsal root ganglia.[66]

Immunologic studies demonstrate that healthy individuals have memory $CD4^+$ and $CD8^+$ T cells specific for VZV antigens in circulating PBMC populations and in lymphoid tissue for decades after their recovery from varicella. By proliferation assay, the precursor frequencies of memory T cells that recognize VZV antigen are approximately 1:40,000 PBMCV in immune adults.[40] When cytokine responses were tested under limiting dilution conditions, up to 85% of proliferating T cells released IFN-γ while 10% or fewer produced IL-4 in response to VZV antigen.[67] Delayed hypersensitive responses to VZV skin test antigens also persist for years after primary infection.[45]

VZV glycoproteins, including gE, gB, gC and gH, and the IE62 protein are recognized by memory T cells from immune individuals, as demonstrated in proliferation and cytokine release assays (Table II). In recent experiments, we found that memory immunity is also maintained against the IE63 protein, the only viral protein detected thus far in latently infected cells of the dorsal root ganglia.[68] When T cells were stimulated with panels of synthetic peptides that mimicked residues of gE and the IE62 protein and contained sequences that fit algorithms for T cell motifs, immune donors had circulating T cells specific for an average of seven of ten IE62 peptides and six of ten gE peptides.[37] T cell clones that recognize amphipathic regions of gB and gI, some of which exhibit B cell helper function, are also detected in memory $CD4^+$ T cell populations from immune individuals.[38] Recognition of several regions on these viral proteins are demonstrated despite differences in the MHC phenotype of the individuals who were tested. Although the glycoproteins and the IE62 protein play different roles in viral replication and infectivity, these experiments with synthetic peptides indicate that VZV infection *in vivo* results in processing them to short amino acid sequences that can be presented by various MHC class II molecules.

Memory T cells that are cytotoxic against target cells expressing VZV proteins are detected in immune adults for at least 20 years after primary VZV infection.[41,43] Among these individuals, the mean precursor frequency for T cells that recognized the IE62 protein was 1:105,000 ± 85,000 SD (range 1:13,000–1:231,000) whereas the mean frequency of CTL precursors specific for gE was 1:121,000 ± 86,000 SD (range 1:15,000–1:228,000).[41] The frequencies of CTL that recognized the IE62 protein and gE were equivalent. The VZV glycoproteins gI and gc are also recognized by memory CTLs, and $CD4^+$ T cell clones specific for gE, gB, gH, or gI can be recovered from PBMC of VZV immune donors.[58,69] Helper and CTL functions of VZV memory T cells are not mutually exclusive because some $CD4^+$ T cell clones that recognize gB or gI mediate B cell help and lysis of VZV-infected targets.[69] Memory CTL recognition of the IE62 protein and gE persists in both $CD4^+$ and $CD8^+$ subpopulations.[41] In the CD^+

population, the mean frequency of CTLs specific for IE62 protein was 1:108,000 (range 1:58,000–180,000) compared to 1:74,000 (range 1:20,000–1:140,000) in the CD8$^+$ population (nonsignificant [NS], paired t test). The mean number of memory CD4$^+$ CTLs that recognized gE targets was 1:119,000 (range 1:30,000–1:220,000) compared to 1:31,000 (range 1:11,000–1:69,000) in CD8$^+$ T cell cultures (NS). Although glycoproteins are less prominent targets of CTL immunity against some viruses, CD4$^+$ and CD8$^+$ CTL recognition of IE62 protein and gE was equivalent. The persistence of CD4$^+$ memory CTLs that recognize VZV proteins in the context of class II MHC expression is likely to be important because class II expression increases during infection of epithelial cells and skin fibroblasts, which are major targets for VZV replication *in vivo*. The CD4$^+$ T cells that infiltrate cutaneous lesions may mediate cytotoxicity directly and also by releasing Th-1 cytokines to recruit CD8$^+$ CTLs specific for VZV.

6. ALTERATIONS IN MEMORY CELL-MEDIATED IMMUNITY AND SUSCEPTIBILITY TO VZV REACTIVATION

The increased incidence of herpes zoster in elderly and immunocompromised patients provided early clinical evidence that immune responses modulate VZV reactivation from latency, influencing either its occurrence or whether the individual develops symptoms of herpes zoster. Although diminished humoral immunity is not a major predisposing factor, decreases in VZV specific cellular immunity alter the host–virus equilibrium. The risk of herpes zoster correlates with declining T cell recognition of VZV antigens in otherwise healthy elderly adults and patients receiving immunosuppressive therapy.[70–72] Because symptoms of herpes zoster do not occur in all cancer patients with decreased T cell proliferation to VZV antigen, impaired virus-specific cellular immunity to VZV appears to be a necessary but not sufficient condition for symptomatic reactivation. Nevertheless, patients who have the most severe and prolonged suppression of cellular immunity, such as bone marrow transplant recipients, are at highest risk for herpes zoster and for the occurrence of cell-associated VZV viremia, which results in hematogenous dissemination and life-threatening complications.[71,73] These clinical observations suggest that the loss of cell-mediated immunity permits the transfer of infectious virus into PBMC from sites of virus replication in dorsal root ganglia or skin and that it prevents the efficient clearance of virally infected T cells, or both. That memory CD4$^+$ T cell responses are important for preserving VZV latency *in vivo* is suggested by the observation that herpes zoster is associated with the decline in CD4$^+$ T cells in patients who have human immunodeficiency viral infection. The clonal expansion of VZV specific CD4$^+$ CTL from T cells of bone marrow transplant recipients, who are at high risk of recurrent VZV infection, is also deficient.[73]

Otherwise healthy elderly adults have a decrease in the frequency of circulating memory T cells specific for VZV antigen and diminished delayed hypersensitive responses to VZV skin test antigen. Immune individuals who were more than 55 years old had fewer T cells producing IFN-γ, and the amount of IFN-γ produced decreased significantly, but the frequency of T cells that released IL-4-making cells was not altered, indicating that immunosenescence interferes with the Th1 more than the Th2 T cell response to VZV.[67]

Other clinical examples relating inadequate cell-mediated immunity and susceptibility to VZV reactivation include the occurrence of herpes zoster in young children who have intrauterine or early postnatal varicella and the short interval between primary and recurrent VZV infections in children with human immunodeficiency viral infection. In one study, 81% of infants who had varicella from 12–24 months of age had T cell proliferation to VZV antigen compared with 59% of those less than one year old ($p < 0.02$).[74]

7. MECHANISMS FOR PRESERVING MEMORY T CELL IMMUNITY TO VZV

Whether virus-specific T cells persist for the life of the host following primary sensitization without requiring antigenic restimulation is an important question in viral immunology. This issue is less problematic in VZV immunity because cellular immunity may be enhanced intermittently when immune individuals are exposed to the virus during the annual epidemics of varicella. Increases in T cell proliferation to VZV antigen can be demonstrated in immune individuals who have household contact with varicella.[75,76] Among mothers of children with varicella, 71% had an increase in T cell proliferation from a mean SI of 7.8 ± 1.3 SE within four days after the onset of varicella in the child compared with 15.3 ± 2.56 SE three to four weeks later.[75] A significant increase in delayed hypersensitivity to VZV skin test antigen was also observed among immune children after varicella exposures.[77] None of these immune individuals who were exposed to varicella developed symptoms, which is consistent with the observation that natural immunity is highly protective against symptomatic reinfection with VZV. The enhanced cellular immunity among close contacts suggests that exposures result in subclinical reinfection or that VZV proteins in the inoculum are processed by antigen-presenting cells at local sites of inoculation without viral replication. Protection against disease caused by new VZV strains occurs even though viral DNA is detected by polymerase chain reaction (PCR) in nasopharyngeal secretions from immune household contacts of children with varicella.[78]

Intermittent subclinical reactivations of VZV from latency may provide a second mechanism for sustaining memory T cell immunity to VZV. Although virological methods can be used to document the reactivation of other herpes viruses, it has been difficult to prove that VZV reactivates without causing symptoms. Nevertheless, in our experience, 19% of bone marrow transplant recipients had subclinical cell-associated VZV viremia.[73] None of the patients who were tested before VZV reactivation had T cell recognition of VZV antigen and five of eight patients (63%) who recovered T cell proliferation had subclinical or clinical VZV reactivation compared to none of six patients who continued to lack VZV immunity. VZV antigens have also been detected in PBMC from some elderly individuals without signs of herpes zoster, indicating that these individuals may experience subclinical reactivations that can boost immunity to VZV.[79]

Both exogenous reexposures to varicella and subclinical reactivation of latent virus may be important for maintaining memory immunity to VZV. Immunity to some late viral proteins, such as the glycoproteins, may depend on reexposure to the intact virus in contacts with varicella whereas reactivation from latency could enhance T cell recognition of viral proteins, such as the IE62 protein, that are made early in replication, even when reactivation is abortive. Recent evidence indicates that cell-mediated immunity to some proteins, such as the IE63 protein, may be preserved as a result of their expression during latency.[68]

8. CELL-MEDIATED IMMUNITY IN THE CONTROL OF VZV REACTIVATION

During VZV reactivation, cytopathic changes are observed in many cells of the involved dorsal root ganglia in association with an extensive local inflammatory response.[63] The reactivation of VZV results in a cutaneous vesicular eruption which follows the dermatomal distribution of a single sensory nerve. Histologically, the vesicles resemble varicella skin lesions.[1] VZV reaches the skin by transport along sensory neurons. Viral proteins are detected within peripheral axons. Replication of the virus at cutaneous sites causes reexposure of the host immune system to VZV antigens because it is associated with abundant expression of viral proteins in epidermal and dermal cells and triggers an intense mononuclear cell inflammatory response.

Cell-mediated immunity to VZV is enhanced significantly between the initial appearance of the cutaneous rash and its resolution.[80] Most healthy individuals have detectable T cell proliferation to VZV antigens at the onset of herpes zoster, but the number of VZV-specific T cells increases immediately as a consequence of restimulation of memory immunity *in vivo*. In contrast, the cell-mediated response is delayed in immunocompromised individuals with herpes zoster, correlating with their risk of severe cutaneous disease, cell-associated viremia, and visceral dissemination. T cell proliferation to VZV antigen in immunocompromised patients with herpes zoster increased from 1.8 ± 0.85 SD to 5.7 ± 3.03 SD by 2–4 weeks whereas the mean peak stimulation index (SI) was >10 in healthy patients and occurred within 1–2 weeks.[70] IFN-α is released by inflammatory cells at cutaneous lesions, and its production correlates with resolution of the rash.[81] this local IFN-α response is impaired in immunocompromised patients, which is likely to be important in pathogenesis because the administration of IFN-α to immunocompromised patients with herpes zoster reduced its severity.[82] Enhanced VZV specific cell-mediated immunity persists for a prolonged period after VZV reactivation, even among immunocompromised patients, and may explain why second episodes of herpes zoster are rare. Antigenic stimulation may continue to occur after the rash resolves because VZV proteins have been detected in PBMC from some patients with postherpetic neuralgia.[83]

9. CELL-MEDIATED IMMUNE RESPONSES AGAINST VZV ELICITED BY PRIMARY IMMUNIZATION WITH VARICELLA VACCINE

Clinical investigations of live attenuated vaccines made from the Oka strain demonstrate that protective immunity against primary VZV infection can be elicited by inoculation with infectious virus grown in tissue culture without causing clinical signs of varicella in the susceptible host.[7,84–86] The live attenuated varicella vaccine is licensed in the United States (VARIVAX™, Merck & Co., Inc.) and is recommended for universal administration in early childhood. As is characteristic of VZV strains, the Oka strain shares antigen cross-reactivity with unrelated VZV isolates. The molecular basis for attenuation of the varicella vaccine is not certain, but it may be related in part to variable production of gC.[28,29] Although no genetic mutations have been defined, the Oka strain can be distinguished from geographically distinct isolates by restriction enzyme analysis of viral DNA.[13]

The attenuation of varicella vaccine preparations made from Oka strain has been demonstrated clinically. Whether it is caused by virological or immunologic factors or a combination of the two is not certain. The varicella vaccine is manufactured to contain a standard inoculum of infectious virus. Because the vaccine is made from infected tissue culture cells, it contains viral proteins that are not incorporated into virions and infectious virus particles. Therefore, vaccine preparations may differ in viral antigen content and in infectious virus titer. Susceptible children who are inoculated with vesicular fluid from cutaneous VZV lesions develop varicella whereas healthy children given the vaccine do not, even when the preparation contains 27,000 plaque-forming units (pfu) of infectious virus.[87] This observation is not unequivocal evidence of virological attenuation of the Oka strain because vesicular fluid is likely to contain predominantly cell-free virus. The attenuation of the Oka vaccine virus observed clinically could be caused by initial priming of the host response by viral proteins, which limits the replication of the infectious viral component. The antigen concentration and the infectious viral content of the vaccine may combine to optimize vaccine immunogenicity because restricted replication of infectious virus in the vaccine may boost the antigen-primed response. In contrast to infection with wild-type VZV, cell-associated viremia, which is essential to varicella pathogenesis, is rarely detectable by culture or PCR following immunization of healthy children.[88] Nevertheless, this modification in virus–host interactions is not attributable to virological factors because the Oka strain retains infectivity for humans $CD4^+$ and $CD8^+$ T cells in the SCID-hu mouse model.[2] In addition, although the vaccine prevents severe varicella in children with leukemia, the occurrence of a varicella-like illness characterized by recovery of Oka strain from scattered cutaneous lesions indicates that the vaccine virus can infect T cells. Mild symptoms induced in secondary cases when the vaccine strain was transmitted from leukemic vaccines with rash to their healthy siblings is the best evidence of a virological basis for attenuation of the vaccine virus.

Factors that have been evaluated in relation to the capacity of varicella vaccine to induce cell-mediated immunity include the infectious viral dose, the relative viral antigen content, the route of administration, and the dosage regimen.[87,89–91] The effects of age and underlying immunosuppressive conditions on vaccine immunogenicity have also been assessed. Among children given vaccine with 1140 pfu/1.6–1.7 relative antigen content, 96% had VZV-specific T cell proliferation at six weeks compared with 58% of vaccines who were tested eight weeks after receiving vaccine with 950 pfu/1.0 relative antigen content ($p < 0.001$, X^2).[91] The mean SI was 28.0 ± 5.5 SE in the first cohort compared to 6.0 ± 1.0 SE in the second group ($p < 0.001$, t test).

Investigations of the kinetics of primary cellular immunity show that more than 95% of varicella vaccine recipients have T cell recognition of VZV antigens by 2–6 weeks.[87,89–91] At six weeks, the mean SI was 15.2 in vaccinees given 1400 pfu/dose (95% ≥ 2.0), and 10.4 in vaccinees given 4350 pfu/dose (100% ≥ 2.0). Among children tested at 0, 2, and 6 weeks, the proportions of vaccinees with positive SI were 80, 89, and 95%, respectively. VZV-specific T cells are detected before IgG antibodies to VZV appear. In one cohort of children, 80% had SI ≥ 3.0 to VZV antigen by two weeks whereas only 40% had antibodies detectable by gpELISA. By six weeks, 97% had VZV IgG antibodies and 95% had T cell recognition of VZV antigen.[90] Skin test reactivity to VZV antigen occurs as early as four days after immunization whereas VZV antibodies are not detected consistently until 10–14 days. This early induction of cell-mediated immunity probably explains the protection conferred by vaccinating susceptible household contacts immediately after exposure to varicella.[92] Because vaccination followed presumed inoculation of mucosal site, T cells specific for VZV antigens must have blocked replication and spread of the virus at other sites, such as the regional lymph nodes, that are important early in the pathogenesis of active infection.

Immunization with varicella vaccine elicits cell-mediated immunity against the VZV proteins that are known targets of natural immunity. In our experiments, T cell recognition of gE was elicited in 91% of vaccinees, and 82% of vaccinees responded to the IE62 protein.[91] The initial mean SIs to gE and the IE62 protein were 25.3 ± 5.1 SE and 11.3 ± 3.1 SE at six weeks and were significantly higher than responses among naturally immune individuals with no recent VZV infection. A comparative analysis of the initial T cell recognition of VZV glycoproteins demonstrated responses to gE in 81%, gB in 65%, and gH in 77% of vaccines, with mean SIs of 9.7 ± 1.8 SE, 6.9 ± 1.4 SE, and 7.6 ± 1.4 SE, respectively.[93] The mean SI was significantly lower for gB compared to gE, which may be a consequence of the fact that gE is the most abundant VZV glycoprotein produced in VZV-infected cells. Overall, 56.3% of vaccine recipients had T cell recognition of all three glycoproteins, and 87.5% of vaccines responded to two of the three glycoproteins by six weeks after immunization. The kinetics and protein specificity of the primary CTL response to the IE62 protein and gE were evaluated in ten vaccines.[58] At 3–4 weeks, the mean precursor frequency for T cells that recognized IE62 protein was 1:156,000 ± 50,000 SE and the mean frequency of CTLs specific for gE was 1:175,000 ± 63,000 SE. CTL recognition of VZV proteins by $CD4^+$ and $CD8^+$ T cells was detected consistently by twelve weeks after immunization. The kinetics of the acquisition of cytotoxic T cell responses paralleled the induction of VZV-specific T cell proliferation in healthy adults who were tested one and two months after immunization.[94]

The assessment of vaccine-induced cellular immunity in susceptible children and adults revealed age-related differences in their capacity to develop primary T cell recognition of VZV antigens.[95] Assays of cellular immunity in two cohorts of children, who were tested after one dose of varicella vaccine, showed mean SIs of 28.6 ± 6.21 SE and 21.1 ± 3.81 SE whereas adult vaccinees had a mean SI of 9.6 ± 1.16 SE ($p = 0.04$). Vaccinees in both age groups had significant boosts in cellular immunity when tested 6–8 weeks after a second dose of vaccine, with increases in mean SI to 36.9 ± 9.13 SE in children and 29.9 ± 8.19 SE in adults. Responses after the second dose were equivalent in children and adults, indicating that a two-dose regimen is effective for overcoming the age-related deficiency in primary cellular immune responses to VZV.

Although more than 90% of children given one dose of varicella vaccine acquire cellular immunity, enhanced initial responses were documented in vaccines given a two-dose regimen.[89,95] In our experience, the mean SI was 21.1 using unfractionated VZV antigen to stimulate PBMC, and 92% of vaccinees had SI ≥ 3.0 at six weeks after the first dose, whereas the mean SI was 36.9 and 100% of vaccinees had positive SIs after the second dose. The mean SI increased from 47.7 to 62.0 in vaccinees who were given the two-dose regimen and tested for T cell proliferation using VZV antigen enriched for glycoproteins.[89]

The first evaluations of the immunogenicity and safety of live varicella (Oka) vaccine in the United States were done with children with leukemia in remission because of their high risk of life-threatening varicella.[96] The experience gained from these studies indicates that children who have moderate immunologic impairment respond to immunization with effective control of the replication of infectious viral component of the vaccine and development of memory immunity to VZV. In a study of 23 immunocompromised children, the mean SI to VZV antigen was 13.3 in patients who were off chemotherapy and 18.0 in those receiving chemotherapy when vaccinated.[97] Vaccine-associated rashes occurred only in the two vaccinees who had the lowest T cell proliferation to VZV at one month, with SIs of 0.65 and 1.35, respectively. The relationship between the capacity of the host to respond to the vaccine and inhibit replication of Oka strain is suggested by the observation of vaccine-related rashes in 40% of those who were receiving chemotherapy compared with 11% of children who were in remission and off chemotherapy.[96]

10. MEMORY T CELL IMMUNITY TO VZV FOLLOWING IMMUNIZATION AND PROTECTION AGAINST VARICELLA

Memory T cell responses to VZV are maintained in most children with vaccine-induced immunity. T cell proliferation was documented in 91–100% of vaccinees who were evaluated at 9–12 months after immunization with vaccine lots containing 1400 or 4350 pfu. The mean SI were 8.9 and 12.5.[87] A subgroup of vaccinees tested at 3–4 years had mean SIs of 27.4, and 95% of recipients had an SI $\geq$ 2.0. Persistence of skin test reactivity was documented in 97% of children immunized at a mean age of 4 years and in 97% of those who had natural infection when both groups were evaluated at 12–13 years of age.[98] Persistence of cellular immunity was evaluated in 214 vaccinees who were tested at a mean of 55 months after immunization with vaccine containing 8700 pfu/dose. The average age at evaluation was 10.5 years.[99] The mean SI was 8.9 $\pm$ 0.6 SE, and 94% of the vaccinees had an SI $\geq$ 3.0. Responder cells that recognized VZV were significantly higher in varicella vaccines who had been immunized an average of 3.5 years before evaluation, compared to naturally immune adults who had varicella in childhood. The frequencies were 1:18,000 $\pm$ 2,000 SE compared with 1:39,000 $\pm$ 3,000 SE ($p = 0.001$, t test for two means).[58]

With regard to protein specificity, 77% of vaccines tested at one year had an SI $\geq$ 2.0 to gE which is comparable to the response rate of 83% among naturally immune individuals.[91] The percentage of subjects with SI $\geq$ 2.0 to IE62 protein was 83% at one year after immunization, which is similar to the detection of responses in 89% of naturally immune adults. IFN-γ production by T lymphocytes stimulated with gE was detected at one year in 60% of vaccinees compared to 56% of naturally immune adults. Forty-four percent of vaccinees and 56% of naturally immune adults had IFN-γ production to IE62 protein. The frequencies of IE62 protein-specific CTL were compared in vaccinees and naturally immune adults. Means were 1:131,000 ≥ 40,000 SE and 1:104,000 ± 19,000 SE, respectively.[58] The frequencies of memory CTL specific for gI were also equivalent. The means were 1:155,000 ± 60,000 SE in vaccinees and 1:170,000 ± 71,000 SE in individuals with natural immunity. The mean frequency of CTL-recognizing gC was 1:208,000 ± 53,000 SE in subjects with vaccine-induced immunity compared to 1:108,000 ± 27,000 SE in naturally immune subjects ($p = 0.15$, t test for two means). The gC-specific CTL frequencies were <1:200,000 in all of the five naturally immune subjects but were >1:200,000 in three of six vaccines, which may be related to the possible variability in its production by Oka strain.[29]

Persistent cellular immunity was observed after immunization with vaccines that have a viral antigen content of ≥1.6, which is characteristic of the licensed vaccine. At one year, the mean SI in children immunized with 1140 pfu/1.6–1.7 relative viral antigen content was 16.3, and 98% had T cell recognition of VZV antigen, whereas the mean SI was 4.0 in children given vaccine with 950 pfu/1.0 relative viral antigen content and only 43% of vaccinees had an SI > 3.0.[37] At one year, only 40% of the children immunized with 950 pfu/1.0 relative viral antigen content vaccine had T cell recognition of gE compared with 77% of those given 1140 pfu/1.6–1.7 relative viral antigen content tested at one year. ($p = 0.03$, X^2). T cell recognition of IE62 protein was also diminished at one year in those who received the 950 pfu/1.0 relative viral antigen content vaccine. Among children who were given vaccine with the same relative viral antigen content of ~2.0 and decreasing infectious viral content of 1,770, 400–500, and 80–160, the mean SIs were 19.6, 18.7 and 15.7 at one year, and 73%, 83%, and 80% of the vaccines had an SI > 3.0.[90] The age-related differences observed in the initial responses of children and adults also affect memory T cell immunity after vaccination.[95] At one year, VZV-specific T cell proliferation was significantly lower in adults who had been given two doses of vaccine compared to children given one dose: the mean SIs were 7.7 ± 1.30 and 16.4 ± 2.21 [$p = 0.02$]. Although 18 of 19 adults (95%) had detectable cellular immunity to VZV at one year, the lower mean SI suggests that the higher incidence of modified varicella after exposure in adult vaccinees may be related to limited T cell recognition of VZV antigens. The dosage regimen does not alter the percentage of children with detectable T cell recognition of VZV antigen, but the mean SI at one year was 23.1 among children given a single dose compared to 34.7 for those given two doses. At one year 93% of vaccinees had SI ≥ 3.0.[93] In a second cohort, the mean SI to VZV antigen was 9.3 at one year among vaccinees given one dose and 22.2 in those who received two doses; 95% of vaccinees in each group had SI ≥ 3.0.

The clinical experience with vaccine-induced immunity to VZV demonstrates that it is highly protective against varicella following close contact with wild-type virus.[85,86,100,101] Some differences in cellular immunity, however, correlate with susceptibility to mild breakthrough infections among vaccinees. These breakthrough infections have been termed modified varicella-like syndrome because most cases are associated with fewer than 50 lesions and no fever or other systemic symptoms. During the first three years after vaccination, breakthrough infections occurred in 19% of children who had low T cell proliferation responses to VZV antigen initially and at one year after immunization with vaccine containing 950 pfu/1.0 relative antigen content compared to 2.2% for vaccinees given vaccine with 1140 pfu/1.6–1.7 relative antigen content.[91] Breakthrough varicella is also more common in adults, correlating with the evidence that cellular immunity to VZV is lower in older vaccines.[102] Cellular immunity was maintained more effectively than antibody titers in leukemic vaccinees, which may explain the modification of varicella severity despite the decline in VZV IgG titers below detectable levels in some vaccines.[103]

As in the case of natural immunity to VZV, vaccine-induced immunity may be maintained by periodic restimulation resulting from exposures to varicella. In our experience, cell-mediated immunity to VZV increased over time in children and adults who participated in clinical trials of varicella vaccine and were tested five years later.[104] Whether this phenomenon will persist when immunization rates are sufficient to reduce the occurrence of annual varicella epidemics is not known.

The attenuated Oka strains also retains some capacity to infect dorsal root ganglia cells and to reactivate because a few cases of herpes zoster have occurred in leukemic and healthy children after immunization. The incidence of symptomatic reactivation is lower in leukemic children who are vaccinated than in those with natural VZV infection. However, subclinical reactivation of vaccine virus could boost and preserve cellular immunity.

11. ENHANCEMENT OF MEMORY T CELL IMMUNITY BY VACCINATION OF NATURALLY IMMUNE INDIVIDUALS

The immunogenicity of live varicella (Oka) vaccine in individuals who are immune to VZV is an important issue because revaccination may be necessary to maintain memory immunity over the lifetime of the individual. The potential of the vaccine to boost natural immunity is also of interest, given the relationship between waning cellular immunity and herpes zoster. The varicella vaccine may be useful as a therapeutic vaccine to reverse declining T cell responses to VZV associated with immunosenescence or immunosuppression. In a cohort of naturally immune vaccinees, the mean SI was 4.8 at baseline and increased to 18.7 at 6–8 weeks. The mean SI declined to 6.2 by 9–12 months.[105] Nine immune individuals who were tested from 2–12 weeks after vaccination had a marked increase in T cell recognition of VZV antigen to a mean SI of 36.3 ± 12.22 SE.[95] By one year, the mean SI among these vaccinees was 12.0 ± 3.12 SE, which was not statistically different from the prevaccine baseline SI of 7.9 ± 2.18 SE. High frequencies of VZV-specific CTL were demonstrated in two vaccinees with immunity to VZV. The CTL frequencies were 1:5,000 for IE62 protein and 1:8,000 for gE in one case and 1:14,000 for IE62 protein and 1:40,000 for gE in the second vaccinee. Clinical observations of naturally immune individuals given varicella vaccine indicate that restimulation of VZV-specific cellular immunity does not induce adverse effects.

When healthy older adults with natural immunity were vaccinated, responder cell frequencies of T cells specific for VZV and IFN-γ release increase significantly.[67,106–108] The vaccinees were 55–87 years old and received vaccine containing 1,110–12,000 pfu. The frequency of VZV-specific T cells rose to 1:40,000 after immunization, which was equivalent to the frequencies detected among immune adults who were 35–40 years old. A persistent increase in cellular immunity to VZV was documented for five years after immunization.[108] The average responder cell frequency was 1:42,000 at four years in vaccinees given 12,000 pfu/dose compared to 1:61,000 among those who received 3,000 pfu/dose. Immunosenescence impaired the response to reexposure to VZV in some older individuals because 10–15% of the population showed no enhancement of T cell proliferation. Placebo-controlled studies are planned to assess whether immunization reduces the incidence or severity of herpes zoster in healthy older adults.

Impairment of cell-mediated immunity to VZV correlates with a high risk of reactivation after bone marrow transplantation. We documented early reconstitution of cellular immunity to VZV by immunization with a heat-inactivated preparation of the varicella vaccine.[109] Among 14 of 28 patients given a single dose one month after BMT, T cell proliferation of VZV antigen was higher in vaccinees after three months. The mean SI was 12.20 ± 3.13 SE in vaccinees compared to 4.83 ± 2.74 SE ($p = 0.036$) in unvaccinated patients, but immunization with a single dose did not affect the severity of herpes zoster. Then twenty-four of 47 patients were randomized to receive vaccine at one, two, and three months after BMT. The mean T cell response was 8.43 ± 3.89 SE in vaccinees compared to 2.00 ± 0.33 SE ($p = 0.014$) in unvaccinated patients at four months. At five months, the mean SIs were 8.56 ± 2.81 SE and 5.30 ± 2.47 SE ($p = 0.043$). IFN-γ and IL-10 release increased with recovery of T-cell proliferation. Although VZV reactivation occurred in 23% of vaccinees and 22% of unvaccinated patients, a marked decrease in disease severity was documented with clinical severity scores of 6.4 ± 1.0 SE compared to 11.8 ± 1.1 SE, respectively ($p = 0.007$). This study of inactivated varicella vaccine in BMT patients provides the first evidence that active immunization can restore cellular immunity and reduce morbidity due to herpesviral reactivation in high-risk populations.

12. SUMMARY

The acquisition of cell-mediated immunity to VZV is essential to resolve viral replication during primary VZV infection. Memory T-cell immunity is necessary to protect against symptoms when the host experiences new exposures to the virus. Because VZV establishes latency, memory immunity is also required to prevent herpes zoster, caused by the reactivation of endogenous virus. Cellular immunity is elicited by immunization with live attenuated varicella vaccine. Although there are dosage effects and age-related differences in immunogenicity, protective immunity is elicited that prevents varicella with most vaccines and reduces the pathogenic effects of the virus in those who experience breakthrough infections. Immunization boosts cellular immunity to VZV in elderly adults who have had natural VZV infections. Reconstruction of VZV immunity in bone marrow transplant patients by immunization with inactivated varicella vaccine modifies the clinical course of herpes zoster. Virus-specific cellular immunity plays a critical role in achieving a balance between the virus and the host during varicella and maintaining this equilibrium through the lifelong persistence of the virus that follows primary VZV infection.

ACKNOWLEDGMENTS. Studies described in this review were supported by Public Health Service grants AI 20459, AI 22280, and AI 18449 from the National Institute of Allergy and Infectious Disease, by Merck Research Laboratories, and by the National Cancer Institute (California 49605). Investigations of VZV immunity in our laboratory were done by Celine Koropchak, Margaret Sharp, Alec E. Wittek, Pamela S. Diaz, Randy Bergen, Sonia Nader, and Rebecca Redman, with the collaboration of John Hay, William Ruyechan, and Paul Kinchington.

REFERENCES

1. Arvin, A. M., 1995, Varicella-zoster virus, in: *Virology*, (B. N. Fields, ed.), Raven Press, New York, pp. 2547–2586.
2. Moffat, J. F., Stein, M. D., Kaneshima, H., and Arvin, A. M., 1995, Tropism of varicella-zoster virus for human CD4+ and CD8+ T lymphocytes and epidermal cells in SCID-hu mice, *J. Virol.* **69**(9)**:**5236–5242.
3. Arvin, A. M., 1992, Cell-mediated immunity to varicella-zoster virus, *J. Infect. Dis.* S35–41.
4. Cohen, J., and Straus, S. E., 1995, Varicella zoster virus and its replication. In *Virology* (B. N. Fields, ed.), Raven Press, New York, pp. 2525–2546.
5. Gershon, A. A., 1995, Varicella vaccine: Its past, present and future, *Pediatr. Infect. Dis. J.* **14**(9)**:**742–744.
6. Arvin, A. M., and Gershon, A. A., 1996, Live attenuated varicella vaccine, *Ann. Rev. Microbiol.* **50:**59–100.
7. Takahashi, M., and Gershon, A., Live attenuated varicella vaccine, in: *Vaccines* (M. Levine, ed.).

8. Clarke, P., Beer, T., Cohrs, R., and Gilden, D. H., 1995, Configuration of latent varicella-zoster virus DNA, *J. Virol.* **69**(12)**:**8151–8154.
9. Davison, A. J., and Scott, J. E., 1986, The complete DNA sequence of varicella-zoster virus, *J. Gen. Virol.* **67:**1759–1816.
10. Gray, W. L., Pumphrey, C. Y., Ruyechan, W. T., and Fletcher, T. M., 1992, The simian varicella virus and varicella-zoster virus genomes are similar in size and structure, *Virology* **186**(2)**:**562–572.
11. Moriuchi, H., Moriuchi, M., Dean, H., Cheung, A. K., and Cohen, J. I., 1995, Pseudorabies virus EPO is functionally homologous to varicella-zoster virus ORF61 protein and herpes simplex virus type 1, *Virology* **209:**281–283.
12. Takada, M., Suzutani, T., Yoshida, I., Matoba, M., and Azuma, M., 1995, Identification of varicella-zoster virus strains by PCR analysis of three repeat elements and a PstI-site less region, *J. Clin. Microbiol.* **33**(3)**:**658–660.
13. LaRussa, P., Lungu, O., Hardy, I., Gershon, A., Steinberg, S. P., and Silverstein, S., 1994, Restriction fragment length polymorphism of polymerase chain reaction products from vaccine and wild-type varicella-zoster virus isolates, *J. Virol.* **66**(2)**:**1016–1020.
14. Forghani, B., Yu, G. J., and Hurst, J. W., 1991, Comparison of biotinylated DNA and RNA probes for rapid detection of varicella-zoster virus genome by *in situ* hybridization, *J. Clin. Microbiol.* **29**(3)**:**583–591.
15. Koropchak, C. M., Solem, S. M., Diaz, P. S., and Arvin, A. M., 1989, Investigation of varicella-zoster virus infection of lymphocytes by *in situ* hybridization, *J. Virol.* **63**(5)**:**2392–2395.
16. Koropchak, C. M., Graham, G., Palmer, J., *et al.* 1991, Investigation of varicella-zoster virus infection by polymerase chain reaction in the immunocompetent host with acute varicella [published erratum appears in *J. Infect. Dis.* **165**(1)**:**188, 1992]. *J. Infect. Dis.* **163**(5)**:**1016–1022.
17. Sawyer, M. H., Wu, Y. N., Chamberlain, C. J., *et al.*, 1992, Detection of varicella-zoster virus DNA in the oropharynx and blood of patients with varicella, *J. Infect. Dis.* **166**(4)**:**885–888.
18. Cohen, J. I., and Seidel, K. E., 1995, Varicella-zoster virus open reading frame 1 encodes a membrane protein that is dispensable for growth of VZV *in vitro, Virology* **206**(2)**:**835–842.
19. Kemble, 1996, personal communication.
20. Grose, C., 1991, Glycoproteins of varicella-zoster virus and their herpes simplex virus homologs, *Rev. Infect. Dis.* **16:**S960–S963.
21. Davison, A., Edson, C., Ellis, R., *et al.*, 1986, New common nomenclature for glycoprotein genes of varicella-zoster virus and their products, *J. Virol.* **57:**1195–1197.
22. Forghani, B., Ni, L., and Grose, C., 1994, Neutralization epitope of the varicella-zoster virus gH:gL glycoprotein complex, *Virology* **199**(2)**:**458–462.
23. Yao, Z., Jackson, W., Forghani, B., and Grose, C., 1993, Varicella-zoster virus glycoprotein gpI/gpIV receptor: Expression, complex formation, and antigenicity within the vaccinia virus-T7 RNA polymerase transfection system, *J. Virol.* **67**(1)**:**305–314.
24. Zhu, Z., Gershon, M. D., Ambron, R., Gabel, C., and Gershon, A. A., 1995, Infection of cells by varicella-zoster virus: Inhibition of viral entry by mannose 6-phosphate and heparin, *Proc. Natl. Acad. Sci. USA* **92**(8)**:**3546–3550.
25. Gabel, C. A., Dubey, L., Steinberg, S. P., Sherman, D., Gershon, M. D., and Gershon, A. A., 1989, Varicella-zoster virus glycoprotein oligosaccharides are phosphorylated during post-translational maturation, *J. Virol.* **63**(10)**:**4264–4276.
26. Gershon, A. A., Sherman, D. L., Zhu, Z., Gabel, C. A., Ambron, R. T., and Gershon, M. D., 1994, Intracellular transport of newly synthesized varicella-zoster virus: Final envelopment in the *trans*-Golgi network, *J. Virol.* **68**(10)**:**6372–6390.
27. Mallory, S., and Arvin, A. M., 1997, Mutational analysis of varicella-zoster virus (VZV) glycoproteins, gI and gE, in recombinant VZV strains, in preparation.

28. Cohen, J. I., and Seidel, K. E., 1994, Absence of varicella-zoster virus (VZV) glycoprotein V does not alter growth of VZV *in vitro* or sensitivity to heparin, *J. Gen. Virol.* **59:**3087–3093.
29. Kinchington, P. R., Ling, P., Pensiero, M., Moss, B., Ruyechan, W. T., and Hay, J., 1990, The glycoprotein products of varicella-zoster virus gene 14 and their defective accumulation in a vaccine strain (Oka), *J. Virol.* **64**(9)**:**4540–4548.
30. Rodriguez, J. E., Moninger, T., and Grose, C., 1993, Entry and egress of varicella virus blocked by same anti-gH monoclonal antibody, *Virology* **196**(2)**:**840–844.
31. Cohen, J. I., and Seidel, K., 1994, Varicella-zoster virus (VZV) open reading frame 10 protein, the homolog of the essential herpes simplex virus protein VP16, is dispensable for VZV replication *in vitro, J. Virol.* **68**(12)**:**7850–7858.
32. Debrus, S., Sadzot, D. C., Nikkels, A. F., Piette, J., and Rentier, B., 1995, Varicella-zoster virus gene 63 encodes an immediate-early protein that is abundantly expressed during latency, *J. Virol.* **69**(5)**:**3240–3245.
33. Kinchington, P. R., Bookey, D., and Turse, S. E., 1995, The transcriptional regulatory proteins encoded by varicella-zoster virus open reading frames (ORFs) 4 and 63, but not ORF 61, are associated with purified virus particles, *J. Virol.* **69**(7)**:**4274–4282.
34. Takahashi, M., 1986, Clinical overview of varicella vaccine: Development and early studies, *Pediatrics* **78S:**736–741.
35. Arvin, A., Kinney-Thomas, E., Shriver, K., *et al.*, 1986, Immunity to varicella-zoster glycoproteins, gpI (gp 90/58) and gp III (gp 118), and to a nonglycosulated protein, p170. *J. Immunol.* **137:**1346–1351.
36. Giller, R. H., Winistorfer, S., and Grose, C., 1989, Cellular and humoral immunity to varicella-zoster virus glycoproteins in immune and susceptible human subjects, *J. Infect. Dis.* **160**(6)**:**919–928.
37. Bergen, R. E., Sharp, M., Sanchez, A., Judd, A. K., and Arvin, A. M., 1991, Human T cells recognize multiple epitopes of an immediate early/tegument protein (IE62) and glycoprotein I of varicella-zoster virus, *Viral Immunol.* **4**(3)**:**151–166.
38. Hayward, A. R., 1990, T-cell responses to predicted amphipathic peptides of varicella-zoster virus glycoproteins II and IV, *J. Virol.* **64**(2)**:**651–655.
39. Arvin, A. M., 1996, Immune responses to varicella-zoster virus, in: *Infectious Disease Clinics of North America,* Vol. 10, (R. W. Ellis, C. W., ed.), W. B. Saunders, Philadelphia, pp. 529–570.
40. Hayward, A., and Herberger, M., 1987, Lymphocyte responses to varicella-zoster in the elderly, *J. Clin. Immunol.* **7:**174–178.
41. Arvin, A. M., Sharp, M., Smith, S., *et al.,* 1991, Equivalent recognition of a varicella-zoster virus immediate early protein (IE62) and glycoprotein I by cytotoxic T lymphocytes of either $CD4^+$ or $CD8^+$ phenotype, *J. Immunol.* **146**(1)**:**257–264.
42. Diaz, P. S., Smith, S., Hunter, E., and Arvin, A. M., 1989, T lymphocyte cytotoxicity with natural varicella-zoster virus infection and after immunization with live attenuated varicella vaccine, *J. Immunol.,* **142**(2)**:**636–641.
43. Hayward, A., Pontesilli, O., Herberger, M., Laszlo, M., and Levin, M., 1986, Specific lysis of varicella-zoster virus-infected B lymphoblasts by human T cells, *J. Virol.* **58:**179–184.
44. Hickling, J. K., Borysiewicz, L. K., and Sissons, J. G., 1987, Varicella-zoster virus-specific cytotoxic T lymphocytes (Tc): Detection and frequency analysis of HLA class I-restricted Tc in human peripheral blood, *J. Infect. Dis.* **156**(5)**:**3463–3469.
45. La Russa, P., Steinberg, S., Seeman, M. D., and Gershon, A. A., 1985, Determination of immunity to varicella by means of an intradermal skin test, *J. Infect. Dis.* **152:**869–875.
46. Asano, Y., Shiraki, K., Takahashi, M., Nagai, T., *et al.,* 1981, Soluble skin test antigen of varicella-zoster virus prepared from the fluid of infected cultures, *J. Infect. Dis.* **143:**684–692.
47. Jenkins, D., and Arvin, A. M., 1997, Primary T-cell responses to the varicella-zoster virus immediate early protein, IE62 elicited *in vitro* by dendritic cell presentation, in preparation.

48. Grose, C. H., 1984, Variation on a theme by Fenner, *Pediatrics,* **68:**735–737.
49. Ozaki, T., Masuda, S., Asano, Y., Kondo, K., Namazue, J., and Yamanishi, K., 1994, Investigation of varicella-zoster virus DNA by the polymerase chain reaction in healthy children with varicella vaccination, *J. Med. Virol.* **42**(1)**:**47–51.
50. Perera, L. P., Mosca, J. D., Ruyechan, W. T., and Hay, J., 1992, Regulation of varicella-zoster virus gene expression in human T lymphocytes [published erratum appears in *J. Virol.,* **69**(4)**:**2723, 1995]. *J. Virol.* **66**(9)**:**5298–5304.
51. Ross, A. H., Lencher, E., and Reitman, G., 1962, Modification of chicken pox in family contacts by administration of gamma globulin, *N. Engl. J. Med.* **267:**369–376.
52. Bowden, R. A., Levin, M. J., Giller, R. H., Tubergen, D. G., and Hayward, A. R., 1985, Lysis of varicella-zoster virus infected cells by lymphocytes from normal humans and immunosuppressed pediatric leukaemic patients, *Clin. Exp. Immunol.* **60:**387–395.
53. Ihara, T., Starr, S., Ito, M., Douglas, S., and Arbeter, A., 1984, Human polymorphonuclear leukocyte-mediated cytotoxicity against varicella-zoster virus-infected fiborblasts, *J. Virol.* **51:**110–116.
54. Arvin, A. M., Kushner, J. H., Feldman, S., Baehner, R. L., Hammond, D., and Merigan, T. C., 1982, Human leukocyte interferon for the treatment of varicella in children with cancer, *N. Engl. J. Med.* **306:**761–765.
55. Arvin, A. M., Koropchak, C. M., Williams, B. R., Grumet, F. C., and Foung, S. K., 1986, Early immune response in healthy and immunocompromised subjects with primary varicella-zoster virus infection, *J. Infect. Dist.* **154:**422–429.
56. Wallace, M. R., Woelfl, I., Bowler, W. A., *et al.,* 1994, Tumor necrosis factor, interleukin-2, and interferon-gamma in adult varicella, *J. Med. Virol.* **43**(1)**:**69–71.
57. Kinchington, P. R., Hougland, J. K., Arvin, A. M., Ruyechan, W. T., and Hay, J., 1992, The varicella-zoster virus immediate-early protein IE62 is a major component of virus particle, *J. Virol.* **66**(1)**:**359–366.
58. Sharp, M., Terada, K., Wilson, A., *et al.,* 1992, Kinetics and viral protein specificity of the cytotoxic T lymphocyte response in healthy adults immunized with live attenuate varicella vaccine, *J. Infect. Dis.* **165**(5)**:**852–858.
59. Gershon, A., and Steinberg, S., 1982, Cellular and humoral immune responses to VZV in immunocompromised patients during and after VZV infections, *Infect. Immun. 1979;***25:**828.
60. Patel, P. A., Yoonessi, S. O'Malley, J., Freeman, A., Gershon, A., and Ogra, P. L., 1979, Cell-mediated immunity to varicella-zoster virus in subjects with lymphoma or leukemia, *J. Pediatr.* **94:**223–230.
61. Straus, S. E., Ostrove, J. M., Inchauspe G., *et al.* NIH conference. Varicella-zoster virus infections. Biology, natural history, treatment, and prevention. *Ann Intern Med 1988;***109:**438–439.
62. Gilden, D. H., Mahalingam, R., Dueland, A. N., Cohrs, R., 1992, Herpes zoster: pathogenesis and latency. *Prog. Med. Virol.* **39**(19)**:**19–75.
63. Lungu, O., Annunziato, P. W., Gershon, A., *et al.,* 1995, Reactivated and latent varicella-zoster virus in human dorsal root ganglia, *Proc. Natl. Acad. Sci. USA* **92**(24)**:**10980–10984.
64. Straus, S. E., 1994, Overview: The biology of varicella-zoster virus infection, *Ann. Neurol.* **56:** S4–8.
65. Gilden, D. H., Rozenman, Y., Murray, R., Devlin, M., and Vafai, A., 1987, Detection of varicella-zoster virus nucleic acid in neurons of normal human thoracic ganglia, *Ann. Emerging Med.* **16**(9)**:**377–380.
66. Sadzot, D. C., Debrus, S., Nikkels, A., Piette, J., and Rentier, B., 1995, Varicella-zoster virus latency in the adult rat is a useful model for human latent infection, *Neurology* **45:**S18–20.
67. Zhang, Y., Cosyns, M., Levin, M. J., and Hayward, A. R., 1994, Cytokine production in varicella-zoster virus-stimulated limiting dilution lymphocyte cultures, *Clin. Exp. Immunol.* **98**(1)**:**128–133.

68. Sadzot-Delvaux, C., Kinchington, P., Debrus, S., Rentier, B., and Arvin, A. M., 1997, Recognition of the latency-associated immediate early protein IE63 of varicella-zoster virus by human memory T-lymphocytes, *J. Immunol.* **150:**2802–2806.
69. Huang, Z., Vafai, A., Lee, J., Mahalingam, R., and Hayward, A. R., 1992, Specific lysis of targets expressing varicella-zoster virus gpI or gpIV by $CD4^+$ human T-cell clones, *J. Virol.* **66**(5)**:**2664–2669.
70. Arvin, A. M., Pollard, R. B., Rasmussen, L., and Merigan, T., 1980, Cellular and humoral immunity in the pathogenesis of recurrent herpes viral infections in patients with lymphoma, *J. Clin. Invest.* **65:**869–878.
71. Meyers, J. D., Flurnoy, N., and Thomas, E. D., 1980, Cell-mediated immunity to varicella-zoster virus after allogeneic bone marrow transplantation, *J. Infect. Dis.* **141:**479–487.
72. Berger, R., Florent, G., and Just, M., 1981, Decrease of the lympho-proliferative response to varicella-zoster virus antigen in the aged, *Infect. Immun.* **32:**24–27.
73. Wilson, A., Sharp, M., Koropchak, C. M., Ting, S. F., and Arvin, A. M., 1992, Subclinical varicella-zoster virus viremia, herpes zoster, and T lymphocyte immunity to varicella-zoster viral antigens after bone marrow transplantation, *J. Infect. Dis.* **165**(1)**:**119–126.
74. Terada, K., Kawano, S., Yoshihiro, K., and Morita, T., 1994, Varicella-zoster virus (VZV) reactivation is related to the low response of VZV-specific immunity afte chicken pox in infancy, *J. Infect. Dis.* **169**(3)**:**650–652.
75. Arvin, A., Koropchak, C. M., and Wittek, A. E., 1983, Immunologic evidence of reinfection with varicella-zoster virus, *J. Infect. Dis.* **148:**200–205.
76. Gershon, A. A., Steinberg, S., and Gelb, L., 1984, NIAID-Collaborative-Varicella-Vaccine-Study-Group. Clinical reinfection with varicella-zoster virus, *J. Infect. Dis.* **149:**137–142.
77. Shiraki, K., Yamanishi, K., and Takahashi, M., 1983, Biological and immunological characterization of the soluble skin test antigen of varicella-zoster virus, *J. Infect. Dis.* **149:**501–504.
78. Connelly, B. L., Stanberry, L. R., and Bernstein, D. I., 1993, Detection of varicella-zoster virus DNA in nasopharyngeal secretions of immune household contacts of varicella, *J. Infect. Dis.* **168**(5)**:**1253–1255.
79. Gilden, D., Mahlingham, R., Dueland, N., and Cohrs, R., 1992, Herpes zoster: Pathogenesis and latency, *Prog. Med. Virol.* **39:**19–75.
80. Hayward, A., Levin, M., Wolf, W., Angelova, G., and Gilden, D., 1991, Varicella-zoster virus-specific immunity after herpes zoster, *J. Infect. Dis.* **163**(4)**:**873–875.
81. Stevens, D., Ferrington, R., Jordan, G., and Merigan, T., 1975, Cellular events in zoster vesicles: Relation to clinical course and immune parameters, *J. Infect. Dis.* **131:**509–515.
82. Merigan, T. C., Rand, K., Pollard, R., Abdallah, P., Jordan, G. W., and Fried, R. P., 1978, Human leukocyte interferon for the treatment of herpes zoster in patients with cancer, *N. Engl. J. Med.* **298:**981–987.
83. Devlin, M. E., Gilden, D. H., Mahalingam, R., Dueland, A. N., and Cohrs, R., 1992, Peripheral blood mononuclear cells of the elderly contain varicella-zoster virus DNA, *J. Infect. Dis.* **165**(4)**:**619–622.
84. Kuter, B. J., Ngai, A., Patterson, C. M., *et al.*, 1995, Safety, tolerability, and immunogenicity of two regimens of Oka/Merck varicella vaccine (Varivax) in healthy adolescents and adults. Oka/Merck Varicella Vaccine Study Group, *Vaccine* **13**(11)**:**967–972.
85. Kuter, B. J., Weibel, R. E., Guess, H. A., *et al.*, 1991, Oka/Merck varicella vaccine in healthy children: Final report of a 2-year efficacy study and 7-year follow-up studies, *Vaccine* **9**(9)**:**643–647.
86. Weibel, R., Kuter, B., Neff, B., *et al.*, 1985, Live Oka/Merck varicella vaccine in healthy children: Further clinical and laboratory assessment, *JAMA* **245:**2435–2439.
87. Arbeter, A., Starr, S.E., and Plotkin, S. A., 1986, Varicella vaccine studies in healthy children and adults, *Pediatrics* **78**(suppl)**:**748–756.

88. Asano, Y., Suga, S., Yoshikawa, T., *et al.*, 1994, Experience and reason: Twenty year follow up of protective immunity of the Oka live varicella vaccine, *Pediatrics* **94:**524–526.
89. Watson, B., Boardman, C., Laufer, D., *et al.*, 1995, Humoral and cell-mediated immune responses in healthy children after one or two doses of varicella vaccine, *Clin. Infect. Dis.* **20**(2)**:**316–319.
90. Watson, B., Piercy, S., Soppas, D., *et al.*, 1993, The effect of decreasing amounts of live virus, while antigen content remains constant, on immunogenicity of Oka/Merck varicella vaccine, *J. Infect. Dis.* **168**(6)**:**1356–1360.
91. Bergen, R. E., Diaz, P. S., and Arvin, A. M., 1990, The immunogenicity of the Oka/Merck varicella vaccine in relation to infectious varicella-zoster virus and relative viral antigen content, *J. Infect. Dis.* **162**(5)**:**1049–1054.
92. Asano, Y., Yoshikawa, T., Suga, S., *et al.*, 1993, Postexposure prophylaxis of varicella in family contact by oral acyclovir, *Pediatrics* **92**(2)**:**219–222.
93. Watson, B., Keller, P. M., Ellis, R. W., Starr, S. E., 1994, Cell-mediated immune responses after immunization of healthy seronegative children with varicella vaccine: Kinetics and specificity, *J. Infect. Dis.* **162**(4)**:**794–799.
94. Hayward, A., Villanueba, E., Cosyns, M., and Levin, M., 1992, Varicella-zoster virus (VZV)-specific cytotoxicity after immunization of nonimmune adults with Oka strain attenuated VZV vaccine, *J. Infect. Dis.* **166**(2)**:**260–264.
95. Nader, S., Bergen, R., Sharp, M., and Arvin, A. M., 1995, Age-related differences in cell-mediated immunity to varicella-zoster virus among children and adults immunized with live attenuated varicella vaccine, *J. Infect. Dis.* **171**(1)**:**13–17.
96. Gershon, A., LaRussa, P., and Steinberg, S., 1995, Clinical trials in immunocompromised individuals, in: *Infectious Disease Clinics of North America,* Vol. 10, (R. E., White, ed.), Saunders, Philadelphia, pp. 583–594.
97. Brunell, P. A., Shehab, Z., Geiser, C., and Waugh, J. E., 1982, Administration of live varicella vaccine to children with leukemia, *Lancet* **2:**1069–1073.
98. Asano, Y., Itakura, N., Hiroishi, Y., *et al.*, 1985, Viral replication and immunologic responses in children naturally infected with varicella-zoster virus and in varicella vaccine, *J. Infect. Dist.* **152:**863–868.
99. Watson, B., Gupta, R., Randall, T., and Starr, S., 1993, Persistence of cell-mediated and humoral immune responses in healthy children immunized with live attenuated varicella vaccine, *J. Infect. Dis.* **169:**197–199.
100. Gershon, A. A., La Russa, P., Hardy, I., Steinberg, S., and Silverstein, S., 1992, Varicella vaccine: The American experience, *J. Infect. Dis.* **166:**S63–68.
101. White, C. J., Kuter, B. J., Hildebrand, C. S., *et al.*, 1991, Varicella vaccine (VARIVAX) in healthy children and adolescents: Results from clinical trials, 1987 to 1989 [see comments], *Pediatrics* **87**(5)**:**604–610.
102. Gershon, A. A., and Steinberg, S. P., 1990, Live attenuated varicella vaccine: Protection in healthy adults compared with leukemic children. National Institute of Allergy and Infectious Diseases Varicella Vaccine Collaborative Study Group, *J. Infect. Dis.* **161**(4)**:**661–666.
103. Gershon, A., Steinberg, S., and Gelb, L., 1986, NIAID-Collaborative-Varicella-Vaccine-Study-Group. Live attenuated varicella vaccine: Use in immunocompromised children and adults, *Pediatrics* **78**(S)**:**757–762.
104. Zerboni, L., Nader, S., and Arvin, A. M., in press, Age-related differences in the persistence of immunity induced by varicella vaccine, *J. Infect. Dis.*
105. Gershon, A. A., Steinberg, S., and Gelb, L., 1984, NIAID-Collaborative-Varicella-Vaccine-Study-Group. Live attenuated varicella vaccine: Efficacy for children with leukemia in remission, *JAMA* **252:**355–362.
106. Levin, M. J., Murray, M., Zerbe, G. O., White, C. J., and Hayward, A. R., 1994, Immune

responses of elderly persons 4 years after receiving a live attenuated varicella vaccine, *J. Infect. Dis.* **170**(3)**:**522–526.

107. Levin, M. J., Murray, M., Rotbart, H. A., Zerbe, G. O., White, C. J., and Hayward, A. R., 1992, Immune response of elderly individuals to a live attenuated varicella vaccine, *J. Infect. Dis.* **166**(2)**:**253–259.
108. Levin, M. J., Rotbart, H. A., and Hayward, A. R., 1995, Immune response to varicella-zoster virus 5 years after acyclovir therapy of childhood varicella [letter], *J. Infect. Dis.* **171**(5)**:**1383–1384.
109. Redman, R. L., Nader, S., Zerboni, L., *et al.,* Early reconstitution of immunity and decreased severity of herpes zoster in bone marrow transplant recipients immunized with inactivated varicella vaccine, submitted.

13

Complement Control Proteins of Rhadinoviruses

JENS-CHRISTIAN ALBRECHT, FRANK NEIPEL, and BERNHARD FLECKENSTEIN

1. INTRODUCTION

Rhadinoviruses, also termed γ2-herpesviruses, are a group of agents that naturally occur in primates, ungulates, rabbits, and rodents. The first human rhadinovirus identified was Kaposi's sarcoma-associated herpesvirus HHV-8 (human herpesvirus type 8). This agent was found by representational difference analysis (RDA) in AIDS-related Kaposi's sarcoma,[1] and subsequent molecular epidemiology has shown that it consistently present in classical, African and AIDS-associated Kaposi's sarcoma, in body cavity-based B cell lymphomas, and in multifocal Castleman's disease.[2] Most rhadinoviruses have a pronounced tropism for T lymphocytes and their precursors. They are usually not pathogenic in their natural host but often oncogenic in distinct but related species.

Herpesvirus saimiri, the prototype of rhadinoviruses, is a common agent of squirrel monkeys (*Saimiri sciureus*), which are native to large areas of South and Central America. The virus persists in T lymphocytes of its natural host over years or for its lifetime. It is estimated that at least one out of 10^6 T lymphocytes of normal healthy squirrel monkeys carry the virus, as monitored by cocultivation

JENS-CHRISTIAN ALBRECHT, FRANK NEIPEL, and BERNHARD FLECKENSTEIN • Institut für Klinische und Molekulare Virologie, Friedrich-Alexander Universität Erlangen-Nürnberg, 91054 Erlangen, Germany.

Herpesviruses and Immunity, edited by Medveczky *et al.* Plenum Press, New York, 1998.

with permissive monolayer cultures.[3] Although *H. saimiri* is not known to be pathogenic in its natural host, it has a pronounced transforming and tumorigenic potential in numerous other New World primates. Tamarin marmosets (*Saguinus* spp.) develop acute lymphoproliferative syndromes, lymphocytic leukemias, and fulminant lymphomas, if parenterally infected with the virus. Various isolates of *H. saimiri* can also transform primary lymphocyte cultures of Old and New World primates. Human T lymphocytes are transformed to continuous growth in cell culture, preserving antigen specificity for long growth periods.[4,5]

Herpesvirus ateles is a rhadinovirus of spider monkeys (*Ateles* spp.) devoid of known pathogenic properties in its natural host. Like *H. saimiri,* it is a potent inducer of lymphatic neoplasias in numerous species of New World primates. The overall genomic structure of *H. ateles* is closely related to *H. saimiri,* and both are quite distinct from the genomic organization of the other herpesvirus genera. An ungulate herpesvirus, the alcelaphine herpesvirus type I (AHV-1) is classified as a γ2-herpesvirus by analysis of the physical properties of its viral genome.[6] AHV-1 naturally occurs in wildebeest and causes malignant catarrhal fever, an acute fatal disease in cattle. A related herpesvirus, bovine herpesvirus type 4 (BHV-4), has a similar genomic organization but does not reveal indications for pathogenic properties.[7] Equine herpesvirus type 2 (EHV-2), a rhadinovirus of horses, is not pathogenic in the adult animal but induces moderate lymphoproliferation in foals.[8]

2. GENOMIC ORGANIZATION OF RHADINOVIRUSES

The rhadinoviruses are defined primarily by distinct structural features of their genomes. They are different in genomic organization from all other herpesvirus genera characterized previously. The intact infectious viral genome (M-genome) of *H. saimiri* is linear double-stranded DNA in the size range from 145 to 165 kbp. Viral genomes persist as episomes in tumors and transformed T cells as covalently closed circles.[3,9] *H. saimiri* DNA molecules are composed of an internal unique 112.9 kbp segment of low G + C (34.5%) content (L-DNA) that is flanked by terminal stretches of high G + C (70.8%) H-DNA repeats. Each H-DNA repeat unit has 1.44 kbp. The repeats are arranged in tandem, and the two H-DNA termini in the M-genomes have the same orientation. Protein coding functions are restricted to the unique L-segment of the M-genomes. The L-DNA has 76 major open reading frames and a set of seven U-RNA genes for a total of 83 potential genes. Homologous sequences are found in herpesviruses of other genera for 60 of the predicted proteins. Genes conserved between *H. saimiri* and the Epstein–Barr virus (EBV) show that their genomes are generally collinear, although conserved gene blocks are separated by unique genes that determine the particular phenotype of the viruses. Unlike other known herpesviruses, *H. saimiri* has numerous genes for viral proteins with significant homologies to cellular proteins of known function. These include two enzymes of the nucleotide metabolism, thymidylate synthase and dihydrofolate reductase, and numerous proteins that are possibly involved in growth regulation of lymphocytes. These include the transformation-associated proteins, STP, that have local similarity to collagen;[10,11] a polypeptide with striking homology to interleukin-17;[12] a protein related to the superantigen encoded by the open reading frame of mouse mammary tumor virus 3′-LTR; a D-type cyclin homologue; and a protein similar to the interleukin-8 receptor.[13] In addition, homology searches have identified two open reading frames related to known cellular complement control proteins[14,15] (Fig. 1). The genomic organization of *H. ateles* is essentially identical to that of *H. saimiri,* although four open reading frames are missing. *H. ateles* has 73 protein encoding genes, among them a structural homologue of complement component C3 convertase inhibitors. However, there is no indication of a second complement regulatory protein in *H. ateles.*[16] The open reading frame number 4 of HHV-8 has structural features of a gene coding for a complement controlling protein. As in *H. ateles,* there is no other gene product interfering with complement action.[2,17] The particular genomic structure of rhadinoviruses is predisposed to the acquisition of cellular genes that, most likely, exert the equivalent function in the context of the virus, providing a selective advantage for the evolution of the viruses. In this review, we summarize the available data on the structures and functions of the known rhadinoviral complement control proteins.

3. THE C3 CONVERTASE INHIBITOR OF *H. SAIMIRI*

The complement system mediates the primary humoral defense mechanism against microbial infection. It is an enzyme cascade leading to destruction of nonself microorganisms and cells. Furthermore, complement activation products initiate most of the inflammatory processes involved in clearing a pathogen. Complement activation can take place by either the classical or the alternative pathway. Initiating the classical pathway of complement activation (CPCA) are antibodies that mark a nonself target for lysis. In contrast, the alternative pathway of complement activation (APCA) is independent of an initiating factor. Consequently, the APCA does not discriminate between self and nonself. Because the complement system has an enormous destructive potential, an effective regulatory system has coevolved to prevent damage to autologous tissues. The key role of complement action is played by the complement component C3 that becomes activated by enzymatic cleavage, forms a C5 converting complex with other components from the CPCA or APCA, and initiates the formation of the terminal

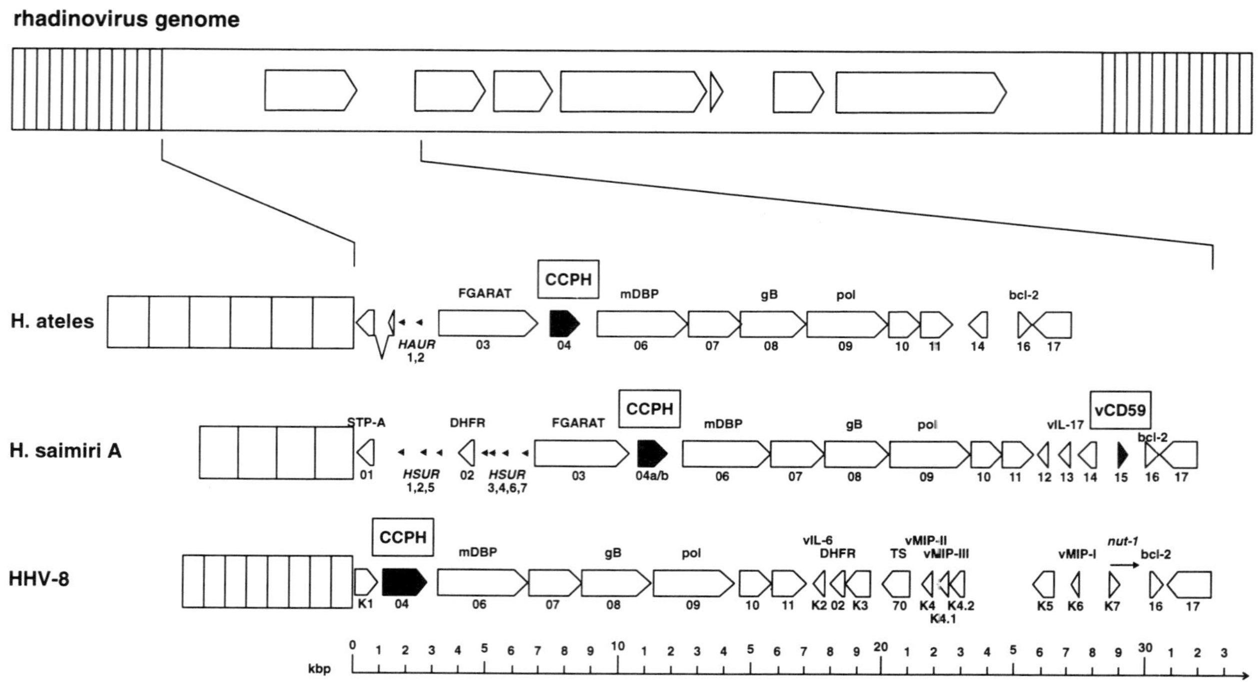

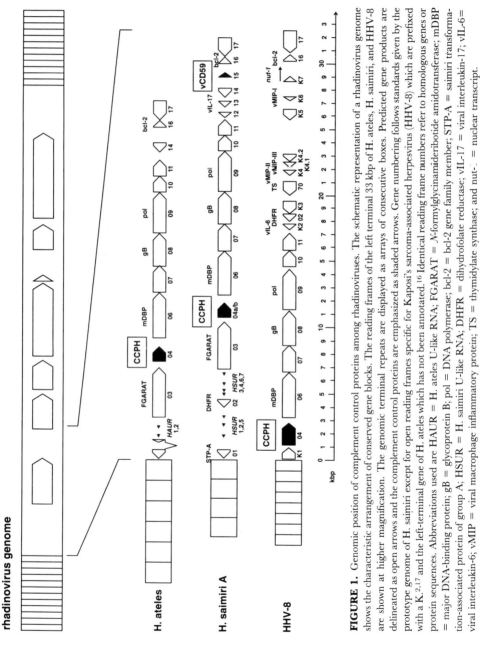

FIGURE 1. Genomic position of complement control proteins among rhadinoviruses. The schematic representation of a rhadinovirus genome shows the characteristic arrangement of conserved gene blocks. The reading frames of the left terminal 33 kbp of H. ateles, H. saimiri, and HHV-8 are shown at higher magnification. The genomic terminal repeats are displayed as arrays of consecutive boxes. Predicted gene products are delineated as open arrows and the complement control proteins are emphasized as shaded arrows. Gene numbering follows standards given by the prototype genome of H. saimiri except for open reading frames specific for Kaposi's sarcoma-associated herpesvirus (HHV-8) which are prefixed with a K.[2,17] and the left-terminal gene of H. ateles which has not been annotated.[16] Identical reading frame numbers refer to homologous genes or protein sequences. Abbreviations used are HAUR = H. ateles U-like RNA; FGARAT = *N*-formylglycinamideribotide amidotransferase; mDBP = major DNA-binding protein; gB = glycoprotein B; pol = DNA polymerase; bcl-2 = bcl-2 gene family member; STP-A = saimiri transformation-associated protein of group A; HSUR = H. saimiri U-like RNA; DHFR = dihydrofolate reductase; vIL-17 = viral interleukin-17; vIL-6= viral interleukin-6; vMIP = viral macrophage inflammatory protein; TS = thymidylate synthase; and nut-1 = nuclear transcript.

membrane attack complex (MAC) which leads to pore formation on a targeted surface. Consequently, complement regulatory activity is concentrated on the breakdown of C3/C5 convertases and inhibiting the MAC.

The continuous turnover of C3 is regulated by the endogenous regulatory protein factor H. Accidental activation of the classical pathway is controlled by the C4b-binding protein (C4bp). On cell surfaces, C3 and C5 convertases are regulated by the membrane-bound glycoprotein complement receptor type I (CR1), complement receptor type II (CR2), membrane cofactor protein (MCP), and decay-accelerating factor (DAF). The coding sequences for C4bp, CR1, CR2, MCP, and DAF form a gene family on human chromosome 1q32, designated as the regulators of complement activation (RCA) complex. The factor H gene is also locoated on chromosome 1, but not within a 1500 kbp fragment carrying the RCA gene cluster. These complement control proteins are related to each other by structural commonalities, most notably, the short consensus repeat motif (SCR) found in all polypeptides that bind to C4 to C3.[18] This repeat motif consists of about 58 to 66 amino acids and is composed of a constant framework of four cysteine residues and of proline, tryptophan, and several hydrophobic residues that are conserved (Fig. 2).

The left termnal *H. saimiri* DNA fragment has revealed a gene for a polypeptide with striking similarities to complement control proteins that interact with C3b or C4b.[14] The open reading frame consists of 1080 nucleotides (nt) which code for a polypeptide of 360 amino acids (aa). The predicted protein has a hydrophobic amino terminus of 20 aa, most likely representing a signal peptide for processing seven Asn-linked glycosylation sites (NxT/S), four units of a repeated amino acid sequence pattern of about 61 aa each, a 62-aa region of unknown significance, a transmembrane domain of 23 aa, and a short stretch of 10 aa at the C-terminus. The four repeat units (33–37% identity to each other) share the typical pattern of the SCR motif known to be the main constituent of members of the human complement control protein family encoded by the RCA complex. This includes C4bp, MCP, DAF, CR1, CR2, and factor H. The global structural layout of the *H. saimiri* complement control protein homologue (CCPH) is most similar to those of the cell-membrane-associated complement regulators DAF and MCP (Fig. 2). Homology searches further reveal a high similarity to the major 35-kDa secretory protein of the vaccinia virus (VCP, ORF C3L) that binds C4b and down-regulates both CPCA and APCA.[19,20] Northern blot hybridization with a probe specific for the CCPH reading frame and second-strand synthesis of cDNA with a transcript specific primer detect transcripts of about 1.5 and 1.7 kb from RNA of owl monkey kidney (OMK) cells that are lytically infected with *H. saimiri*. Sequence analysis of cDNAs show that CCPH transcripts occur as unspliced or single-spliced mRNA. The unspliced mRNA directs translation of a membrane-bound glycoprotein (mCCPH), whereas splicing removes a 193-bp intron which encodes the transmembrane domain and

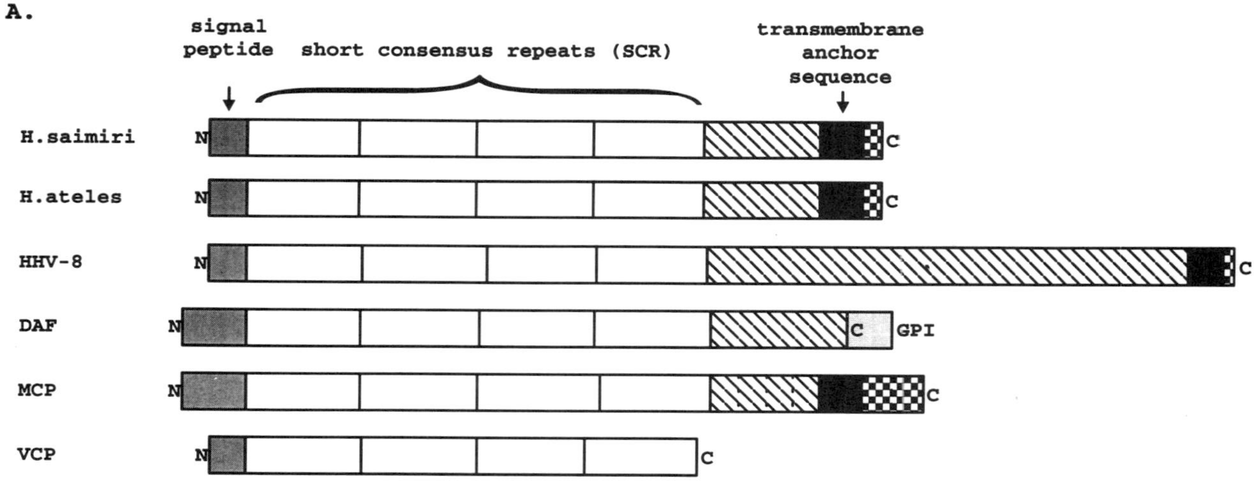

B.

```
                        1                                                            67
H.ateles      SCR2  KrrCstPadllNGwytvtgn...lyyGsvItytCNtGYqLlGsp.tssCllgpdgrvnWtprpPiCe
H.saimiri A   SCR2  KkkCstPqdllNGrytvtgn...lyyGsvItytCNsGYsLiGst.tsaCllkrggrvdWtprpPiCd
H.saimiri C   SCR2  KrrCstPkellhGrydipgd...lyyGsnItykCdkGYmLiGkn.tstCllneegqmqWipkpPlCe
H.saimiri B   SCR2  KkkCpnPadilNGqthiikd...lyfGssItykCNiGYvLlGdk.kstCllgenskvqWipkpPiCe
VCP           SCR2  KrrCpsPrdidNGqldi.gg...vdfGssItysCNsGYhLiGes.ksyCelgstgsmvWnpeaPiCe
DAF           SCR3  KksCpnPgeirNGqidvpgg...ilfGatIsfsCNtGYkLfGst.ssfCli.sgssvqWsdplPeCr
HHV-8         SCR2  KksCpnPgeiqNGkvifhggqdalkyGanIsyvCNeGYfLvGreyvryCmigasgqmaWssspPfCe
MCP           SCR2  retCpyirdplNGqavpang..tyefGyqmhfiCNeGYyLiGee.ilyCel.kgsvaiWsgkpPiCe
      Consensus     K--C--P----NG------------G--I---CN-GY-L-G------C---------W----P-C-
```

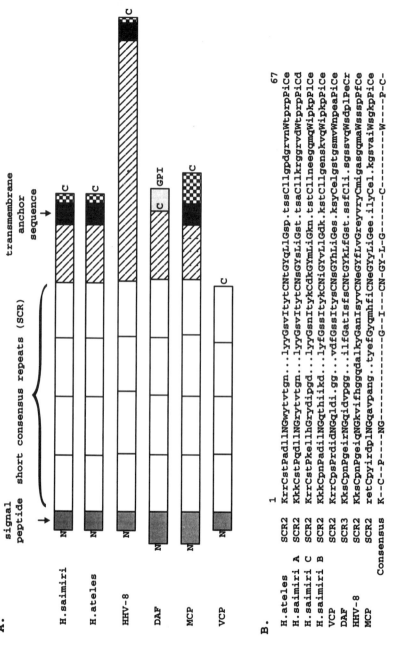

FIGURE 2. Structural similarities among complement control proteins. All known complement regulatory protein sequences carrying four contiguous SCRs are represented in (a). At the N-terminal end, predicted signal sequences are shaded, whereas transmembrane anchor sequences near the C-terminal end are black. A region of unknown significance is marked by hatched lines, and the cytoplasmic tail is shown as pattern-filled box. In the case of DAF, the hydrophobic amino acid sequence at the C-terminus, which is replaced by a GPI anchor, is shaded. A comparison of the best conserved short consensus repeat (SCR) is shown in (b). In the case of DAF and MCP, the SCR corresponds to a split exon. H. saimiri A refers to the prototype sequence of strain A11. H. saimiri B and C refer to group B strain SMHI, and group C strain C-488,[16,41] respectively.

results in a secretory protein (sCCPH) with a different C-terminus.[14] A similar mechanism exists in the human DAF gene, in which the majority of transcripts specifying the membrane-bound form of DAF is spliced and in which unspliced transcripts containing an intron encoding contiguous protein sequence result in a secreted form of DAF.[21] However, no antigen related to the hydrophilic form of DAF has been detected yet, although soluble DAF molecules have been found in plasma, urine, saliva, and synovial fluid, and in HeLa cell culture supernatants. This can be explained by a report demonstrating the release of glycosylphosphatidylinositol (GPI)-anchored DAF by cell-associated GPI-specific phospholipase D.[22]

The *H. saimiri* gene for CCPH has basically an intronless structure in contrast to all mammalian genes for complement regulators. The genes for the RCA proteins exhibit several splicing characertistics. Each SCR is generally encoded by a separate exon. The exon/intron junction of exons containing one or two SCRs occurs after the first nt of the codon triplet. This is believed to have facilitated the process of genetic duplication which have led to the repeated SCR structural motif.[18,23] Additionally, one of the SCRs is encoded by two exons in genes of the RCA complex. Interestingly, this 'split' exon occurs after the second nt of a conserved glycine codon. This pattern is specific for split exons of complement regulatory proteins, but no other SCR-containing gene carries such an exon.[24] This suggests that complement control proteins that bind C3b/C4b may have evolved from a gene that contained an ancestral split exon or that there is an RCA-specific activity encoded by the split exon. Most notably, the SCR encoded by the split exon is the best conserved SCR among mammalian and viral complement regulatory proteins. Searching current database entries (Genbank 97.0, 289,329 sequences, 387,914,422 nt) with a consensus sequence corresponding to the exon/exon boundary of split exons and conserved viral SCR coding sequences did not reveal other occurrences than that of split exons from DAF, MCP, CR2, factor H, and their rodent homologues (Table I). This indicates that the CCPH gene and the VCP gene of vaccinia virus may contain a remnant of a split exon. This suggests that the viral genes have evolved from RCA genes or a common ancestor or have been captured by acquisition of a cDNA structure derived from the RCA locus of their natural hosts.

Two neutralizing mouse monoclonal antibodies (SD, SE) were described[25] that both recognize two glycoproteins of different sizes from extracts of cells infected with *H. saimiri*. The larger glycoprotein, which has the mobility of a 65–75 kDa protein in denaturing polyacrylamide gels, was precipitated from detergent extracts of purified virions. The smaller glycoprotein has a molecular mass of 45–52 kDa in infected cells. The recognition pattern of these monoclonal antibodies suggests that the two proteins share an identical epitope which is assumed to be realized by translation of two proteins from a single gene. Complementary DNA clones representing both mRNA species of the CCPH gene were transiently

TABLE 1
Consensus Sequence of the Exon/Exon Boundary of Split Exons of Complement Control Proteins

Organism protein	SCR	Codon −3	−2	−1	0	+1	+2	+3	+4	+5
H. saimiri	2	TGT	AAT	TCA	GG|C	TAC	AGC	TTA	ATT	GGA
CCPH		C	N	S	G	Y	S	L	I	G
H. ateles	2	TGT	AAT	ACA	GG|A	TAT	CAG	TTG	TTA	GGG
CCPH		C	N	T	G	Y	Q	L	L	G
HHV-8	2	TGT	AAT	GAA	GG|A	TAT	TTT	TTG	GTT	GGT
CCPH		C	N	E	G	Y	F	L	V	G
Human	3	TGT	AAC	GAG	GG|G	TAC	AAA	TTA	TTT	GGC
DAF		C	N	T	G	Y	K	L	F	G
Human	2	TGT	AAT	GAG	GG|T	TAT	TAC	TTA	ATT	GGT
MCP		C	N	E	G	Y	Y	L	I	G
Vaccinia	2	TGT	AAT	AGC	GG|A	TAT	CAT	TTG	ATC	GGT
VCP		C	N	S	G	Y	H	L	I	G
Consensus[a]		TGT	AAY	NNN	GG|N	TAY	NNN	TTR	NNN	GGN

[a]Ambiguous codes used are Y = pYrimidine, R = puRin, and N = aNy.

expressed in COS-7 cells. Cells transfected with either cDNA expression clone were analyzed by an indirect immunofluorescence assay allowing antibody internalization. This procedure showed a broad cytoplasmic fluorescence with COS-7 cells transfected with either of the clones, whereas preventing antibody diffusion revealed only staining of cells transfected with the mCCPH cDNA clone. This has proven that the monoclonal antibodies are directed against the two forms of CCPH, and it confirmed surface expression of mCCPH. Neither protein was detectable by immunoprecipitation of COS-7 cell extracts. However, the sCCPH protein secreted from COS-7 cells was precipitated by monoclonal antibody SE, and a glycoprotein of comparable size also was precipitated from OMK cell culture supernatants lytically infected with *H. saimiri*. Precipitation of the 65–75 kDa glycoprotein from detergent extracts of purified virions identified this glycoprotein as mCCPH. In summary, it was demonstrated that the membrane-bound CCPH is a viral structural component, whereas the alternative form sCCPH is secreted by productively infected cells.[14]

Stable cell lines expressing mCCPH or an amino-terminal, FLAG-epitope tagged derivative (5′FLAGmCCPH) were generated for functional studies of the viral complement regulator. The cell lines conferred resistance to complement-mediated cell damage by inhibiting the lytic activity of both human and rat whole serum complement. The mechanism by which mCCPH regulates complement was further defined by showing that the mCCPH and 5′FLAGmCCPH transfectants exhibit reduced cell-surface deposition of C3b as detected by surface-bound C3d, another cleavage product of activated C3. This demonstrated that mCCPH inhibits the CPCA at the level of C3 activation, thereby preventing serum complement-mediated lysis of cells from expressing this virally encoded protein.[26] The localization of mCCPH on the surface of the *H. saimiri* virion particle, taken together with the complement-regulatory activity of the molecule, suggest that the virus evolved a humoral immuno-evasive mechanism by inhibiting complement-mediated clearance, resulting in increased survival of the virus. Host primate cells susceptible to viral infection express endogenous complement control proteins that protect the cell from autologous complement. However, it might be possible that gene expression of these cellular proteins is decreased during viral replication. Nevertheless, increased cell-surface expression of complement inhibitors has been demonstrated to be more efficient in protecting cells against complement-mediated cell lysis.[27] Another potential role for CCPH during viral infection involves the formation of C3 degradation products which are known to be ligands of cellular receptors. The binding of these ligands to the viral particle following C3 degradation may facilitate cellular entry. Numerous examples of complement-mediated and complement-dependent cellular entry involving complement proteins or complement protein/cellular receptor-interactions are documented, including the Epstein-Barr virus (γ-1 herpesvirus), the West Nile virus[28] (flavivirus), echoviruses[29] (enterovirus), and measles virus[30] (morbillivirus). The use of complement receptors as a mode of cellular entry is remarkable considering that these receptors are predominantly expressed on lymphoid cells. These are also the main target and reservoir for *H. saimiri.*

4. THE COMPLEMENT CONTROL PROTEIN OF *H. ATELES*

The reading frame four of *H. ateles,* structurally similar to the *H. saimiri* CCPH gene, encodes a protein of 360 amino acids.[16] It is at an identical genomic position (Fig. 1). The structure of the putative gene product is almost identical with regard to signal sequence, the arrangement of short consensus repeats, transmembrane sequence, and cytoplasmic tail (Fig. 2). The degree of overall conservation is 73%. Table II summarizes the degree of sequence conservation among herpesvirus ateles and three representative strains of *H. saimiri* subgroups A, B, and C. The genetic distance between *H. ateles* and the three *H. saimiri* stains is not higher than among the three *H. saimiri* strains. Until now, there are no functional data on the CCPH gene product of *H. ateles* although its overall structural conservation strongly suggests that it functions like CCPH of *H. saimiri.*

TABLE II
Conservation of Complement Control Proteins among *H. ateles* and *H. saimiri* Strains of Subgroups A, B, and C

	H. ateles #73	*H. saimiri* #11 (A)	*H. saimiri* SMHI (B)	*H. saimiri* #488 (C)
H. ateles #73		73.1	59.8	58.9
H. saimiri #11 (A)	81.9		64.0	59.2
H. saimiri SMHI (B)	71.1	72.2		55.0
H. saimiri #488 (C)	70.0	70.0	67.1	

Upper right: % identity. Lower left: % similarity, allowing conservative replacements

5. THE COMPLEMENT REGULATOR OF KAPOSI'S SARCOMA-ASSOCIATED HERPESVIRUS HHV-8

The genomic sequence of HHV-8 has been determined recently by two groups.[2,17] Reading frame four of HHV-8 codes for a protein of 550 amino acids that contains four short consensus repeats in the N-terminal part of the molecule, whereas the C-terminal half is structurally different. The predicted protein has a C-terminal transmembrane anchor sequence, suggesting that the C-terminal half has the spacer function of a transmembrane protein exposing the complement antagonistic short consensus repeats to the cell surface (Fig. 2). No data have been reported with respect to control of gene expression, splicing pattern, or function of the protein.

6. THE TERMINAL COMPLEMENT INHIBITOR OF *H. SAIMIRI*

The reading frame ORF-15, identified within the *H. saimiri* group A strain 11 L-DNA consists of 363 nucleotides (nt), and is located at position 29,231–29,593 of the prototype genome. The nucleotide sequence taken as a query to search through databases for homologies revealed a nucleotide sequence identity of 67% between position 33 and 323 of ORF-15 and the corresponding region of human CD59 cDNA. When the entire reading frames were compared, sequence identity decreased to 64%. These data indicate that ORF-15 is as homologous to the cellular CD59 gene as the *H. saimiri* thymidylate synthase gene is to its cellular counterpart. The amino acid sequence deduced from

ORF-15 consist of 121 residues. A leader peptide of 19 aa of the ORF-15 gene product, termed vCD59, is predicted by the method of Heijne[31] and by comparison with human CD59 (huCD59). It may be processed at the equivalent residue of huCD59 because the environment of the cleavage site is identical. Both the predicted viral polypeptide and the huCD59 have a single consensus sequence for Asn-linked glycosylation of the same context (Asn-Cys-Ser), and it has been demonstrated that huCD59 is glycosylated at this asparagine residue. vCD59 and huCD59 have hydrophobic C-terminal sequences. In the case of huCD59, it is known that this sequence is replaced by a glycosylphosphatidyl-inositol (GPI)-anchor. The processing site for GPI-linkage of huCD59 is not known but is assumed to be Asn-77, based on GPI anchor signal commonalities. This suggests that vCD59 may also be a GPI-anchored protein because there is an asparagine residue at position 77, and the domain surrounding the proposed GPI modification site is highly conserved (Fig. 3). Processing of the vCD59 precursor peptide may result in a mature peptide of 77 aa. The structures of the viral and the cellular protein are expected to be very similar based on the observation that both predicted mature proteins share an overall identity of 52% (65% similarity allowing conservative replacements), both have a single Asn-linked glycosylation site of the same context, and all cysteine residues are at identical positions.[15] To define the molecular relationship between CD59 homologues more precisely, sequences derived from the African green monkey (agmCD59), baboon (babCD59), owl monkey (omCD59), cottontop marmoset (ctmCD59), squirrel monkey (smCD59), and three different isolates of *H. saimiri* (vCD59A-C) were cloned by polymerase chain reaction (PCR), sequenced, and analyzed.[32] The alignment of the corresponding amino acid sequences indicates that the signal peptide and the hydrophobic domain responsible for GPI-anchoring are highly conserved among New World and Old World primates with respect to viral sequences (Fig. 3). This is also reflected by phylogenetic analyses comparing complete amino acid sequences of CD59 homologues, which indicate that each group may have evolved independently. However, comparison of predicted mature peptides indicates that the CD59 homologue of *H. saimiri* is most closely related to sequences of New World monkeys, especially to that of the squirrel monkey. Computing evolutionary relationships of predicted mature peptides confirmed this result, suggesting that *H. saimiri* may have captured the CD59 gene from its natural host, the squirrel monkey. However, the sequence derived from the squirrel monkey encodes three extra amino acids which may be the result of an internal duplication of nine nucleotides. Taken together, it suggests that *H. saimiri* acquired the gene from monkeys that are different from the squirrel monkeys known today.

Human CD59 is an inhibitor of the terminal membrane attack complex of complement. It is an 18 to 20-kDa glycoprotein. The common features of huCD59 and vCD59 suggest that both polypeptides may also share functional properties. To investigate the complement regulatory potential of the ORF-15

```
                      1                                     >L                                      66
vCD59 H.saimiri B     MkrkvmyILFkLvLaFvVFCkSsySLQCYsCsrsT...vqCTTsTsCTSNLDSCLIAKAGsrVYYR
vCD59 H.saimiri C     .....myILFkLMLa.sVFCkSsySLQCYnCshsT...mqCTTsTsCTSNLDSCLIAKAGskVYYR
vCD59 H.saimiri A     .....myILFtLvLtF.VFCkpihSLQCYnCshsT...mqCTTsTsCTSNLDSCLIAKAGsgVYYR
agmCD59               MGIQGGSVLFGLLLaLAVFCHSGhSLQCYnCPnpT...tnCkTaiNCsSgFDtCLIArAGlqVYnq
babCD59               MGIQGGSVLFGLLLvLAVFCHSGhSLQCYnCPnpT...tdCkTaiNCsSgFDtCLIArAGlqVYnq
huCD59                MGIQGGSVLFGLLLvLAVFCHSGhSLQCYnCPnpT...adCkTavNCsSdFDaCLItKAGlqVYnk
omCD59                MGIQGGSVLFGLLLvLAVFCHSGnSLQCYsCPypT...tqCTmtTNCTSNLDSCLIAKAGsrVYYR
ctmCD59               MGIQGGSVLFGLLLiLAVFCHSGhSLQCYsCPysT...arCTTtTNCTSNLDSCLIAKAGlrVYYR
smCD59                MGIQGGSVLFGLLLvLAVFCHSGnSLQCYsCPlpTmesmeCTasTNCTSNLDSCLIAKAGsgVYYR
         Consensus    --------LF-L-L---VFC----SLQCY-C---T-----C-----C-S--D-CLI--AG--VY--
                                          #    # #         #    #     #

                      67                                         N<                        131
vCD59 H.saimiri B     CWKFDDCsFkrISNQLSEtqLKYhCCKnNLCNvNkviENGkrTiSDKa.LLLlalFLvtA*nFpl
vCD59 H.saimiri C     CWKFDDCsFkrISNQLSEtqLKYhCCKKNLCNvNkviENGkrTiSDKa.LLLlalFLvtA*nLsl
vCD59 H.saimiri A     CWKFDDCsFkrISNQLSEtqLKYhCCKKNLCNvNkgiENikrTiSDKa.LLLlalFLvtAWnFpl
agmCD59               CWKFanCnFndIStlLkEseLqYfCCKedLC..NeqLENGGTsLSEKTVLLLVtpLLaAAWcLHP
babCD59               CWKFanCnFndIStlLkEseLqYfCCKKdLCNfNeqLENGGTsLSEKTVvLLVtlLLaAAWcLHP
huCD59                CWKFEhCnFndVttrLrEneLtYyCCKKdLCNfNeqLENGGTsLSEKTVLLLVtpFLaAAWsLHP
omCD59                CWKFEDCtFsrVSNQLSEneLKYyCCKKNLCNfNeaLkNGGTTLSkKTVLLLVipFLvAAWsLHP
ctmCD59               CWKFEDCtFrqlSNQLSEneLKYhCCreNLCNfNgiLENGGTTLSkKTVLLLVtpFLaAAWsLHP
smCD59                CWKFDDCsFkrISNQLSEtqLKYhCCKKNLCNvkevLENGGTTLSkKTILLLVtpFLaAAWsrHP
         Consensus    CWKF--C-F------L-E--L-Y-CC---LC-------N-----S-K---LL----L--A-----
                      #   #            ##   #
```

FIGURE 3. Comparison of CD59 homologues. Three viral sequences from H. saimiri subgroups (deduced from ORF-15 of strains A-11, B-SMHI, and C-488) are shown aligned to three sequences from Old World primates (African green monkey = agm, baboon = bab, human = hu) and three sequences from New World monkeys (owl monkey = om, cottontop marmoset = ctm, squirrel monkey = sm). The beginnings of the predicted mature peptides are marked at a lysine residue (>L), and the predicted GPI-attachment site is marked at its respective asparagine residue (<N). Amino acids are given in single letter code with conserved amino acids capitalized. Stars (*) within the vCD59 sequences of H. saimiri B and C denote premature stop codons. Cysteine residues are emphasized by a cross (#) below a consensus sequence.

gene product and to allow a detailed functional comparison, the CD59 homologues from *H. saimiri* and from its natural host, the squirrel monkey (smCD59), were cloned for stable expression in murine BALB/3T3 cells. HuCD59-expressing clones were identified using anti-huCD59 polyclonal serum or monoclonal antibodies, but anti-huCD59 antibodies did not cross-react with vCD59 or smCD59. To identify clones expressing these molecules, PCR analysis of reverse transcribed mRNA was performed, and positive clones were chosen for functional analysis. Stably transfected BALB/3T3 clones were tested in a dye release assay for protection against lysis by human complement. Cells expressing vCD59, smCD59, or huCD59 were effective in inhibiting membrane damage by human complement. To address the question of species restriction, the same assay was performed using rat complement. Cells expressing vCD59 were protected from complement-mediated lysis by whole rat serum, whereas cells expressing smCD59 or huCD59 were complement-sensitive. To rule out complement regulation before the initiation of the terminal complement cascade, cells expressing vCD59 or smCD59 were tested for C3 deposition relative to cells transfected with huCD59 or vector alone. The same cells were subjected to complement-mediated lysis by human complement to examine the degree of complement-mediated cytolysis relative to the level of C3 deposition. C3 deposition on the surface of all transfectants, including the vector alone, was equal. However, cells expressing vCD59, smCD59, or huCD59 released less than 20% of the cytoplasmic dye when subjected to 20% human serum, whereas cells transfected with the vector alone released 60%. This finding indicates that the complement regulatory activity of vCD59 and smCD59 occurs after C3 deposition.[33]

The C-terminal hydrophobic domains of viral and cellular CD59 are quite distinct in sequence. This region is known to be essential for GPI anchoring, the mechanism of membrane attachment demonstrated for huCD59. To investigate the membrane attachment characteristics of vCD59, transfectants were incubated with phosphatidylinositol-specific phospholipase C (PIPLC) prior to exposure to complement. The complement regulatory activities of cells expressing vCD59 or huCD59 were substantially reduced (55 to 75%) by PIPLC treatment, demonstrating GPI-anchoring of these molecules on the cellular surface.[33] Potential complement regulatory activity of vCD59 in the natural host of *H. saimiri* was tested by exposing vCD59-expressing cells to squirrel monkey serum as the source of complement. vCD59 effectively inhibited membrane damage mediated by squirrel monkey complement, and cells expressing smCD59 showed the same level of protection. These data established the efficient complement regulatory activity of vCD59. Complement regulation mediated by vCD59 may be a mechanism for immune evasion during the life cycle of the virus. This hypothesis was supported by examining the transcription of ORF-15. Northern blot hybridization experiments, using a probe specific for the open reading frame encoding vCD59, detected specific transcripts in polyadenylated RNA isolated from owl monkey kidney (OMK) cells that were lytically infected with *H. saimiri*. However, no transcripts were detected in *H. saimiri*-transformed cell lines.[33] These results demonstrate that the gene for vCD59 is transcribed only during lytic reproduction of infectious virus and provide evidence for the potential expression of vCD59 on the surface of infected cells and/or virions.

The importance of complement regulatory molecules is evident from the fact that a variety of viruses express C3/C4-binding proteins. By mimicking human proteins, these viral polypeptides may enhance virulence by preventing complement activation. The complement control protein of the vaccinia virus is most homologous to C4bp. It is the major secretory protein which regulates CPCA and APCA, and it contributes to the virulence of vaccinia virus.[34] Multiple host immune evasion mechanisms are not unprecedented. Herpes simplex virus expresses the glycoprotein gC on the viral envelope and on virally infected cells. It is an inhibitor of the C3/C5 convertases and renders cells more resistant to complement-mediated attack.[35,36] However, gC does not exhibit convincing homologies to members of the RCA family. Herpes simplex virus also expresses a functional Fc-receptor that inhibits complement lysis and possibly Fc-receptor-mediated phagocytosis.[37]

The viral complement counteracting proteins may also be of interest with regard to transplantation immunity and complement-mediated tissue injury. In general, xenografts are poorly tolerated and are rapidly rejected. Human organs do not meet the need. For numerous reasons, a primate source is unlikely to become an acceptable alternative. Usually, xenografts are acutely rejected because heterophilic antibodies activate complement on the endothelium of a transplanted organ's vasculature. In theory it should be possible to prevent acute graft rejection by expressing complement regulatory proteins on graft endothelium. There is evidence that pig endothelial cells coated with DAF and certain cell lines expressing MCP or DAF are protected from this type of injury. Transgenic animals that express human complement regulators provide an experimental system to examine these hypotheses more rigorously. The problem of species restriction of human complement inhibitors may be overcome by introducing complement regulatory proteins from nonhuman species or even viral complement regulators that exhibit less species restriction. Another approach is to infuse soluble complement inhibitors to prevent complement-dependent tissue injury. Initial trials suggest that various complement regulatory proteins may become useful therapeutic reagents.[38–40]

7. CONCLUDING REMARKS

The rhadinovirus genes homologous to known cellular sequences have most likely been sequestered from the host during the viral genome evolution. There is precedent for gene capturing in other DNA viruses. The vaccinia virus complement control protein VCP is a well-known example. The remarkably high number of virus-cell homologous genes in rhadinovirus DNA may be related to the particular genomic structure that is predisposed to the acquisition of foreign DNA. The large proportion of repetitive DNA in the viral genome allows incorporation of cellular DNA sequences without loss of genetic information. As herpesviruses are assumed to replicate according to the rolling circle mechanism after initial circularization in infected cells, foreign DNA stretches could easily be incorporated during the early phase of replication. All known cellular homologous genes of rhadinoviruses have the structure of cDNA copies of the respective cellular transcripts, suggesting that reverse transcription is involved in gene capturing. Most likely, acquired cellular genes provide selective advantage to the viruses. The homologies are limited to the respective open reading frame or to certain functional domains of proteins. The acquisition of two different types of complement inhibitors by the rhadinovirus *H. saimiri* has probably not taken place simultaneously. CCPH widely diverges among different *H. saimiri* strains, and a gene at the same genetic distance is also present in *H. ateles*. The viral CD59 homologues closely reflect the squirrel monkey's CD59 structure. *H. ateles* and HHV-8 do not have a CD59 homologue. This suggests that the vCD59 of *H. saimiri* is an independent and more recent gene acquisition.

H. saimiri complement control proteins are unique in many regards. First, in contrast to herpes simplex or the vaccinia virus, the two complement regulators of *H. saimiri* act at two different levels of complement activation, namely, on the formation and assembly of the classical C3/C5 convertases and on the formation of the terminal membrane attack complex. Unlike CCPH of *H. saimiri*, the herpes simplex glycoprotein gC does not reveal convincing sequence homologies to cellular complement regulators. This makes it rather unlikely that the gC gene of herpes simplex and other alpha herpesviruses is a derivative of cellular DNA. There is no known precedent for the vCD59 homologue in any other virus. The vaccinia protein VCP is expressed only as a secretory molecule. In contrast, *H. saimiri* CCPH exists in two forms, as a membrane-bound functional protein and as a secreted polypeptide which is translated from a spliced mRNA. Both forms may help to counteract destruction of virus-containing cells by the immune system. Unlike poxviruses, herpesviruses characteristically persist for the lifetime of the host. The high viral load of naturally infected animals may indicate that circulating white blood cells are an important reservoir for *H. saimiri* or that they mediate the transport from persistently infected cell types to virus-producing secretory cells. Possibly, the dual endowment of *H. saimiri* with complement regulators relates to the special role of peripheral blood cells as a viral reservoir that, by means of a viral gene product, is protected against elimination through humoral immune responses.

ACKNOWLEDGMENTS. Original work was supported by Dr. Mildred Scheel-Stiftung and Sonderforschungsbereich 466 *Lymphoproliferation und virale Immundefizienz*.

REFERENCES

1. Chang, Y., Cesarman, E., Pessin, M. S., Lee, F., Culpepper, J., Knowles, D. M., and Moore, P. S., 1994, Identification of herpesvirus-like DNA sequences in AIDS—associated Kaposi's sarcoma, *Science* **266:**1865–1869.
2. Neipel, F., Albrecht, J. C., and Fleckenstein, B., 1997, Cell-homologous genes in the Kaposi's sarcoma-associated rhadinovirus HHV-8-determinants of its pathogenicity? *J. Virol.*, **71:**4187–4192.
3. Fleckenstein, B., and Desrosiers, R. C., 1982, *The Herpesviruses*, Vol. 1 (Roizman, B. ed.), Plenum Press, New York, London, pp. 253–332.
4. Biesinger, B., Muller-Fleckenstein, I., Simmer, B., Lang, G., Wittmann, S., Platzer, E., Desrosiers, R. C., and Fleckenstein, B., 1992, Stable growth transformation of human T lymphocytes by herpesvirus saimiri, *Proc. Natl. Acad. Sci. USA* **89:**3116–3119.
5. Weber, F., Meinl, E., Drexler, K., Czlonkowska, A., Huber, S., Fickenscher, H., Müler-Fleckenstein, I., B., Wekerle, H., and Hohlfeld, R., 1993, Herpesvirus saimiri-transformed human T cell lines expressing functional receptor for myelin basic protein, *Proc. Natl. Acad. Sci. USA* **90:**11049–11053.
6. Ensser, A., Pflanz, R., and Fleckenstein, B., 1997, Primary structure of the alcelaphine herpesvirus type I genome, *J. Virol.* **71:**6517–6525.
7. Lomonte, P., Bublot, M., van Santen, V., Keil, G. M., Pastoret, P. P., and Thiry, E., 1995, Analysis of bovine herpesvirus 4 genomic regions located outside the conserved gammaherpesvirus gene blocks, *J. Gen. Virol.* **76:**1835–1841.
8. Telford, E. A., Watson, M. S., Aird, H. C., Perry, J., and Davison, A. J., 1995, The DNA sequence of equine herpesvirus 2, *J. Mol. Biol.* **249:**520–528.
9. Werner, F. J., Bornkamm, G. W., and Fleckenstein, B., 1977, Episomal viral DNA in a herpesvirus saimiri-transformed lymphoid cell line, *J. Virol.* **22:**794–803.
10. Jung, J. U., Trimble, J. J., King, N. W., Biesinger, B., Fleckenstein, B. W., and Desrosiers, R. C., 1991, Identification of transforming genes of subgroup A and C strains of herpesvirus saimiri, *Proc. Natl. Acad. Sci. USA* **88:**7051–7055.
11. Medveczky, M. M., Geck, P., Vassallo, R., and Medveczky, P. G., 1993, Expression of the collagen-like putative oncoprotein of Herpesvirus saimiri in transformed T cells, *Virus Genes* **7:**349–365.
12. Yao, Z., Painter, S. L., Fanslow, W. C., Ulrich, D., Macduff, B. M., Spriggs, M. K., and Armitage, R. J., 1995, Human IL-17: A novel cytokine derived from T cells, *J. Immunol.* **155:**5483–5486.
13. Albrecht, J. C., Nicholas, J., Biller, D., Cameron, K. R., Biesinger, B., Newman, C., Witmann, S., Craxton, M. A., Coleman, H., and Fleckenstein, B., 1992, Primary structure of the herpesvirus saimiri genome, *J. Virol.* **66:**5047–5058.
14. Albrecht, J. C., and Fleckenstein, B., 1992, New member of the multigene family of complement control proteins in herpesvirus saimiri, *J. Virol.* **66:**3937–3940.
15. Albrecht, J. C., Nicholas, J., Cameron, K. R., Newman, C., Fleckenstein, B., and Honess, R. W., 1992, Herpesvirus saimiri has a gene specifying a homologue of the cellular membrane glycoprotein CD59, *Virology* **190:**527–530.
16. Albrecht, J. C., and Fleckenstein, B., 1998, Primary structure of the herpesvirus ateles genome, unpublished.

17. Russo, J. J., Bohenzky, R. A., Chen, M. C., Chen, J., Yan, M., Maddalena, D., Parry, J. P., Peruzzi, D., Edelman, I. S., Chang, Y., and Moore, P. S., 1996, Nucleotide sequence of the Kaposi's sarcoma-associated herpesvirus (HHV8), *Proc. Natl. Sci. USA* **93:**14862–14867.
18. Hourcade, D., Holers, V. M., and Atkinson, J. P., 1989, The regulators of complement activation (RCA) gene cluster, *Adv. Immunol.* **45:**381–415.
19. Kotwal, G. J., Isaacs, S. N., McKenzie, R., Frank, M. M., and Moss, B., 1990, Inhibition of the complement cascade by the major secretory protein of vaccinia virus, *Science* **250:**827–830.
20. McKenzie, R., Kotwal, G. J., Moss, B., Hammer, C. H., and Frank, M. M., 1992, Regulation of complement activity by vaccinia virus complement-control protein, *J. Infect. Dis.* **166:**1245–1250.
21. Caras, I. W., Davitz, M. A., Rhee, L., Weddell, G., Martin, D. W. Jr., and Nussenzweig, V., 1987, Cloning of decay-accelerating factor suggests novel use of splicing to generate two proteins, *Nature* **325:**545–549.
22. Metz, C. N., Brunner, G., Choi-Muira, N. H., Nguyen, H., Gabrilove, J., Caras, I. W., Altszuler, N., Rifkin, D. B., Wilson, E. L., and Davitz, M. A., 1994, Release of GPI-anchored membrane proteins by a cell-associated GPI-specific phospholipase D, *EMBO J.* **13:**1741–1751.
23. Holers, V. M., Kinoshita, T., and Molina, H., 1992, The evolution of mouse and human complement C3-binding proteins: Divergence of form but conservation of function, *Immunol. Today* **13:**231–236.
24. Farries, T. C., and Atkinson, J. P., 1991, Evolution of the complement system, *Immunol. Today* **12:**295–300.
25. Randall, R. E., Newman, C., and Honess, R. W., 1984, Isolation and characterization of monoclonal antibodies to structural and nonstructural herpesvirus saimiri proteins, *J. Virol.* **52:**872–883.
26. Fodor, W. L., Rollins, S. A., Bianco Caron, S., Rother, R. P., Guilmette, E. R., Burton, W. V., Albrecht, J. C., Fleckenstein, B.,, and Squinto, S. P., 1995, The complement control protein homolog of herpesvirus saimiri regulates serum complement by inhibiting C3 convertase activity, *J. Virol.* **69:**3889–3892.
27. Lachmann, P. J., 1991, The control of homologous lysis, *Immunol. Today* **12:**312–315.
28. Cooper, N. R., 1991, Complement evasion strategies of microorganisms, *Immunol. Today* **12:**327–331.
29. Bergelson, J. M., Chan, M., Solomon, K. R., St. John, N. F., Lin, H., and Finberg, R. W., 1994, Decay-accelerating factor (CD55), a glycosylphosphatidylinositol-anchored complement regulatory protein, is a receptor for several echoviruses, *Proc. Natl. Acad. Sci. USA* **91:**6245–6249.
30. Naniche, D., Varior-Krishnan, G., Cervoni, F., Wild, T. F., Rossi, B., Rabourdin-Combe, C., and Gerlier, D., 1993, Human membrane of cofactor protein (CD46) acts as a cellular receptor for measles virus, *J. Virol.* **67:**6025–6032.
31. Nielsen, H., Engelbrecht, J., Brunak, S., and von Heijne, G., 1997, Identification of procaryotic and eukaryotic signal peptides and prediction of their cleavage sites, *Protein Eng.* **10:**1–6.
32. Fodor, W. L., Rollins, S. A., Bianco Caron, S., Burton, W. V., Guilmette, E. R., Rother, R. P., Zavoico, G. B., and Squinto, S. P., 1995, Primate terminal complement inhibitor homologues of human CD59, *Immunogenetics* **41:**51.
33. Rother, R. P., Rollins, S. A., Fodor, W. L., Albrecht, J. C., Setter, E., Fleckenstein, B., and Squinto, S. P., 1994, Inhibition of complement-mediated cytolysis by the terminal complement inhibitor of herpesvirus saimiri, *J. Virol.* **68:**730–737.
34. Isaacs, S. N., Kotwal, G. J., and Moss, B., 1992, Vaccinia virus complement-control protein prevents antibody-dependent complement-enhanced neutralization of infectivity and contributes to virulence, *Proc. Natl. Acad. Sci. USA* **89:**628–632.
35. Harris, S. L., Frank, I., Yee, A., Cohen, G. H., Eisenberg, R. J., and Friedman, H. M., 1990, Glycoprotein C of herpes simplex virus type 1 prevents complement-mediated cell lysis and virus neutralization, *J. Infect. Dis.* **162:**331–337.

36. McNearney, T. A., Odell, C., Holers, V. M., Spear, P. G., and Atkinson, J. P., 1987, Herpes simplex virus glycoproteins gC-1 and gC-2 bind to the third component and provide protection against complement-mediated neutralization of viral infectivity, *J. Exp. Med.* **166:**1525–1535.
37. Bell, S., Cranage, M., Borysiewicz, L., and Ninson, T., 1990, Induction of immunoglobulin G Fc receptors by recombinant vaccinia virus expressing glycoproteins E and I of herpes simplex virus type I, *J. Virol.* **64:**2181–2186.
38. Ryan, U. S., 1995, Complement inhibitory therapeutics and xenotransplantation, *Nat. Med.* **1:**967–968.
39. Kennedy, S. P., Rollins, S. A., Burton, W. V., Sims, P. J., Bothwell, A. L., Squinto, S. P., and Zavoico, G. B., 1994, Protection of porcine aortic endothelial cells from complement-mediated cell lysis and activation by recombinant human CD59, *Transplantation* **57:**1494–1501.
40. Kroshus, T. J., Bolman, R. M., III, Dalmasso, A. P., Rollins, S. A., Guilmette, E. R., Williams, B. L., Squinto, S. P., and Fodor, W. L., 1996, Expression of human CD59 in transgenic pig organs enhances organ survival in an ex vivo xenogeneic perfusion model, *Transplantation* **61:**1513–1521.
41. Medveczky, P., Szomolanyi, E., Desrosiers, R. C., and Mulder, C., 1984, Classification of herpesvirus saimiri into three groups based on extreme variation in a DNA region required for oncogenicity, *J. Virol.* **52:**938–944.

Index

An "*f*" or "*t*" following a page number indicates that the term may be found in a figure or table, respectively, on the page indicated.